Beach-Dune System Morphodynamics

Beach-Dune System Morphodynamics

Editors

Felice D'Alessandro
Giuseppe Roberto Tomasicchio
Ferdinando Frega

MDPI • Basel • Beijing • Wuhan • Barcelona • Belgrade • Manchester • Tokyo • Cluj • Tianjin

Editors
Felice D'Alessandro
University of Milan
Italy

Giuseppe Roberto
Tomasicchio University of Salento
Italy

Ferdinando Frega
University of Calabria
Italy

Editorial Office
MDPI
St. Alban-Anlage 66
4052 Basel, Switzerland

This is a reprint of articles from the Special Issue published online in the open access journal *Journal of Marine Science and Engineering* (ISSN 2077-1312) (available at: https://www.mdpi.com/journal/jmse/special_issues/Beach-Dune).

For citation purposes, cite each article independently as indicated on the article page online and as indicated below:

LastName, A.A.; LastName, B.B.; LastName, C.C. Article Title. *Journal Name* **Year**, *Volume Number*, Page Range.

ISBN 978-3-0365-5027-5 (Hbk)
ISBN 978-3-0365-5028-2 (PDF)

Cover image courtesy of Antonio Francone.

Contents

About the Editors

Felice D'Alessandro

Felice D'Alessandro is Associate Professor of Coastal Engineering at the University of Milan. He was/is partner of national and EU projects (e.g., FP6-Hydralab III, FP7-Hydralab IV). He was a Visiting Research Scientist at Laboratorio Nacional de Engenharia Civil (LNEC), Lisbon, Portugal, and at the Coastal and Hydraulics Laboratory, U.S. Army ERDC, Vicksburg, Mississippi. He is member of the Ph.D. scientific board in Environmental Sciences at University of Milan. His research activity mainly focuses on wave-breaking modelling in Boussinesq-Type Equations (BTE), wave–dune interaction and beach resilience, longshore transport, multivariate design and structural risk in coastal and off-shore engineering, and coastal ecosystem restoration. He is author/co-author of more than 100 papers published in refereed international journals and conference proceedings.

Giuseppe Roberto Tomasicchio

Giuseppe Roberto Tomasicchio is Full Professor of Coastal Engineering and Director of EUMER Laboratory at the University of Salento. He was scientific coordinator of national and EU projects (e.g., FP6-Hydralab III and FP7-Hydralab IV) and member of several PIANC working groups. His research interests are directed towards a better understanding of coastal and offshore engineering topics such as wave breaking and longshore transport, coastal structures such as reshaping or berm breakwaters, armor stone degradation, long-term wave prediction, rehabilitation of eroded/polluted beaches, design and structural risk of coastal structures, wave-induced forces at pipelines, and forces at floating structures. He is author of more than 100 papers published in peer-reviewed international journals and conference proceedings. Since 2004, he has organized the International Short Conference/Course on Applied Coastal Research (a IAHR and ASCE-COPRI promoted event). Since 2018, he has been President of the Italian Association for Offshore and Maritime Engineering.

Ferdinando Frega

Ferdinando Frega is Assistant Professor of Coastal Engineering at the University of Calabria. He was/is partner of national and EU projects. He was a Visiting Research Scientist at the Coastal and Hydraulics Laboratory, U.S. Army ERDC, Vicksburg, Mississippi. He is a member of the Urban Hydraulic and Hydrology Laboratory (LIU) at the University of Calabria. His research activity mainly focuses on wave-breaking modelling in Boussinesq-Type Equations (BTE), wave–dune interaction and beach resilience, longshore transport, multivariate design and structural risk in coastal and off-shore engineering, coastal ecosystem restoration, scaling behavior, physical models and urban hydrology. He is author/co-author of more than 50 papers published in refereed international journals and conference proceedings.

Preface to "Beach-Dune System Morphodynamics"

Coastal dunes are known for their functions in ecological systems in addition to their aesthetic qualities, providing a highly valuable and unique habitat due to their biodiversity of flora and fauna. They also represent the boundary between land and sea, acting as a protective natural barrier against flooding due to storm surges and wave action. Beach–dune systems are highly dynamic features whose evolution is primarily determined by the mutual and complex exchange of sand through hydrodynamic and eolian processes. The sustainable and resilient conservation of beach–dune (eco)systems in a changing climate requires important insights from multidisciplinary studies and approaches. Toward this vision, this Special Issue is dedicated to collecting original scientific contributions based on field observations, laboratory experiments, and/or numerical models.

The Guest Editors would like to thank all the authors, the *JMSE* Editors, and the reviewers for their outstanding contributions and dedication to this Special Issue.

Felice D'Alessandro, Giuseppe Roberto Tomasicchio, and Ferdinando Frega
Editors

Journal of
*Marine Science
and Engineering*

MDPI

Editorial

Beach-Dune System Morphodynamics

Felice D'Alessandro [1,*], Giuseppe Roberto Tomasicchio [2] and Ferdinando Frega [3]

1 Department of Environmental Science and Policy, University of Milan, Via Celoria 2, 20133 Milan, Italy
2 EUMER, Department of Engineering for Innovation, University of Salento, Campus Ecotekne, Via Monteroni, 73100 Lecce, Italy; roberto.tomasicchio@unisalento.it
3 Department of Civil Engineering, University of Calabria, Ponte P. Bucci, 87036 Rende, Italy; ferdinando.frega@unical.it
* Correspondence: felice.dalessandro@unimi.it

Citation: D'Alessandro, F.; Tomasicchio, G.R.; Frega, F. Beach-Dune System Morphodynamics. *J. Mar. Sci. Eng.* **2022**, *10*, 802. https://doi.org/10.3390/jmse10060802

Received: 1 June 2022
Accepted: 8 June 2022
Published: 10 June 2022

Coastal dunes are known for their functions in ecological systems in addition to their aesthetic qualities, providing a highly valuable and unique habitat of due to their biodiversity of flora and fauna. They also represent the boundary between land and sea, acting as a protective natural barrier against flooding due to storm surges and wave action. Beach–dune systems are highly dynamic phenomena whose evolution is primarily determined by the mutual and complex exchange of sand through hydrodynamic and eolian processes. The sustainable and resilient conservation of beach–dune (eco)systems in a changing climate requires important insights from multidisciplinary studies and approaches. Toward this vision, this Special Issue is dedicated to collecting original scientific contributions based on field observations, laboratory experiments, and/or numerical models.

The following is an overview of the contributions to this Special Issue.

The first published paper is entitled "Application of the Saffir-Simpson Hurricane Wind Scale to Assess Sand Dune Response to Tropical Storms" by Ellis et al. [1]. This study investigated the volumetric changes of two dune systems on the Isle of Palms, South Carolina, USA, before and after Hurricane Irma (2017) and Hurricane Florence (2018), which struck the island as tropical storms with different characteristics. Irma had relatively high wave heights and rainfall, resulting in an average volumetric dune loss of 39%. During Florence, a storm with low rainfall and moderate winds, the net volumetric dune loss averaged 3%. The primary driving force for dune change during Irma was water (rainfall and storm surge), while during Florence it was wind (aeolian transport). The authors suggest reconsidering the use of the Saffir–Simpson hurricane wind scale because different geomorphological responses were measured, even though Irma and Florence were both classified as tropical storms. Site-specific studies of dune morphology before and after the storm, as well as site-specific meteorological measurements of the storm (wind characteristics, storm surge, rainfall), are urgently needed.

The second published paper is entitled "Sensitivity of Storm Response to Antecedent Topography in the XBeach Model" by Mickey et al. [2]. In this study, a methodology was developed to investigate the uncertainty of the profile response within the XBeach storm impact model associated with different antecedent topography. A parameterized island Gaussian fit model (PIGF) generated an idealized base profile and a set of idealized profiles with different characteristics based on collected LiDAR data from Dauphin Island, AL, USA. Six synthetic storm scenarios were simulated and analyzed on each of the idealized profiles using XBeach in both one- and two-dimensional configurations to determine the morphological response and uncertainties associated with the different antecedent topography. The morphological response of the profile tends to scale with the strength of the storm, but for different profiles, the uncertainty of the profile response to the medium-strength storm scenarios is greater than for the low- and high-strength storm scenarios. XBeach can be very sensitive to morphological thresholds, both antecedent and time-varying, especially with respect to beach slopes.

The third published paper is entitled "Predicting Morphodynamics for Beach Intertidal Systems in the North Sea: A Space-Time Stochastic Approach" by Bogaert et al. [3]. The authors developed a stochastic model of a macrotidal sandy-barred beach based on cross-shore elevation profiles. Intensive survey campaigns using Real Time Kinematics-GPS (RTK-GPS) and LiDAR approaches were conducted. The results of this study demonstrate how the temporal evolution of an integrated property, such as total displaced sand volume, can be estimated over time. This suggests that a similar stochastic approach could be useful for estimating other properties, as long as one is able to capture the stochastic space–time variability of the underlying processes.

The fourth published paper is entitled "Development and Application of an Empirical Dune Growth Model for Evaluating Barrier Island Recovery from Storms" by Dalyander et al. [4]. In this study, an empirical dune growth model (EDGR) was developed to predict the evolution of the primary foredune of a barrier island. EDGR was assessed against observed dune evolution along the western end of Dauphin Island, Alabama, over the 10 years following Hurricane Katrina (2005). The computational efficiency of EDGR allows for the rapid prediction of dune evolution and enables ensemble predictions to limit the uncertainties that can arise when terminal dune characteristics are unknown. In addition, EDGR can be coupled with an external model to estimate dune erosion and/or the long-term evolution of other subaerial features to enable the prediction of barrier island evolution on a decadal scale.

The fifth published paper is entitled "Investigating Changes in Aeolian Sediment Transport at Coastal Dunes and Sand Trapping Fences: A Field Study on the German Coast" by Eichmanns and Schüttrumpf [5]. In this paper, the authors present the results of field experiments conducted at the beach, on coastal dunes, and at a sand trapping fence on the East Frisian island Langeoog. The vertical sediment flux profile was measured by vertical mesh sand traps, and saltiphones measured instantaneous sediment transport. A meteorological station was set up to obtain wind data. On the beach, dune toe, and dune crest, the stationary wind profile can be effectively described by the law of the wall. Saturated aeolian sediment transport rates on the beach and dune toe were predicted by widely used empirical models. At the sand trapping fence, these empirical transport models could not be applied, as no logarithmic wind profile existed. The upwind sediment supply reduced after each brushwood line of the sand trapping fence, thereby, leading to increased deviation from the saturated conditions.

The sixth published paper is entitled "Use of Nanosilica for Increasing Dune Erosion Resistance during a Sea Storm" by Leone et al. [6]. In this study, the reliability of an innovative non-toxic colloidal silica-based solution for coastal sand dunes was verified for the first time using laboratory experiments. An extensive experimental campaign was carried out in the 2D wave flume of the EUMER laboratory at the University of Salento (Italy) to investigate the effectiveness of the use of nanosilica. The study was initially based on a horizontal seabed, and then a cross-shore beach–dune profile similar to those generally observed in nature was drawn. Detailed measurements of wave characteristics and observed cross-shore beach–dune profiles were analyzed for a wide range of wave conditions. In both cases, two sets of experiments were conducted. After performing the first set of experiments, which resembled the native conditions of the models composed with natural sand, the effects of injecting silica-based mineral colloidal grout were investigated. The observations show that mineral colloidal silica increases the mechanical strength of non-cohesive sediments, thus reducing the volume of dune erosion and improving the resistance and longevity of the beach–dune system.

The seventh published paper is entitled "Role of Storm Erosion Potential and Beach Morphology in Controlling Dune Erosion" by Lemke and Miller [7]. This study investigated the role of storm intensity and the vulnerability of location in controlling dune erosion based on data compiled for eighteen historical events in New Jersey. Storm intensity was characterized by the Storm Erosion Index (SEI) and Peak Erosion Intensity (PEI), factors used to describe a storm's cumulative erosion potential and maximum erosive power, respectively. In this study, a direct relationship between these parameters, beach morphology characteristics, and expected dune response was established through a classification tree ensemble. Of the seven input parameters, PEI was the most important, indicating that peak storm conditions with time scales on the order of hours were the most critical in predicting dune impacts. Results suggested that PEI alone was successful in distinguishing between storms most likely to result in no impacts (PEI < 69) and those likely to result in some impacts (PEI > 102), regardless of beach conditions. For intensities in between, where no consistent behavior was observed, beach conditions must be considered. Because of the propensity for beach conditions to change over short spatial scales, it is important to predict impacts on a local scale. This study established a model with the computational effectiveness to provide such predictions.

The eighth published paper is entitled "Effects of Anthropogenic Pressures on Dune Systems-Case Study: Calabria (Italy)" by Foti et al. [8]. This study aimed to evaluate the effects of anthropogenic pressures on the Calabrian dune systems, especially with regard to the triggering of coastal erosion processes. For this purpose, historical and current cartographic data, such as shapefiles, cartography, and satellite imagery, were analyzed using QGIS. This evaluation was carried out through the comparison between the current extension of the dune systems and their extensions after the Second World War, i.e., before increased anthropogenic pressures. This evaluation was also carried out through the analysis of shoreline changes in coastal areas where dune systems are currently present, and in coastal areas where dune systems have been partially or totally destroyed by anthropogenic causes. These changes were compared to the status of these same coastal areas in the 1950s, thus excluding coastal areas without dune systems in the 1950s and analyzing what was built in place of the destroyed dune systems. Two criteria were defined to identify the levels of destruction of the dune systems and to identify coastal erosion processes. The analysis showed a strong correlation between the destruction of dune systems by anthropogenic causes and the triggering of coastal erosion processes.

The ninth published paper is entitled "A Numerical Study on the Impact of Building Dimensions on Airflow Patterns and Bed Morphology around Buildings at the Beach" by Pourteimouri et al. [9]. In this study, a numerical model was developed using the open-source computational fluid dynamics solver OpenFOAM. First, the model was used to predict the airflow patterns around a single rectangular building. The model predictions were validated with wind-tunnel data, which had a good agreement. Second, a reference beach building was introduced, and then the building dimensions were increased in length, width, and height, each up to three times the size of the reference building dimensions. The impact of each dimensional extent on the near-surface airflow patterns was investigated. The results show that the near-surface airflow patterns are least dependent on the length of the building in the wind direction, and they depend most on the width of the building perpendicular to the wind direction. Third, the convergence of a third-order horizontal near-surface velocity field was calculated to interpret the impact of changes in airflow patterns on potential erosion and deposition patterns around the building. The numerical predictions were compared with the observed erosion and sedimentation patterns around scale models in the field. The comparisons showed satisfactory agreements between numerical results and field measurements.

The tenth published paper is entitled "Beach and Dune Erosion: Causes and Interventions, Case Study: Kaulon Archaeological Site" by Barbaro et al. [10]. The paper, through a case study, analyzed erosive processes of the beaches and dunes, their causes, and the possible interventions. The case study concerns the archaeological site of Kaulon, located on a dune on the Ionian coast of Calabria (Italy). The beach near the site was affected by erosive processes, and during the winter of 2013–2014, the site was damaged by two sea storms. To identify the causes of these processes, three erosive factors were analyzed. These factors are anthropogenic pressure, wave climate and sea storms, and river transport. The effects produced by these factors were assessed in terms of shoreline changes and damage to the beach–dune system, also evaluating the effectiveness of the defense interventions.

The main causes of the erosive processes were identified through a cross analysis of erosive factors and their effects. This analysis highlighted that, in the second half of the 20th century, the erosive processes mainly correlated with anthropogenic pressure, while recently, natural factors have prevailed, especially sea storms. Regarding the interventions, the effects produced by two interventions carried out during the winter of 2013–2014, one built in urgency between the first and second sea storm and the other built a few years after the second sea storm, were analyzed. This analysis highlighted that the latter intervention was more effective in defending the site.

The eleventh published paper is entitled "The Dune Engineering Demand Parameter and Applications to Forecasting Dune Impacts" by Janssen and Miller [11]. The breaching or overtopping of coastal dunes is associated with greater upland damages. Reliable tools are needed to efficiently assess the likelihood of dune erosion during storm events. Existing methods rely on numerical modeling (extensive investment) or insufficiently parameterize the system. To fill this gap, in this paper, a fragility curve model using a newly developed dune Engineering Demand Parameter (EDP) was introduced. Conceptually, the EDP is similar to the Shield's parameter in that it represents the ratio of mobilizing terms to stabilizing terms. Physically, the EDP is a measure of storm intensity over the dune's resilience. To highlight potential applications, the proposed EDP fragility curve models were fit to a spatially and temporally robust dataset and used to predict dune response subjected to varying storm intensities, including both extratropical and tropical storms. This approach allows for the probabilistic prediction of dune impacts through an innovative, computationally efficient model. Several different forms of the EDP were tested to determine the best schematization of the dune resilience. The final recommended EDP is the Peak Erosion Intensity (PEI) raised to the fourth power over the product of the dune volume and berm-width squared. Including both storm intensity and resilience terms in the EDP enables a comparison of different beach configurations in different storm events, directly fulfilling a need that existing vulnerability assessors cannot currently account for.

The twelfth published paper is entitled "Beach–Dune System Morphodynamics" by D'Alessandro et al. [12]. This study reviews traditional issues and design advances, building bridges between potential risks and adaptation measures. The benefits of nature-based and hybrid solutions, as well as the need for multidisciplinary studies and approaches to promote sustainable and resilient conservation of the coastal environment, are emphasized. Considering the importance and complexity of the subject, this research remains incomplete. It is limited in providing a general overview and outlining some important future directions. Instead, it intends to serve as a springboard for further research in the field of beach–dune system morphodynamics.

Author Contributions: Conceptualization, F.D., G.R.T., and F.F.; writing—original draft preparation, F.D., G.R.T., and F.F.; writing—review and editing, F.D., G.R.T., and F.F. All authors have read and agreed to the published version of the manuscript.

Funding: This research received no external funding.

Conflicts of Interest: The authors declare no conflict of interest.

References

1. Ellis, J.T.; Harris, M.E.; Román-Rivera, M.A.; Brianna Ferguson, J.; Tereszkiewicz, P.A.; McGill, P.A. Application of the Saffir-Simpson Hurricane Wind Scale to Assess Sand Dune Response to Tropical Storms. *J. Mar. Sci. Eng.* **2020**, *8*, 670. [CrossRef]
2. Mickey, R.C.; Dalyander, P.S.; McCall, R.; Passeri, D.L. Sensitivity of Storm Response to Antecedent Topography in the XBeach Model. *J. Mar. Sci. Eng.* **2020**, *8*, 829. [CrossRef]
3. Bogaert, P.; Montreuil, A.-L.; Chen, M. Predicting Morphodynamics for Beach Intertidal Systems in the North Sea: A Space-Time Stochastic Approach. *J. Mar. Sci. Eng.* **2020**, *8*, 901. [CrossRef]
4. Dalyander, P.S.; Mickey, R.C.; Passeri, D.L.; Plant, N.G. Development and Application of an Empirical Dune Growth Model for Evaluating Barrier Island Recovery from Storms. *J. Mar. Sci. Eng.* **2020**, *8*, 977. [CrossRef]
5. Eichmanns, C.; Schüttrumpf, H. Investigating Changes in Aeolian Sediment Transport at Coastal Dunes and Sand Trapping Fences: A Field Study on the German Coast. *J. Mar. Sci. Eng.* **2020**, *8*, 1012. [CrossRef]

6. Leone, E.; Kobayashi, N.; Francone, A.; De Bartolo, S.; Strafella, D.; D'Alessandro, F.; Tomasicchio, G.R. Use of Nanosilica for Increasing Dune Erosion Resistance during a Sea Storm. *J. Mar. Sci. Eng.* **2021**, *9*, 620. [CrossRef]
7. Lemke, L.; Miller, J.K. Role of Storm Erosion Potential and Beach Morphology in Controlling Dune Erosion. *J. Mar. Sci. Eng.* **2021**, *9*, 1428. [CrossRef]
8. Foti, G.; Barbaro, G.; Barillà, G.C.; Frega, F. Effects of Anthropogenic Pressures on Dune Systems-Case Study: Calabria (Italy). *J. Mar. Sci. Eng.* **2022**, *10*, 10. [CrossRef]
9. Pourteimouri, P.; Campmans, G.H.P.; Wijnberg, K.M.; Hulscher, S.J.M.H. A Numerical Study on the Impact of Building Dimensions on Airflow Patterns and Bed Morphology around Buildings at the Beach. *J. Mar. Sci. Eng.* **2022**, *10*, 13. [CrossRef]
10. Barbaro, G.; Foti, G.; Barillà, G.C.; Frega, F. Beach and Dune Erosion: Causes and Interventions, Case Study: Kaulon Archaeological Site. *J. Mar. Sci. Eng.* **2022**, *10*, 14. [CrossRef]
11. Janssen, M.S.; Miller, J.K. The Dune Engineering Demand Parameter and Applications to Forecasting Dune Impacts. *J. Mar. Sci. Eng.* **2022**, *10*, 234. [CrossRef]
12. D'Alessandro, F.; Tomasicchio, G.R.; Frega, F.; Leone, E.; Francone, A.; Pantusa, D.; Barbaro, G.; Foti, G. Beach–dune System Morphodynamics. *J. Mar. Sci. Eng.* **2022**, *10*, 627. [CrossRef]

Journal of
Marine Science and Engineering

MDPI

Article

Application of the Saffir-Simpson Hurricane Wind Scale to Assess Sand Dune Response to Tropical Storms

Jean T. Ellis *, Michelle E. Harris, Mayra A. Román-Rivera, J. Brianna Ferguson, Peter A. Tereszkiewicz and Sean P. McGill

Department of Geography, University of South Carolina, Columbia, SC 29208, USA; mh28@email.sc.edu (M.E.H.); mayrar@email.sc.edu (M.A.R.-R.); jbf2@email.sc.edu (J.B.F.); petert@email.sc.edu (P.A.T.); spmcgill@email.sc.edu (S.P.M.)

* Correspondence: jtellis@sc.edu; Tel.: +1-803-777-1593

Received: 24 July 2020; Accepted: 18 August 2020; Published: 1 September 2020

Abstract: Over one-third of the Earth's population resides or works within 200 km of the coast. The increasing threat of coastal hazards with predicted climate change will impact many global citizens. Coastal dune systems serve as a natural first line of defense against rising sea levels and coastal storms. This study investigated the volumetric changes of two dune systems on Isle of Palms, South Carolina, USA prior to and following Hurricanes *Irma* (2017) and *Florence* (2018), which impacted the island as tropical storms with different characteristics. *Irma* had relatively high significant wave heights and precipitation, resulting in an average 39% volumetric dune loss. During *Florence*, a storm where precipitation was low and winds were moderate, net volumetric dune loss averaged 3%. The primary driving force causing dune change during *Irma* was water (precipitation and storm surge), and during *Florence*, it was wind (aeolian transport). We suggest that the application of the Saffir-Simpson Hurricane Wind Scale classifications should be reconsidered because different geomorphic responses were measured, despite *Irma* and *Florence* both being designated as tropical storms. Site-specific pre- and post-storm studies of the dune morphology and site-specific meteorological measurements of the storm (wind characteristics, storm surge, precipitation) are critically needed.

Keywords: coastal storms; tropical cyclones; coastal hazards; incipient foredunes; Saffir-Simpson Hurricane Wind Scale; Isle of Palms; South Carolina

1. Introduction

Approximately 3.2 billion people either live or work within 200 km of a coastline [1]. Additionally, it is estimated that more than 200 million people are under threat of extreme sea-level events resulting from coastal storms [2]. During these storms, the beach-dune system serves as a natural defense mechanism for the coastline, which is unfortunately threatened worldwide by storm-induced erosion [3]. Dune response to storms is controlled by storm characteristics and the pre-existing dune morphology [4–7]. Dunes are also weakened from multiple, frequent storm impacts [8–11]. Maintenance of the dunes is imperative not only to the natural environment, but also to the coastal built environment and surrounding communities that provide provisioning, regulatory, cultural, and support functions [12]. Accordingly, understanding the geomorphic response of dune systems to storm events is vital based on the current population distribution, its anticipated growth, and estimations of more frequent and intensifying tropical cyclones [13].

Dunes are a critical sedimentological component of the beach-dune-bar system [14,15]. Storms can have a range of impacts on the dunes, from minor scarping to major overwash or breach events [16,17]. Foredune height and extent, relative to the tidal stage during the storm event, are controlling factors on dune response [16,18,19], and are also directly related to sediment availability [20–23]. Individual characteristics of the storm or dune system can affect the observed geomorphic responses. Generally, they are summarized as the relationships between land elevations, water levels, and the stages of rising storm waters [24].

Many studies have been conducted in barrier island dune systems to describe the observations and mechanisms of storm-based erosion (described in [14]), here we highlight those specific to hurricanes. Pre- and post-*Dennis* (2005) digital elevation models (DEMs) were generated to calculate a 7% volumetric loss along a 2 km long study site along the Florida Panhandle [25]. Following Hurricane *Ivan* (2004, Florida Panhandle), 70% of the incipient foredunes were destroyed and in several of their field study locations, storm overwash was noted [26]. During *Ivan*, it was suggested that wave set-up and swash were the significant contributors to overwash processes and dune erosion [27]. Specific to South Carolina, a 25% volumetric loss and a 9% volumetric gain at 14th and 56th Avenues, respectively, was measured on Isle of Palms following Hurricane *Hugo* (1989) [28]. These surveys were approximately 250 m in shore-perpendicular length (to −1.5 m MSL), and therefore covered the dune, beach, and a portion of the nearshore bar system [28]. Another study investigated *Hugo* along the coast of South Carolina and concluded there was severe beach and dune erosion [29]. They also found that high and continuous dunes served as a solid barrier to coastal inundation [29].

The Saffir-Simpson scale is the industry standard to classify tropical cyclones formed in the Atlantic and northern Pacific (East of the International Date Line). A form of this scale has been used for a half a century since Saffir aimed to provide a similar estimation to the Richter earthquake magnitude scale for hurricane property damage [30]. Simpson enhanced this concept with storm surge estimates, which resulted in the Saffir-Simpson Hurricane Scale (SSHS) [31]. Despite its pervasive use, the SSHS has received substantial criticisms, including saturation at its higher end and issues related to the 'hard transitions' between categories, which then have substantial impacts on decision-makers and/or evacuation orders [32]. Following Hurricane *Katrina* (2005), many have proposed alternatives to the SSHS that consider adding maximum storm wind velocity, storm size, storm surge, and/or offshore bathymetry [32–34]. In response, the original SSHS that considered barometric pressure, storm surge, and maximum sustained wind velocity for 1 min was revised in 2010. The new scale, coined the Saffir-Simpson Hurricane Wind Scale (SSHWS), only considers a 1–5 categorical designation of the hurricane, based on the maximum 1 min sustained wind velocity. Unfortunately, the aforementioned 'hard transition' issue between the categories was not rectified. However, the SSHWS is used as an indicator to predict coastal geomorphic change [6].

Scant research has acknowledged that variability of coastal morphologic change is related to and/or explained by SSHWS or its predecessor [16], and additional research is needed. However, it has been demonstrated that the interactions between the shoreline, storm, and subsequent coastal response are complex, and that site-specific studies are needed (e.g., [34]). Therefore, this study reports field-based geomorphic assessments of two dune systems on Isle of Palms, SC, USA, before and after Hurricanes *Irma* (2017) and *Florence* (2018). These data are used to determine the storm-based volumetric changes. We also compare these changes to the characteristics of the two tropical storms upon impact with Isle of Palms.

2. Study Area

Isle of Palms (IOP, Figure 1) is a barrier island located 15.5 km northeast of Charleston, SC, USA, and is bordered by Dewees Inlet to the northeast and Breach Inlet to the southwest. It is a drumstick barrier island and a mixed energy coast [35]. The tides are semidiurnal, with an average range of 1.5 m and an average spring tidal range of 2.5 m [36]. Onshore wave heights average 0.6 m, and fine-grained quartz sand transports alongshore from northeast to southwest at average rates of 120,000 m^3/yr [37]. The island's beaches have been altered by humans since the 1970s [38]. Most notably, there were substantial nourishments to the NE portion of the island towards Dewees Inlet (NE of 53rd Avenue, Figure 1) in 2010 and 2018 [38]. The 2018 nourishment cost $13.5 million, spanned

~3.8 km, and comprised 1.282 million m^3 of sand [38]. Following hurricanes, it is common on this island to scrape the beach, which is moving sand from the foreshore to the pre-storm foredune line (c.f., [39]).

Figure 1. Study area located on Isle of Palms, SC, USA. The blue and red dots indicate Site A at Pavilion Drive and Site B at 53rd Avenue, respectively. The inset map shows the location of Isle of Palms within SC (red box). Base imagery obtained from Google Earth from September 2018.

Historically, the dunes on the SW end of the island had heights that nearly doubled those found in the NE, at approximately 1.5 and 0.7 m, respectively [35]. The NE end of the island was historically described as unstable and erosional, while the SW end was accretional [40]. Accretion in this area was reinforced by alongshore sediment transport patterns [35] and average low wave energy conditions [41]. However, more recently, the island's SW portion (especially southwest of Site A at Pavilion Drive, Figure 1) is highly erosional (personal observation). There have been high energy events impacting South Carolina annually from 2015–2020. Furthermore, the influence of king tides is increasing (c.f., [39]).

Two dune systems along IOP were investigated. The selected locations are part of a longitudinal study that has been in place since Hurricane *Matthew* (2016) (c.f., [42]). Both sites are located adjacent to public beach access points, one at Pavilion Drive along Front Beach (Site A; Figure 2a) and a second at 53rd Avenue (Site B; Figure 2b). Site A has higher relative pedestrian foot traffic, with 50–150 m of beach in front of the current setback line, depending on the tidal stage [43]. Site B is a wider beach-dune system with 125–190 m of beach-dune complex between the setback line and waterline [43], depending on the tidal stage. This site receives minimal impact from beachgoers because of its proximity to a private, gated community that is located just NE of the study site and extends to the end of the island. Additional details about the dune characteristics are discussed later.

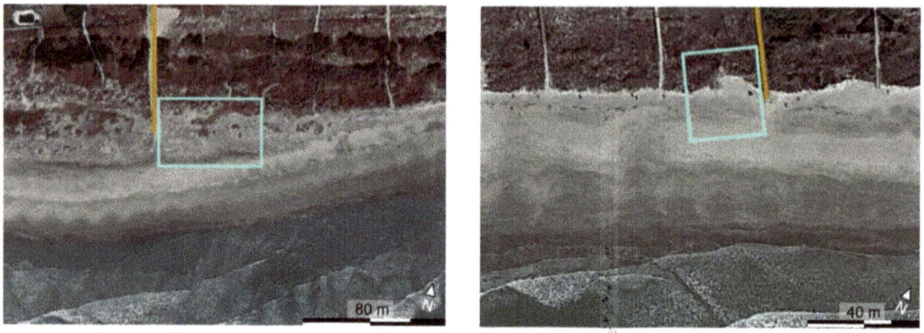

(**a**) (**b**)

Figure 2. Aerial images of (**a**) Sites A and (**b**) B. The yellow colored lines show the location of the pedestrian boardwalk at (**a**) and the 53rd Avenue beach access (**b**). Google Earth imagery from March 2018.

3. Hurricanes *Irma* and *Florence*

This study considers Hurricanes *Irma* and *Florence* (Figure 3). At peak intensity, *Irma* was a Category 5 [44] using the SSHWS. *Irma* first made landfall on U.S. soil at Cudjoe Key, Florida as a Category 3 storm. It traveled up the Florida peninsula and dissipated over Missouri. *Florence* was a Category 4 hurricane at its peak intensity according to the SSHWS [45]. *Florence* made landfall at Wrightsville Beach, North Carolina, approximately 240 km from IOP, as a Category 1 hurricane, using the SSHWS. Tropical storm conditions were measured in South Carolina during *Irma* and *Florence*, despite *Irma*'s eye not traveling into the state [44,45].

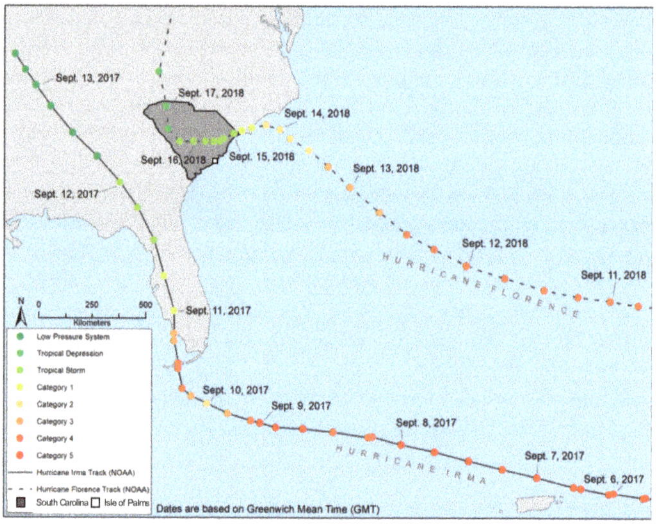

Figure 3. Hurricanes *Irma* and *Florence* path [46,47].

Wind and wave data were obtained from the National Oceanic and Atmospheric Administration's (NOAA) National Data Buoy Center (NDBC) (Station 41004) to temporally encompass the South Carolina-issued hurricane watches for *Irma* and *Florence* and the field surveys conducted for this study. The hurricane watches were 30 August 2017 to 12 September 2017 for *Irma*, and 30 August 2018 to 18 September 2018 for *Florence*. The field survey protocols are described in Section 3. This NOAA

NDBC station is located 73 km southeast of Breach Inlet (Figure 1) and is the closest instrument suite that simultaneously measures wind and waves. NOAA reports a 20 min average significant wave height every hour. Maximum and average 8 s wind gust speeds, measured during 2 min periods, are reported every 10 min from an anemometer 4.1 m above the surface. The maximum wind gust is from the 10 min time interval. All wind and wave data obtained from the NOAA station are averaged to 3 h time blocks for data presentation purposes. Wind and wave data were used to identify possible forcing events, herein defined as >2σ of the hurricane watch temporal duration. The >2σ threshold has been used by other researchers, including those studying dunes at IOP [39,48,49]. Figure 4 shows the NOAA Station 41004 data where the 2σ threshold significant wave heights, wind speeds, and wind gusts during the hurricane watch are denoted by arrows. The temporal duration of the hurricane watch and the date of storm impact on IOP are also noted.

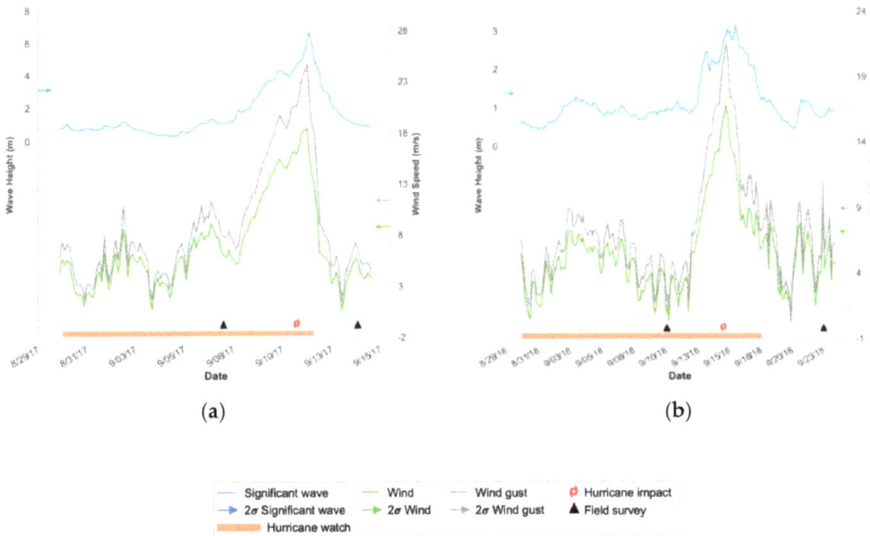

Figure 4. Significant wave heights, wave speeds, and wind gusts during the (**a**) *Irma* and (**b**) *Florence* study periods.

Irma weakened to a tropical storm as it approached the U.S. Southeast. It did not make landfall on the coast of South Carolina, but impacted IOP on 11 September 2017. Mean wind speeds were 7.3 m/s and bidirectional, with peak wind gusts of 30.1 m/s from the NE (Table 1). The Charleston Community Collaborative Rain, Hail & Snow Network (CoCoRaHS) sites recorded an average of 184.25 mm rainfall during the storm [44]. The peak storm surge (storm tide minus the astronomical tide) at the Charleston NOAA National Ocean Service gauge was 1.28 m MHHW [44]. The peak storm tide observed at Charleston Harbor was the third-highest on record at that time, as the storm impacted IOP approximately at high tide [50]. Table 1 provides a summary of storm conditions for Hurricanes *Irma* and *Florence*, as they pertain to conditions on, or nearest available to, IOP.

Table 1. Summary storm conditions during *Irma* and *Florence* obtained from the closest instrument(s) to Isle of Palms.

		Irma	*Florence*
SSHWS		Tropical storm	Tropical Storm
Wind Speed (m/s)	Average	7.3 *	6.0 *
	2σ	8.9 *	7.2 *
Wind Gust (m/s)	Average	9.4 *	7.6 *
	2σ	11.5 *	9.0 *
	Maximum	30.1 *	23.6 *
Dominant Storm Wind Direction		NE *	SSW *
Significant Wave Height (m)	Average	1.8 *	3.1 *
	2σ	1.3 *	1.4 *
Storm Surge (m) MHHW		1.28 °	0.45 #
Storm Precipitation Total (mm)		184.25 °	31.75 #

* NOAA Station 41004; ° [44]; # [45].

Similarly to *Irma*, *Florence* weakened from a hurricane to a tropical storm before impacting IOP on 15 September 2018 when the tides were low. Mean wind speeds were 6.0 m/s (predominantly from the SSW), with peak wind gusts measured at 23.6 m/s that were predominantly from the SSW (NOAA Station 41004; Table 1). *Florence* was a slow, forward-moving storm along its path. This resulted in large precipitation totals and devastating floods, which were evident in north South Carolina and North Carolina [45]. However, closer to the IOP study site (at the Charleston airport), less rain was measured (31.75 mm) during the storm [45]. The peak storm surge at the NOAA's National Ocean Service Charleston gauge was 0.45 m MHHW [45].

4. Assessing Dune Change

Field data collection was executed prior to and after each storm at Sites A and B. Pre- and post-*Irma* data collection was conducted on 7 and 14 September 2017. The pre-hurricane *Florence* survey took place on 10 September 2018, and the post-*Florence* survey was conducted on 23 September 2018.

Topographic data were obtained using a Sokkia Series 30R Total Station, which has an instrument accuracy of +/−2 mm. Points were recorded approximately every meter and at geomorphic breaks along shore-perpendicular transects. At Site A, eight transects were spaced 6.5 m apart over a shore parallel length of 44.5 m (Figure 2a). At Site B, ten transects were spaced 5.5 m apart over a shore parallel length of 55 m (Figure 2b). There was an average of 165 points at Site A and an average of 265 points at Site B recorded during each survey. Site B is a wider beach-dune system in the cross-shore and alongshore directions. Benchmarks and the total station locations were established using an X90-OPUS Static GPS receiver that has a +/−5 mm instrument accuracy.

This study focuses on the dunes, however, measurements were also obtained (and presented) for the beach to understand the erosion and accretion of both systems. The onshore extents of Sites A and B are landward of the pre-Hurricane *Matthew* primary foredune, and are the same baselines that were used by others at the same field sites [42]. The offshore extents of the dunes are defined according to the dune toes, which are dynamic and defined as where there is a distinct change in slope between the dune and backshore. The offshore extents of the beaches are static. The intervening space between the dune toes and the offshore beach lines comprise a narrow zone of the upper backshore that is not inundated by tides. It is recognized that the beach definition used in this paper is not traditional, and it is not the same that is used by others [14,51]. However, there is precedent when studying the same IOP system for using these same beach and dune definitions [42].

The survey points were used to create digital elevation models (DEMs) and DEM-based change maps to estimate beach and dune volumetric change using the above-mentioned beach and dune definitions. DEMs were generated using the ordinary kriging method, with cell and lag sizes of 0.2 m. The semivariogram model was spherical. The average RMSEs for Sites A and B were 0.03 m and 0.02 m, respectively. These averages do not include the Site A post-*Irma* survey because changing the backsite resulted in an RMSE of 0.15 m. The DEM plane height was −1.0 m for both sites. Normalized DEM volumes were calculated by dividing the post-storm by pre-storm volumes. DEM change maps are valuable to identify regions of erosion and accretion over time, which in this case is pre- to post-storm conditions. However, exact survey points and field notes were used to identify the dune toe line.

5. Results

5.1. Site A Beach-Dune System

Before Hurricane *Irma* at Site A, the dune system was generally characterized as a foredune that was recovering following Hurricane *Matthew* (2016) (c.f., [42]). It was a rounded dune ridge with no defined scarp. Vegetation was abundant onshore of the dune crest and portions of the offshore dune slope (Figure 5a–c). Hurricane *Irma* scarped the dune resulting in a near-vertical slope with exposed roots (Figure 5d–f). The storm deposited beach wrack offshore of the post-storm dune toe (Figure 5d–f). The DEM change map (Figure 6a) shows almost consistent spatial volumetric gain onshore of the post-storm dune toe line. Substantial volumetric loss was observed directly onshore of the pre-storm dune toe line. The lateral onshore dune toe erosion from *Irma* ranged from 2.0 to 8.5 m. The total range of change from *Irma*, using a DEM change map was 2.51 m (−1.19 to 1.32 m).

The pre-*Florence* photographs show that the Irma storm scarp is still present and that slumping occurred during the last year (Figure 5g–i). The details of the intervening time between *Irma* and *Florence* are beyond the scope of this paper, but addressed elsewhere (c.f., [42]). Pioneering stabilizing species, such as *Hydrocotyle bonariensis*, were present along the pre-storm dune toe. The most noticeable geomorphic change post-Hurricane *Florence* was the depositional lobe at the scarp base (Figure 5j–l). This depositional lobe is best illustrated in Figure 5i,l that compare the pre- and post-storm conditions and show the post-storm vegetation burial. The *Florence* DEM change map reveals a complex pattern of erosion and deposition ranging from 0.44 to −0.59 m. The post-storm depositional lobe on the NE side of the site (noted above) is captured by the DEM change map (Figure 6b, green box). The DEM change map also reveals erosion along the SW portion of the site. From a quantitative perspective, Site A had a greater loss in dune volume following *Irma* (28%), compared to the volume loss following *Florence* that was statistically negligible (Table 2). There was minimal lateral dune toe change when comparing pre- and post-storm conditions, the maximum retreat approximated 0.5 m.

Table 2. Dune volumes (m^3 and normalized volumes (*nv*)) for pre- and post-*Irma* and *Florence* for Sites A and B.

	Irma		*Florence*	
	Pre-	**Post-**	**Pre-**	**Post-**
SITE A				
Volume (m^3)	1762.2	1437.1	1488.9	1482.1
nv	1.00	0.82	1.00	1.00
SITE B				
Volume (m^3)	4934.6	2017.3	2953.5	2772.7
nv	1.00	0.41	1.00	0.94

Figure 5. Site A field photographs showing conditions (**a–c**) pre-*Irma*, (**d–f**) post-*Irma*, (**g–i**) pre-*Florence*, and (j–l) post-*Florence*. Symbology denotes anthropogenic or natural locations that are consistent to more than one photograph.

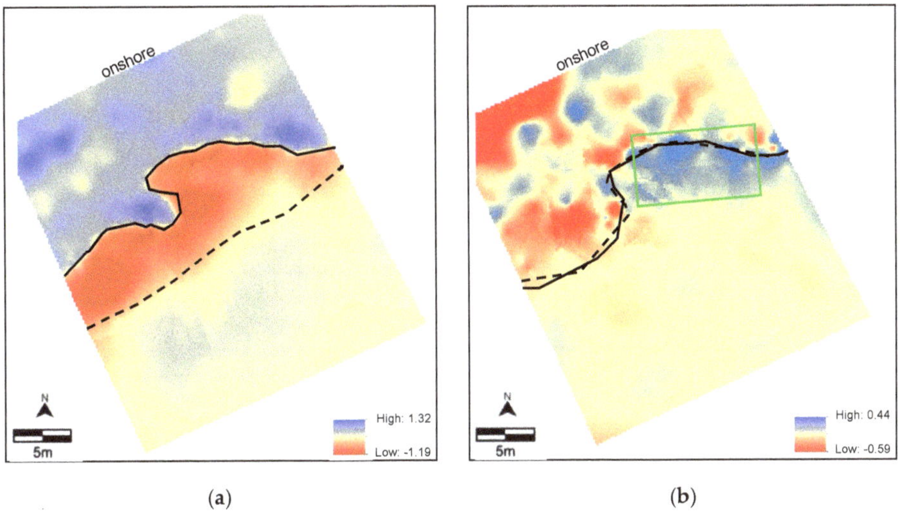

Figure 6. Site A digital elevation models (DEM) change maps for Hurricanes (**a**) *Irma* and (**b**) *Florence*. The dashed and solid lines show the pre- and post-storm dune toe lines, respectively.

5.2. Site B Beach-Dune System

Before Hurricane *Irma* at Site B, there was an incipient dune field with new-growth vegetation (Figure 7a,b). Hurricane *Irma* eroded the offshore portion of the incipient dune field, leaving behind storm wrack (Figure 7c,d). The DEM change map (Figure 8a) confirms the substantial dune erosion. Four shore-parallel incipient foredunes were eliminated, resulting in a maximum of 1.41 m of erosion (Figure 8a, green box). The onshore portion of the study site experienced accretion (up to 0.40 m), with that sand concentrated around a few mounds toward the NE and SE extremes (Figure 8a). Table 2 shows that the normalized volume decreased substantially from 1.00 to 0.41 from the pre- to post-*Irma* surveys. The onshore, lateral dune retreat from *Irma* ranged from 12.0 to 20.0 m.

Prior to *Florence*, the incipient foredune system was visually similar to the pre-*Irma* condition, with new-growth vegetation present (Figure 7e,f). Field observations and the corresponding photographs (Figure 7g,h) suggested that the site was minimally impacted by *Florence*. The solitary vegetation patch visible before and after the storm (Figure 7e,g) remained intact. However, the DEM change map (Figure 8b) reveals a complex spatial pattern of volumetric gains and losses. The magnitude of change ranges from 0.10 to –0.54 m, which is substantially smaller compared to the other location and scenarios considered. Table 2 codifies this low magnitude change; the normalized volume was 0.94. The onshore dune toe retreat from *Florence* was minimal, with an average onshore regression of approximately 1.0 m. The normalized post-storm volume changes at this site were 0.41 for *Irma* and 0.94 for *Florence*.

Figure 7. Site B field photographs showing conditions (**a**,**b**) pre-*Irma*, (**c**,**d**) post-*Irma*, (**e**,**f**) pre-*Florence*, and (**g**,**h**) post-*Florence*. Symbology denotes anthropogenic or natural locations that are consistent to more than one photograph.

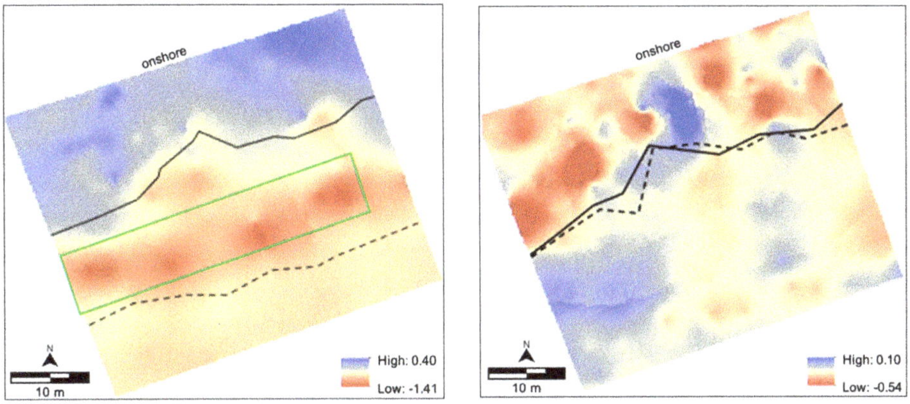

(**a**) (**b**)

Figure 8. Site B DEM change maps for Hurricanes (**a**) *Irma* and (**b**) *Florence*. The dashed and solid lines show the pre- and post-storm dune toe lines, respectively.

6. Discussion

Hurricanes *Irma* and *Florence* impacted the IOP dunes as tropical storms with similar intensities based on SSHWS. However, analysis of the storm characteristics (Table 1 and Figure 4) suggests that the overall system energetics were substantially less for *Florence* compared to *Irma*, except for significant wave height (despite the closer geographic proximity of *Florence* to the field site). For example, during *Irma*, the maximum and average wind gust was 28% and 24% higher, respectively (Table 1). In addition, the precipitation and storm surge was 574% and 284% higher during *Irma* (Table 1). Figure 4 strongly suggests that the storm energetics were limited to the duration of *Irma* and *Florence* and that no substantial geomorphic changes were likely outside the temporal duration of either storm. These data also exemplify the variability of conditions observed in situ, even within the 'tropical storm' classification. Accordingly, even though both systems impacted the IOP dunes as tropical storms (i.e., the same SSHWS designation), *Irma* and *Florence* resulted in different geomorphic responses. This suggests that the geomorphic changes resulting from *Irma* are most indicative of water-based processes and geomorphic changes from *Florence* are largely related to wind-driven, or aeolian processes. The following paragraphs justify this statement and discuss each storm to highlight the different responses with an emphasis on the relationships between the geomorphic change and the dominant forcing factor(s).

Hurricane *Irma* impacted IOP at approximately high tide. A substantial amount of rain fell (184 mm, Table 1), which prohibited aeolian transport, despite the high wind speeds (c.f., [52] for review on the negative impact of moisture on aeolian transport potential). Both sites experienced dune erosion and volumetric loss (averaging 39%), with patterns suggestive of rising water levels and associated storm surge. However, the geomorphic response was different at Site A compared to Site B. The scarping at Site A was likely caused by rising water levels, as evidenced by the high angled slopes (Figure 5d–f). At this site, there was extensive dune scarping (up to 8.5 m); this erosion is the main contributor to the minimum DEM value of −1.2 m and a volumetric loss of 18%. At Site B, *Irma* substantially decreased the total volume of sediment (pre–and post–storm normalized values of 1.00 and 0.41, respectively). The storm-induced erosion was concentrated on four incipient foredunes that decreased in elevation by approximately 1.4 m, from what we believe was the storm surge. The dune toe regressed by up to 20 m. The deposition on the onshore portion of the site was distributed (with a maximum of 0.40 m elevation change) and concentrated to previously low-lying areas. Based on our observations and the associated data analyses, we surmise that during *Irma*, the IOP dunes were most strongly influenced by storm-associated water damage, which was a combination of high tides, storm surge, and precipitation.

When Hurricane *Florence* impacted IOP, the tide was low with obliquely onshore winds (SSW dominant and gusts peaking at 23.6 m/s), and precipitation was limited to ~32 mm (Table 1). The post-storm geomorphic response showed that the dune toe line minimally changed at both sites. At Site A, an accumulation of dry sand was observed offshore of the post-storm dune toe (Figure 6b (grey box)), suggesting that aeolian processes were paramount. A visible gap between the scarp and this accumulation of sand (visible on Figure 5l) strongly suggests that the sand source was offshore of the scarp. However, the NE side of the dune scarp (the right side of the DEM) showed evidence of post-storm deposition, which is consistent with onshore winds. Holistic analysis of this site reveals no net volumetric change when comparing pre- and post-storm conditions (normalized volume change of 1.00). Hurricane *Florence* had minimal impact on Site B. A majority of the post-storm deposition was observed onshore of the dune toe line (within the incipient foredunes), which suggests aeolian transport. We therefore surmise that resulting from *Florence*, the IOP dunes were most strongly influenced by aeolian transport, which also explains the minimal net geomorphic change.

The immediate morphologic response dunes of IOP, following *Irma* and *Florence,* can be compared to other dune systems following hurricanes. At Site B (*Irma*) the incipient foredunes were devastated, which is similar to the response of the incipient foredunes along the Florida coast after *Ivan* [26]. Similar to *Ivan*, we surmise that the *Irma* damage is largely from water [27]. *Hugo* is considered the 'benchmark' storm of the South Carolina coast because of its devastation to IOP and the entire South Carolina coast [28,29]. A previous study [28] reported beach-dune volumetric changes for ~250 m transects (~10x longer than those in this study). Net erosion and accretion were found along the transects closest to our study sites [28]. The dune volumetric loss reported here was 39% for *Irma* and 3% for *Florence*.

7. Conclusions

General assumptions are made about tropical storm intensities and the associated, potential dune damage. For example, the United States Geologic Survey (USGS) Coastal Change Hazards Portal [53] predicts dune damage according to SSHWS. The pre- and post-storm morphologic data (Table 2, Figures 5–8) from two dune sites on Isle of Palms, SC related to two tropical storms (*Irma* and *Florence*) suggest that the general assumptions are misguided because dramatically different geomorphic responses were measured, and these were the same storm category. Our findings are similar to previous research that concluded that stronger tropical cyclones do not equate with greater morphologic coastal change [16].

The findings from this study reveal that the precipitation-dominant *Irma* (as observed on IOP), which also had relatively high significant wave heights, and impacted the Island during high tide, resulted in an average 39% volumetric loss of the IOP dunes. During *Florence*, a storm where precipitation and the tides were low, winds were moderate, and aeolian transport geomorphically impacted the dunes, the net volumetric change was inconsequential (average of 3% volumetric loss). We therefore attest that not only is the magnitude of the tropical storm important, which is summarized by SSHWS, but also the characteristics of the storm, such as wave height, storm surge, wind speed, and precipitation. The latter are not components of SSHWS, but should be in future iterations. In the meantime, the SSHWS should be applied with caution when predicting or characterizing morphologic change of beach-dune systems.

This research emphasizes the importance of not overlooking site-specific geomorphic and meteorological measurements. *Florence,* because of its slow-forward moving speed, resulted in catastrophic flooding over parts of North and South Carolina (c.f., [54]). However, IOP was spared from this devastation. Interestingly, *Irma* produced more precipitation on the island in comparison, affirming the need for the site-specific meteorological measurements. We attest that site-specific measurements of coastal topography, wind characteristics, waves, storm surge, and precipitation should be ascertained.

The research presented herein exhibits that when investigating dune morphology immediately prior to, and following storms, generalizations based on the application of the SSHWS classification are over-simplified. Specifically, during *Irma*, the IOP dunes were most strongly influenced by high tides, storm surge, and precipitation. During *Florence*, the IOP dunes were most strongly influenced by aeolian transport, which was possible due to adequate wind speeds and minimal precipitation. This research demonstrated that two tropical storms impacting Isle of Palms, SC resulted in systematically different geomorphic impacts to the coastal dune system (*Irma* = –39%; *Florence* = –3%). We aspire that future studies will reconsider the application of SSHWS and will include site-specific geomorphic and meteorological measurements, all of which will not only benefit academic research, but also will support and improve coastal management decisions.

Author Contributions: Conceptualization, J.T.E.; methodology, J.T.E., M.A.R.-R.; formal analysis, J.T.E., M.E.H.; field-based data collection, J.T.E., M.A.R.-R., S.P.M., M.E.H., P.A.T., J.B.F; data curation, M.E.H., M.A.R.-R.; writing—original draft preparation, J.T.E.; writing—review and editing, J.T.E.; M.E.H.; M.A.R.-R.; P.A.T.; J.B.F.; S.P.M.; visualization, M.E.H.; J.T.E.; J.B.F.; supervision and project administration, J.T.E. All authors have read and agreed to the published version of the manuscript.

Funding: This research was partially supported by the Department of Geography at the University of South Carolina and HVRI.

Acknowledgments: J.T.E would like to thank IOP Fire Chief Ann Graham, IOP Mayor Jimmy Carroll, and Linda Tucker for their cooperation and support of this research. The research team appreciates the many that helped us in the field: Erika Chin, Holly Gould, April Hiscox, and Alexandria McCombs. J.T.E. and M.E.H. thank ALE and PPC for their unconditional support during this study.

Conflicts of Interest: The authors declare no conflict of interest. The funders had no role in the design of the study; in the collection, analyses, or interpretation of data; in the writing of the manuscript, or in the decision to publish the results.

References

1. The Coastal Population Explosion. Available online: http://oceanservice.noaa.gov/websites/retiredsites/natdia_pdf/3hinrichsen.pdf (accessed on 8 September 2018).
2. Nicholls, R.J. Planning for the impacts of sea level rise. *Oceanography* **2011**, *24*, 144–157. [CrossRef]
3. Harley, M.D.; Ciavola, P. Managing local coastal inundation risk using real-time forecasts and artificial dune placements. *Coast. Eng.* **2013**, *77*, 77–90. [CrossRef]
4. Hall, M.J.; Halsey, S.D. Comparison of overwash penetration from Hurricane Hugo and pre-storm erosion rates for Myrtle Beach and North Myrtle Beach, South Carolina, USA. *J. Coast. Res.* **1991**, *SI8*, 229–235.
5. Claudino-Sales, V.; Wang, P.; Horwitz, M.H. Factors controlling the survival of coastal dunes during multiple hurricane impacts in 2004 and 2005: Santa Rosa Barrier Island, Florida. *Geomorphology* **2008**, *95*, 295–315. [CrossRef]
6. Plant, N.G.; Stockdon, H.F. Probabilistic prediction of barrier-island response to hurricanes. *J. Geophys. Res. Earth Surf.* **2012**, *117*. [CrossRef]
7. Houser, C.; Wernette, P.; Rentschlar, E.; Jones, H.; Hammond, B.; Trimble, S. Post-storm beach and dune recovery: Implications for barrier island resilience. *Geomorphology* **2015**, *234*, 54–63. [CrossRef]
8. Coco, G.; Senechal, N.; Rejas, A.; Bryan, K.R.; Capo, S.; Parisot, J.P.; Brown, J.A.; MacMahan, J.H.M. Beach response to a sequence of extreme storms. *Geomorphology* **2014**, *204*, 493–501. [CrossRef]
9. Karunarathna, H.; Pender, D.; Ranasinghe, R.; Short, A.D.; Reeve, D.E. The effects of storm clustering on beach profile variability. *Mar. Geol.* **2014**, *348*, 103–112. [CrossRef]
10. Dissanayake, P.; Brown, J.; Wisse, P.; Karunarathna, H. Effect of storm clustering on beach / dune erosion. *J. Geol.* **2015**, *370*, 63–75.
11. Angnuureng, D.B.; Almar, R.; Senechal, N.; Castelle, B.; Addo, K.A.; Marieu, V.; Ranasinghe, R. Shoreline resilience to individual storms and storm clusters on a meso-macrotidal barred beach. *Geomorphology* **2017**, *290*, 265–276. [CrossRef]
12. Everard, M.; Jones, J.; Watts, B. Have we neglected the societal importance of sand dunes? An ecosystem services perspective. *Aquat. Conserv.* **2010**, *20*, 476–487. [CrossRef]

13. Webster, P.J.; Holland, G.J.; Curry, J.A.; Chang, H.R. Changes in tropical cyclone number, duration, and intensity in a warming environment. *Science* **2005**, *5742*, 1844–1846. [CrossRef]
14. Houser, C.; Ellis, J. Morphodynamic systems: Beach and dune interaction. In *Treatise on Geomorphology*; Schroder, J., Sherman, D.J., Eds.; Academic Press: San Diego, CA, USA, 2013; pp. 267–288.
15. Román-Rivera, M.A. Innovative Approaches Using Multispectral Imagery to Detect Nearshore Bars and Elucidate Beach System Dynamics. Ph.D. Thesis, University of South Carolina, Columbia, SC, USA, 2019.
16. Sallenger, A.H. Storm impact scale for barrier islands. *J. Coast. Res.* **2000**, *16*, 890–895. [CrossRef]
17. Hesp, P. Foredunes and blowouts: Initiation, geomorphology and dynamics. *Geomorphology* **2002**, *48*, 245–268. [CrossRef]
18. Thieler, E.R.; Young, R.S. Quantitative evaluation of coastal geomorphological changes in South Carolina after Hurricane Hugo. *J. Coast. Res. Spec. Issue* **1991**, *8*, 187–200. [CrossRef]
19. Houser, C.; Hapke, C.; Hamilton, S. Controls on coastal dune morphology, shoreline erosion and barrier island response to extreme storms. *Geomorphology* **2008**, *100*, 223–240. [CrossRef]
20. Leatherman, S.P. Barrier island dynamics: Overwash processes and eolian transport. In *Coastal Engineering 1976, Proceedings of the 15th International Conference*; American Society of Civil Engineers: Reston, VA, USA, 1976; pp. 1958–1974.
21. Leatherman, S.P. Barrier dune systems: A reassessment. *Sediment. Geol.* **1979**, *24*, 1–16. [CrossRef]
22. Armon, J.W. Dune erosion and recovery on a northern barrier. In *Coastal Zone '80, Proceedings of the 2nd Symposium on Coastal and Ocean Management*; American Society of Civil Engineers: Reston, VA, USA, 1980; pp. 1233–1250.
23. Psuty, N.P. Spatial variation in coastal foredune development. In *Coastal Dunes: Geomorphology, Ecology and Management for Conservation*; Carter, R.W.G., Curtis, T.G.F., Sheehy-Skeffington, M.J., Eds.; Balkema: Rotterdam, The Netherlands, 1992; pp. 3–13.
24. Morton, R.A. Factors controlling storm impacts on coastal barriers and beaches—A preliminary basis for near real-time forecasting. *J. Coast. Res.* **2002**, *18*, 486–501.
25. Priestas, A.M.; Fagherazzi, S. Morphological barrier island changes and recovery of dunes after Hurricane Dennis, St. George Island, Florida. *Geomorphology* **2010**, *114*, 614–626. [CrossRef]
26. Claudino-Sales, V.; Wang, P.; Horwitz, M.H. Effect of Hurricane Ivan on coastal dunes of Santa Rosa Barrier Island, Florida: Characterized on the basis of pre- and poststorm LIDAR surveys. *J. Coast. Res.* **2010**, *26*, 470–484. [CrossRef]
27. Wang, P.; Kirby, J.H.; Haber, J.D.; Horwitz, M.H.; Knorr, P.O.; Krock, J.R. Morphological and sedimentological impacts of Hurricane Ivan and immediate poststorm beach Recovery along the Northwestern Florida barrier-island coasts. *J. Coast. Res.* **2006**, *226*, 1382–1402. [CrossRef]
28. Katuna, M.P. The effect of Hurricane Hugo on the Isle of Palms, South Carolina: From destruction to recovery. *J. Coast. Res. Spec. Issue* **1991**, *8*, 263–273.
29. Coch, N.K.; Wolff, M.P. Effects of Hurricane Hugo storm surge in coastal South Carolina. *J. Coast. Res.* **1991**, *SI*, 201–226.
30. Saffir, H.S. Hurricane wind and storm surge. *Military Eng.* **1973**, *423*, 4–5.
31. Simpson, R.H. The hurricane disaster-potential scale. *Weatherwise* **1974**, *27*, 169–186.
32. Kantha, L. Time to replace the Saffir-Simpson Hurricane Scale? *Eos Trans. Am. Geophys. Union* **2006**, *87*, 3–6. [CrossRef]
33. Powell, M.D.; Reinhold, T.A. Tropical cyclone destructive potential by integrated kinetic energy. *Bull. Am. Meteorol. Soc.* **2007**, *88*, 513–526. [CrossRef]
34. Anthony, E.J. Geomorphology storms, shoreface morphodynamics, sand supply, and the accretion and erosion of coastal dune barriers in the southern North Sea. *Geomorphology* **2013**, *199*. [CrossRef]
35. Hayes, M.O.; Moslow, T.F.; Hubbard, D.K. *Beach Erosion in South Carolina*; Coastal Research Division, Department of Geology, University of South Carolina: Colombia, SC, USA, 1978; p. 99.
36. Tides & Currents: Isle of Palms Pier, SC–Station ID 8665494. Available online: https://tidesandcurrents.noaa.gov/stationhome.html?id=8665494 (accessed on 24 July 2020).
37. Kana, T.W. Beach erosion during minor storms. *J. Waterway Port Coast. Ocean Division* **1977**, *103*, 505–518.
38. CSC (Coastal Science & Engineering). *Final Report 2018 Beach Restoration Project City of Isle of Palms*; Technical Report #CSE–2453FR; CSE: Columbia, SC, USA, 2018; p. 57.

39. Ellis, J.T.; Román-Rivera, M.A. Assessing natural and mechanical dune performance in a post-hurricane environment. *J. Mar. Sci. Eng.* **2019**, *7*, 126. [CrossRef]
40. Stephen, M.F.; Brown, P.J.; Fitzgerald, D.M.; Hubbard, M.K.; Hayes, M.O. *Beach Erosion Inventory of Charleston County, South Carolina: A Preliminary Report*; Technical Report No. 4; S.C.; Sea Grant: Charleston, SC, USA, 1975; p. 79.
41. Fico, C. Influence of Wave Refraction on Coastal Geomorphology - Bull Island to Isle of Palms, South Carolina. Master's Thesis, Coastal Research Division, Department of Geology, University of South Carolina, Columbia, SC, USA, 1978; p. 190.
42. Ellis, J.T.; Román-Rivera, M.A.; Harris, M.E.; Tereszkiewicz, P.A. Two years and two hurricanes later: Did the dunes recover? *Shore & Beach* **2020**. (in print).
43. South Carolina Department of Health and Environmental Control-Ocean & Coastal Management (SCDHEC-OCRM). SC Beachfront Jurisdictional Lines. Available online: https://gis.dhec.sc.gov/shoreline/ (accessed on 5 May 2020).
44. Cangialosi, J.P.; Latto, A.S.; Berg, R. National Hurricane Center Tropical Cyclone Report: Hurricane Florence (AL112017). National Hurricane Center. 30 June 2018. p. 111. Available online: https://www.nhc.noaa.gov/data/tcr/AL112017_Irma.pdf (accessed on 21 December 2019).
45. Stewart, S.R.; Berg, R. National Hurricane Center Tropical Cyclone Report: Hurricane Florence (AL062018). National Hurricane Center. 30 May 2019. 2019; p. 98. Available online: https://www.nhc.noaa.gov/data/tcr/AL062018_Florence.pdf (accessed on 21 December 2019).
46. National Hurricane Center and Central Pacific Hurricane Venter. NHC GIS Archive – Tropical Cyclone Best Track for AL 11207. Available online: https://www.nhc.noaa.gov/gis/archive_besttrack_results.php?id=al11&year=2017&name=Hurricane%20IRMA (accessed on 1 July 2020).
47. National Hurricane Center and Central Pacific Hurricane Venter. NHC GIS Archive – Tropical Cyclone Best Track for AL 062018. Available online: https://www.nhc.noaa.gov/gis/archive_besttrack_results.php?id=al06&year=2018&name=Hurricane%20FLORENCE (accessed on 1 July 2020).
48. Harris, M.E.; Ellis, J.T.; Barrineau, C.P. Evaluating the geomorphic response from sand fences on dunes impacted by hurricanes. *Ocean Coast. Manage* **2020**. (in print). [CrossRef]
49. Zhang, K.; Douglas, B.; Leatherman, S.; The, S.; July, N.; Zhang, K.; Douglas, B.; Leatherman, S. Do storms cause long-term beach erosion along the U.S. east barrier coast? *J. Geol.* **2002**, *110*, 493–502. [CrossRef]
50. National Weather Service (NWS). Tropical Storm Irma–September 10-11, 2017. Available online: https://www.weather.gov/chs/TropicalStormIrma-Sep2017f (accessed on 9 March 2020).
51. Masselink, G.; Hughes, M.G. *Introduction to Coastal Geomorphology and Processes*; Arnold: London, UK, 2003; p. 354.
52. Ellis, J.T.; Sherman, D.J. Fundamentals of aeolian sediment transport: Wind blown sand. In *Treatise on Geomorphology*; Schroder, J., Lancaster, N., Sherman, D.J., Baas, A.C.W., Eds.; Academic Press: San Diego, CA, USA, 2013; Volume 11, pp. 85–108.
53. USGS Coastal Change Hazards. Available online: https://marine.usgs.gov/coastalchangehazardsportal/ (accessed on 9 March 2020).
54. Griffen, M.; Malsick, M.; Mizzell, H.; Moore, L. Historic rainfall and record-breaking flooding from Hurricane Florence in the Pee Dee Watershed. *J. South Carolina Water Res.* **2019**, *6*, 28–35. [CrossRef]

Journal of
Marine Science and Engineering

MDPI

Article

Sensitivity of Storm Response to Antecedent Topography in the XBeach Model

Rangley C. Mickey [1,*], Patricia S. Dalyander [2], Robert McCall [3] and Davina L. Passeri [1]

[1] U.S. Geological Survey St. Petersburg Coastal and Marine Science Center; St. Petersburg, FL 33701, USA; dpasseri@usgs.gov
[2] The Water Institute of the Gulf, Baton Rouge, LA 70802, USA; sdalyander@thewaterinstitute.org
[3] Deltares, 2629 Delft, The Netherlands; Robert.mccall@deltares.nl
* Correspondence: rmickey@usgs.gov

Received: 16 September 2020; Accepted: 19 October 2020; Published: 21 October 2020

Abstract: Antecedent topography is an important aspect of coastal morphology when studying and forecasting coastal change hazards. The uncertainty in morphologic response of storm-impact models and their use in short-term hazard forecasting and decadal forecasting is important to account for when considering a coupled model framework. This study provided a methodology to investigate uncertainty of profile response within the storm impact model XBeach related to varying antecedent topographies. A parameterized island Gaussian fit (PIGF) model generated an idealized baseline profile and a suite of idealized profiles that vary specific characteristics based on collated observed LiDAR data from Dauphin Island, AL, USA. Six synthetic storm scenarios were simulated on each of the idealized profiles with XBeach in both 1- and 2-dimensional setups and analyzed to determine the morphological response and uncertainty related to the varied antecedent topographies. Profile morphologic response tends to scale with storm magnitude but among the varied profiles there is greater uncertainty in profile response to the medium range storm scenarios than to the low and high magnitude storm scenarios. XBeach can be highly sensitive to morphologic thresholds, both antecedent and time-varying, especially with regards to beach slope.

Keywords: numerical modeling; XBeach; morphodynamic; dune; beach; synthetic storm; antecedent topography; uncertainty

1. Introduction

On barrier coasts, the morphodynamic response of barrier islands to major storms is critical in the assessment of coastal resiliency and vulnerability [1]. While storm characteristics are inherently important, antecedent morphology has been shown to be a primary factor controlling beach response to extreme events [2,3]. A variety of models have been developed to aid coastal impact assessment [2,4–7]. The degree of sensitivity of each model to the antecedent morphology (e.g., fore-dune height and width, beach slope and width, etc.) varies depending on model complexity and use. Specifically, simple parameterized models, such as the Sallenger storm impact scale [2], and the Stockdon et al., hurricane response model [8], are sensitive to morphologic thresholds related to antecedent topography since they rely on present or historic data for model inputs of beach slope, as well as dune crest and dune base elevations for impact classification. More complex morphodynamic process-based models, such as XBeach [6], can also be highly sensitive to the initial morphologic conditions [9–11] and simulation results may vary depending on specific characteristics of the pre-storm morphology.

Due to this dependence of modeled beach response to antecedent conditions, efforts to model the response of a barrier island or mainland beach to a storm event would ideally incorporate accurate pre-storm morphology rather than approximate or outdated morphology. The prohibitive cost of acquiring elevation data through LiDAR (Light Detecting and Ranging) or other methods immediately preceding a storm event typically means that there is considerable uncertainty in the antecedent morphology. This uncertainty tends to be largest for the beach and nearshore regions, which can evolve rapidly over timescales ranging from days to weeks and under low magnitude conditions. Although dunes are more stable features that recover on timescales of years following higher magnitude storm events [12,13], antecedent dune morphology for a storm may still have uncertainty associated with accretion over time or erosion during storm events that occur following the most recent elevation survey. There can also be systematic errors or offsets in elevation data that do not result from real-world change, which need to be accounted for and corrected before the data can be used in any analytical method [14].

Recently, there has been a push towards the use of storm impact models to operationally predict coastal hazards [10,15–17], and the use of storm impact models in coupled modeling systems to predict long-term barrier island response [18–21]. Both applications enhance the need to understand the sensitivities of the storm-impact models to antecedent conditions to evaluate how this uncertainty influences the accuracy of predicted storm response, and ultimately the implications for a given application. For example, if the primary objective of the coupled model system is to estimate the vulnerability of a restoration project to storms over a 10-year project lifetime, a recovery model only needs to resolve the pre-storm profile to the level of accuracy that influences the response predicted by the storm-impact model. Additionally, the same concept could be applied to existing but out-of-date morphologic data for regions forecasted to be impacted by storms within an operational framework. Quantifying how uncertainty in initial model elevations propagates through storm impact models will enhance understanding of the predictive skill of model frameworks applied over storm- or decadal-time scales.

In the current study, a methodology was developed to systematically assess the sensitivity and uncertainty of morphologic change resulting from the storm-impact model XBeach [6]. XBeach is a process-based numerical storm-impact model that solves for wind/swell waves, infragravity waves, flow, sediment transport and morphological change, and has been well-validated for storm impacts on barrier islands [22–25]. Evolution of idealized barrier island profiles with varying characteristics of the antecedent topography (beach width, beach slope, fore-dune width and height, fore-dune cross shore position, island base height, and the presence of a berm feature) were simulated under varying hydrodynamic conditions to identify what their effect is on the overall predicted morphological response.

An outline of the paper is as follows. Section 2 (Materials and Methods) describes the methods related to (i) extracting real-world morphologic profile characteristics from LiDAR data; (ii) generation of idealized baseline and idealized modified profiles; (iii) description of simulated storm scenarios and predicted storm impact regime classification; and (iv) numerical model setup. Section 3 (Results) presents the morphological response for all profile scenarios and uncertainty derived from comparison between modified and baseline profiles. Section 4 (Discussion) discusses the morphologic impacts associated with storm magnitude and the identification of uncertainties related to antecedent topography and morphologic response. Section 5 (Conclusions and Future Work) discusses the main conclusions along with future work and expansion of the proposed methodology.

2. Materials and Methods

2.1. Study Area and LiDAR Analysis

This methodology was tested at Dauphin Island, AL, USA, a low-lying barrier island off the coast of Alabama in the northern Gulf of Mexico (Figure 1). Dauphin Island has been impacted by and recovered from several major storms over the past two decades; the storms have historically overwashed and inundated portions of the island [26]. The elevation data from Dauphin Island used in this study provides a range of profile characteristics that describe various island states (pre-storm, post-storm, recovered). Eleven LiDAR datasets were analyzed to determine the variability of the island dune and beach characteristics. This study focused on island features from the western portion of Dauphin Island (west of the post-Hurricane Katrina breach of 2005, commonly referred to as Katrina Cut; Figure 1), which is uninhabited and not as anthropogenically influenced as the area east of Katrina Cut [27]. Dauphin Island represents a composite barrier island formed from Pleistocene barrier ridges on the eastern portion and Holocene sediments comprising the narrow western portion [28]. This western portion of Holocene sediments is highly influenced by the dominant westward littoral drift and is extremely vulnerable to island overwash and breaching during storms [29]. This vulnerability leads to major variations in island morphology throughout the time span of the provided LiDAR dataset due to storm impacts which provide a range of antecedent morphologic conditions for this study.

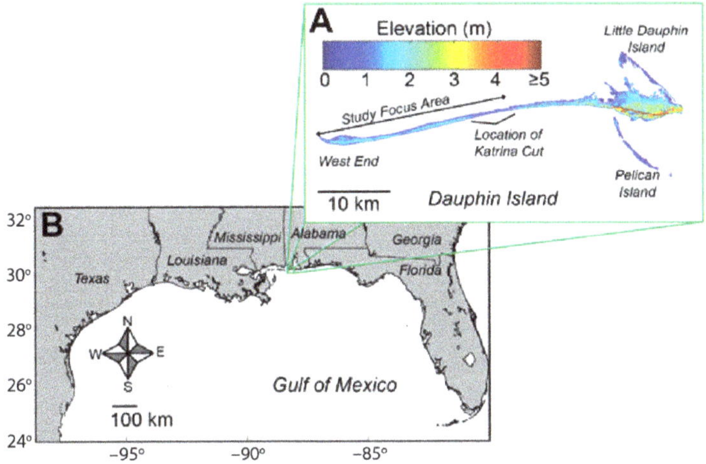

Figure 1. Location of Dauphin Island (**A**) within the northern Gulf of Mexico (**B**) (modified from [21]).

The LiDAR datasets span approximately 10 years from September 2005 to January 2015 [30]. These datasets cover different morphological stages of island configurations, including a post-Hurricane Katrina survey that characterized a degraded island and a recovered configuration of island features represented by the latest survey collected in 2015 [31]. Features already extracted from the LiDAR dataset [30], include fore-dune position, fore-dune height (relative to NAVD 88), fore-dune toe (fore-dune base derived from the location of maximum slope change from fore-dune crest to shoreline identified using methods of [32], and shoreline position taken as the mean high water (MHW) level position at 0.23 m (m; NAVD88) obtained from the National Oceanic and Atmospheric Association's (NOAA) National Data Buoy Center (NDBC) tide gauge at Dauphin Island (8735180). Shore-normal profiles were extracted from gridded LiDAR data with an alongshore resolution of 10 m and a cross-shore resolution of 2.5 m [33]. In some cases, fore-dune height, fore-dune toe, or shoreline position were not identifiable in LiDAR-derived profiles. For example, if measurements were taken at high tide, the MHW line may be obscured; these profiles were excluded from the analyses. Beach width was calculated as the horizontal distance from the fore-dune toe to the shoreline (as illustrated in Figure 2A), and beach slope was calculated using an endpoint method following methods from [33].

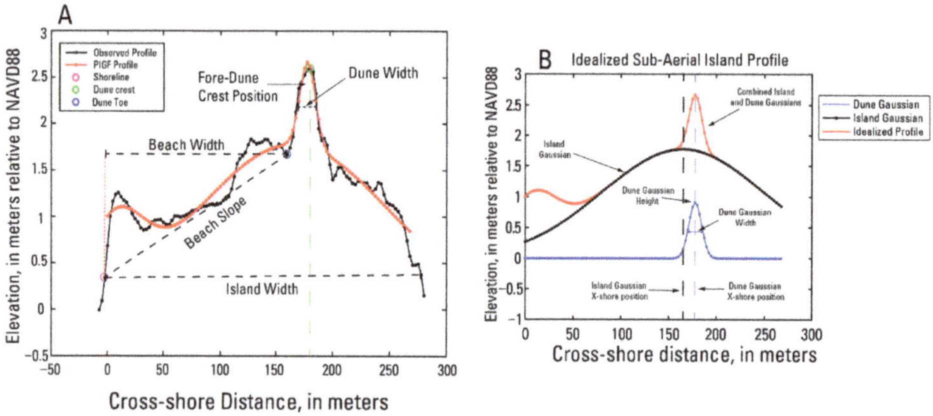

Figure 2. (**A**) Example of parameterized island gaussian fit (PIGF) profile (red) fitted to an observed profile (black) from the June 2008 LIDAR survey, with location indicators where profile features were extracted or calculated. (**B**) Sum of Gaussians with labeled individual parts that make up the final idealized profile in (**A**).

While many characteristics were derived directly from the LiDAR profile, others were obtained by applying a parameterized island Gaussian fit (PIGF) model, developed by [30], which provided a systematic estimation of other characteristics that were not defined previously. The advantage of the PIGF model is it can be applied to large datasets to quickly provide well defined features that robustly characterize a cross-shore profile. The PIGF model fits Gaussian curves to specific features of the observed profiles to generate a smoothed island profile consisting of the sum of multiple Gaussian curves with the profile beginning at the 0 m elevation (NAVD88), designated the 0 m cross-shore position, and terminating at various lengths landward of that depending on the input variable thresholds (set to 0 m elevation). One Gaussian curve represents the sub-aerial island centroid, and one or more curves represent dunes and berms, if present (Figure 2B). Applying the PIGF model provides an estimate of relative fore-dune height above the island centroid Gaussian curve at that dune position compared to the fore-dune height typically extracted that is measured as an absolute elevation above relative datum. Having this relative dune height is important since the absolute height would vary across the profile depending on where the island centroid Gaussian is positioned versus the dune feature position. The PIGF model also provides a more consistent estimation of fore-dune width, calculated as the full width at half maximum of dune Gaussian height versus the dune width based on position of dune toe. Fore-dune and island centroid Gaussian positions were measured relative to the 0 m elevation (NAVD88). Occasionally, the best fit of combined Gaussian curves identified with the PIGF model did not properly identify the island centroid and dune features (for example, the solution amalgamated multiple features into one Gaussian rather than resolving the individual features), see [21] Appendix 2 for model details. These profiles were excluded from analysis by specifying a 100 m maximum threshold for fore-dune widths. Elevation features were adjusted to correct for systematic offsets identified in [14] for these LiDAR surveys.

After island features were extracted, they were averaged and interpolated to 200 m in the alongshore through application of a 400 m wide Hanning window, following methods of [33]. The mean, 5th, and 95th percentile of characteristics over all profiles and times were then calculated to establish the range of values for sensitivity testing (Table 1). Since berms were not a persistent feature in time or space, characteristic values for berm width and height were derived by averaging the Gaussian derived berm heights and widths of selected profiles from the 2015 LiDAR survey wherein berm features were identified (Table 1). A doubling of the mean berm height, to 0.5 m, was established arbitrarily as an endmember "high" value for berm crest elevation.

Table 1. Dauphin Island beach, dune, and island feature height and width means, highs (95th percentile), and lows (5th percentile), in meters (elevations relative to NAVD88; slope in m/m).

Characteristic	Mean	High	Low
		95th %	5th %
Beach Width	61.03	73.33	33.68
Beach Slope	0.022	0.028	0.017
Fore-Dune Height	2.06	2.61	1.24
Fore-Dune Width	33.93	49.41	27.11
Fore-Dune Gauss Height	0.95	1.39	0.40
Fore-Dune Gauss cross-shore Position	153.29	170.57	75.97
Berm Height	0.25	0.50 [1]	-
Berm Width	15.8	-	-
Island Gauss Height	1.33	1.74	0.94
Island Gauss Width	274.11	-	-
Island Gauss cross-shore Position	115.41	-	-

[1] High berm height was arbitrarily set as double the mean value.

2.2. Generation of Baseline and Varied Idealized Profiles

The baseline characteristic profile was generated by applying the mean values for each characteristic listed in Table 1 to a modified version of original PIGF model that includes an additional function for a linear beach portion. This linear beach portion is joined to form a piecewise continuous island profile including the sum-of-Gaussians approximation for the combined island base centroid and fore-dune curves.

$$\begin{aligned} z_{island}(x) &= \left\{ (x > x_T) \times G_{high,cent} \times e^{\left[-\left(\frac{x - G_{x,cent}}{0.6005612 \times G_{width,cent}}\right)\right]^2} \right\} \\ &+ \left\{ (x \leq x_T) \times \left[mx + \left(G_{high,cent} \times e^{\left[-\left(\frac{x_T - G_{x,cent}}{0.6005612 \times G_{width,cent}}\right)\right]^2} - m \times x_T \right) \right] \right\} \end{aligned} \quad (1)$$

Equation (1) uses the cross-shore position of the beach and dune transition point (represented by x_T; derived from shoreline position plus the mean beach width), along with the beach slope (m), island Gaussian cross-shore position ($G_{x,cent}$), height ($G_{high,cent}$), and width ($G_{width,cent}$) to generate the base profile of the island (z_{island}) for the length of the cross-shore coordinates (x) (Figure 2B black dotted line). A Gaussian curve representing the fore-dune was derived from mean parameters (fore-dune cross-shore position, $G_{x,dune}$; height, $G_{high,dune}$; and width, $G_{width,dune}$; Figure 2B blue line) and added to the z_{island} profile to generate the full baseline profile in Figure 2B (red line).

A characteristic offshore elevation profile was generated using a pre-Hurricane Ivan (2004) digital elevation model (DEM) developed by [24]. Offshore bathymetry data from a relatively alongshore uniform section of western Dauphin Island (Figure 3A; red box) was averaged alongshore to provide a mean profile of offshore bathymetry, which was connected to the idealized island profile generated from the modified PIGF model. The averaged offshore profile extended from the 0.23 m MHW line to a water depth of approximately 16 m. To preserve the calculated mean beach width of the idealized island profile portion, the offshore profile portion was adjusted positively or negatively in elevations ≤0.8 m depending on the difference in the shoreward end point elevation of each idealized island profile (Figure 3B); this adjustment provided for a seamless connection to each of the idealized island profiles without adjusting island profile elevations. The cross-shore resolution of the profiles is variable and ranges from approximately 12.5 m offshore to 2.5 m from the shoreline to the back barrier. The baseline and modified profiles represent idealized states of the island to inform how XBeach evolves varying antecedent morphologic conditions.

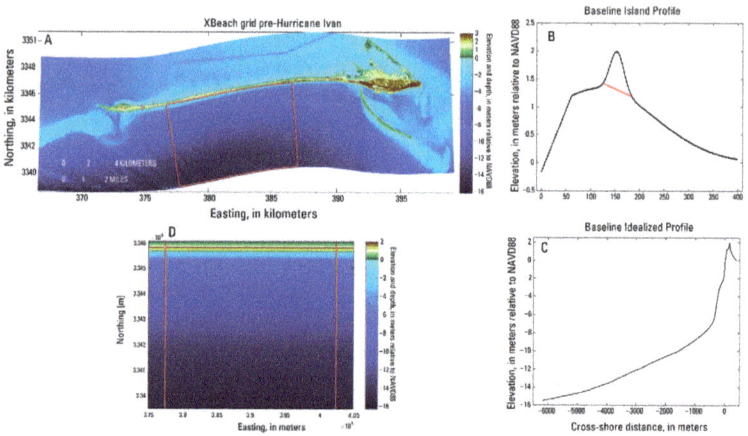

Figure 3. (**A**) Digital elevation model (DEM) of Dauphin Island from 2004, prior to Hurricane Ivan [24]. Red box indicates area used to calculate an alongshore average offshore depth profile to use in the current study. (**B**) Baseline island idealized profile generated using mean values for island, dune, and beach profile characteristics; red line indicates dune footprint area examined in post-storm analyses for dune volume and width changes. (**C**) Seamless topographic and bathymetric idealized profile generating by combining the baseline idealized island profile and alongshore averaged bathymetric profile. (**D**) Alongshore uniform 2D grid generated by replication of the idealized profile in (**C**). Area east and west of the vertical red lines were excluded from analysis because of wave shadowing effects.

The sensitivity of model results to varying initial conditions was tested by individually modifying parameters of the baseline profile by replacement of the mean value with high (95th) and low (5th) percentile values of profile characteristics. To add a berm to the baseline profile, another set of Gaussian values (mean berm height, width, and position) was added to the modified PIGF model. Sensitivity testing was also conducted to evaluate the response of the model to varying heights of the island Gaussian curve, while keeping the absolute height (fore-dune height and island height combined) of the profile fixed. The afore-mentioned modifications resulted in a total of 15 different island profiles: one baseline profile (Figures 3B and 4E); eight profiles based on the 95th and 5th percentile values of beach width, beach slope, fore-dune height, and fore-dune width (Table 1, Figure 4A,B,F,I,J,M,N); two profiles with a berm present seaward of the baseline fore-dune (Figure 4C,D); two profiles with varying island centroid heights but consistent overall island heights (dune plus island centroid; Figure 4K,L), and two profiles with modified dune positions that vary seaward and landward of the baseline position (Figure 4G,H). The modified profile with the low value of fore-dune height was excluded from analysis because the maximum elevation is lower than the elevation of the island centroid at the cross-shore location of the placed fore-dune. Evaluation of observed profiles indicates that this inconsistency results from correlation in island characteristics, specifically that the lowest fore-dune heights correspond to post-storm profiles where the island centroid is also in a deflated state. In general, it should be noted that individual characteristics of realistic profiles will tend to be cross-correlated and defined by environmental forcing (e.g., seasonal steepening and narrowing of the beach, erosion of the island centroid and fore-dune, etc.). For the current study, these parameters were individually modified to allow for sensitivity testing of the model to specific antecedent characteristics. Additionally, the complexity in applying fixed values for inherently correlated features must be noted, with specific regards to varying beach slope while trying to maintain a fixed beach width and vice versa. The modified beach width profiles (Figure 4A,B) illustrate this model complexity through the linear beach portion of the profile generated by the modified PIGF model. This linear portion is wider for the high beach width profile (Figure 4B) compared to the low beach width profile (Figure 4A) while the overall beach width, as described in Figure 2A from shoreline to dune toe, is narrower.

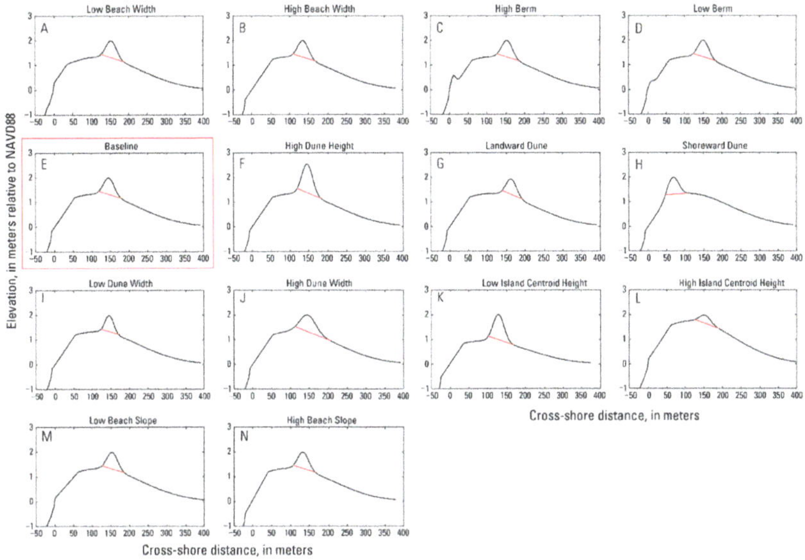

Figure 4. Idealized initial profiles varying island characteristics as described in Table 1; (**A**) low beach width case, (**B**) high beach width case, (**C**) high berm height case, (**D**) low berm height case, (**E**) baseline case, (**F**) high dune height case, (**G**) landward positioned dune case, (H) shoreward positioned dune case, (**I**) low dune width case, (**J**) high dune width case, (**K**) low island centroid height case, (**L**) high island centroid height case, (**M**) low beach slope case, and (**N**) high beach slope case. Red line indicates dune footprint area examined in post-storm analyses for dune volume and width changes (see Supplementary Materials).

2.3. Synthetic Storms for Simulation

Each idealized profile was subjected to six synthetic storm scenarios derived from previous modeling efforts by [21,34], (four from the 2017 study and two from the 2020 study). Each of the six synthetic storms was selected to provide a spread in maximum total water levels (TWL; defined as the combination of wave run-up, tide, and surge water levels relative to NAVD88) and duration of TWL exceeding the height of the foredune toe. The TWL for each synthetic storm was calculated via the parameterization of [5], using the mean fore-shore beach slope from Table 1 along with significant wave height (H_s), peak wave period (T_p), and tide plus surge from the various synthetic storm hydrodynamic time series. The TWL threshold to determine storm duration was set as the exceedance of the average dune toe, 1.47 m (NAVD88), calculated from the LiDAR data set discussed above based on methods from [34]. Four of the synthetic hydrodynamic conditions (open circles in Figure 5A) were derived from historical storm events (1996–2006) that were discretized into 9 bins based on maximum TWL elevation and duration of exceedance above an average dune toe threshold [34]. The wave and tide conditions for all discretized storm events in each of the 9 bins were averaged together to provide a representative scenario of hydrodynamic conditions for each bin. Details for the hydrodynamic conditions of the 9 representative scenarios (wave parameters and water levels) can be found in [34]. Of the 9 representative scenarios developed, 4 were used in this modeling study due to their spread in maximum TWL and duration: storms 1, 2, 3, and 6 (Figure 5A). Based on the spread of duration between storms 3 and 6, two other synthetic storm scenarios (tropical cyclones) were provided to fill this gap since none existed from [34]. The two additional synthetic storms (storms 4 and 5; Figure 5A) were from a suite of 295 synthetic storms from the Federal Emergency Management Agency (FEMA) Risk Mapping Region IV probabilistic coastal storm surge modeling effort for the Alabama and Florida Panhandle coasts [35]. These two storms ranged between the maximum TWL and duration of scenarios

3 and 6 (Figure 5A). It must be noted that the original surge timeseries for the FEMA derived storms did not include astronomical tide. Therefore, a morphological tide sequence, which represents the average tidal conditions based on data from the NOAA NDBC tide gauge at Dauphin Island (8735180), was added to the timeseries so that maximum surge occurred during a high tide phase. Description of the morphological tide sequence applied can be found in [21]. Figure 5A illustrates the relationship between maximum TWL and duration above the dune toe threshold for the six synthetic storms used in this study. Timeseries plots for each synthetic storm are shown in Figure 5B (aligned at the time of maximum TWL) to illustrate how the magnitude and duration of TWL above the 1.47 m dune toe threshold varies between storms.

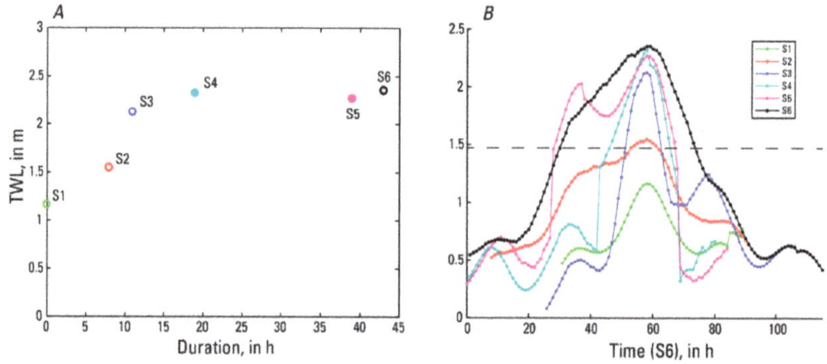

Figure 5. (**A**) Scatter of Total Water Level (TWL) versus duration (hours of TWL above dune toe threshold) for the 6 synthetic storms (S1–S6; open circles represent scenarios from [34], and closed circles from [21]). (**B**) TWL time series of each synthetic storm scenario aligned so that peak TWL coincides with S6 peak TWL; dashed line indicates the 1.47 m dune toe elevation threshold.

Storm 1 (S1), represented a short-lived, low magnitude (maximum TWL = 1.16 m) storm event that did not reach above the dune toe threshold; storm 2 (S2) had slightly higher maximum TWL (1.54 m) and reached above the dune toe threshold for approximately 8 h; storm 3 (S3) had a similar duration (11 h) to S2 with a higher maximum TWL (2.12 m); storm 4 (S4) had a similar maximum (2.32 m) to S3 but with a longer duration above the threshold (19 h); storm 5 (S5), was similar to S3 and S4 in maximum TWL (2.26 m) but had a much longer duration of almost 40 h; and storm 6 (S6) represented another prolonged (43 h), high magnitude (maximum TWL = 2.35 m) storm event. All storms were simulated on each idealized profile using the significant wave height (H_s), peak wave period (T_p), and tide plus surge inputs described in [34] (S1, S2, S3, S6) and [21] (S4, S5). The wave direction for each scenario was fixed to be shore-normal (180 degrees). The predicted storm impact regime based on [2], was determined for each profile for the six storm scenarios using the morphologic threshold of pre-storm dune crest and dune toe for each profile (Table 2) and is presented as a benchmark by which to compare model result variability.

Table 2. Storm impact regime predicted for each idealized profile and storm scenario based on [2]. X marked cells for the low dune height case which was not modeled.

	S = Swash			C = Collison			O = Overwash			I = Inundation								
	Sallenger Regime Classifications																	
	S1			S2			S3			S4			S5			S6		
	Low	Baseline	High	Low	Baseline	High	Low	Baseline	High	Low	Baseline	High	Low	Baseline	High	Low	Baseline	High
Beach Width	S		S	S		S	C		O	O		O	O		O	O		O
Beach Slope	S		S	S		S	C		O	O		O	O		O	O		O
Dune Height	X		S	X		S	X		C	X		C	X		C	X		C
Dune Width	S	S	S	S	S	S	C	C	O	O	O	O	O	O	O	O	O	O
Dune Position	S		S	C		S	O		O	O		O	O		O	O		O
Island Height	S		S	C		S	C		O	O		O	O		O	O		O
Berm present	S		S	S		S	C		C	O		O	O		O	O		O

2.4. Numerical Model Setup: 1D and 2D XBeach

XBeach is a 2D process-based numerical model that solves horizontal equations for wave propagation, flow, sediment transport and bottom changes under varying spectral wave and flow boundary conditions and was developed to study nearshore dune and barrier island response to storm events [6]. Wave group scale equations for roller energy, time-dependent short-wave action balance, nonlinear shallow water mass and momentum, and sediment transport are concurrently solved in XBeach ([6]). XBeach can be set up in a 1D parameter space [17,36–38] or a 2D parameter space [22–25,39] depending on the appropriateness of the application, computational efficiency, and research goals. The 2D setup of the XBeach model has been validated at Dauphin Island by [24] for morphological changes as a result of Hurricanes Ivan and Katrina. Both 1D and 2D setups were utilized in this study using the same model parameters for comparison and evaluation for future study.

The 2D simulations were setup by replicating the 1D idealized profiles in the alongshore with a resolution of 25 m resulting in a rectangular grid 30 km alongshore and approximately 6.5 km cross-shore (Figure 3D). The 2D model setup was used to better approximate the response of the profile considering both alongshore and cross-shore hydrodynamics and sediment transport dynamics [38]. XBeach model parameters for all 2D simulations were setup like those in [34], with the 'single_dir' option in place and a morphological acceleration factor of 10, with all other parameters set to default. For the 2D simulations, a buffer area extending approximately 2.5 km from the lateral boundaries of the model domain was excluded from post-storm topographic analysis to eliminate any wave boundary effects that might impact the profiles (Figure 3D). The 1D setup simulations used the same model parameters, however, preliminary 1D results showed greater erosion compared to the 2D setup. To replicate reasonable corresponding morphological changes from the 2D setup in a 1D setup, sensitivity tests that varied the model bed friction coefficient (BFC) parameter were performed which found that a spatially constant BFC of 44 was better suited than the default BFC parameter of 55.

2.5. Profile Change Metrics

By comparing the variability in profile response between the baseline and modified profiles, the degree to which antecedent morphologic features influence storm-induced morphological change can be identified. Analyses included calculating changes to beach width, beach slope, beach volume, maximum island elevation (dune and island base), island width, shoreline retreat distance, dune width, and dune volume. Post-storm beach characteristics were based on the post-storm derived fore-dune toe and shoreline location. Changes to dune volume and width were calculated based on the changes to the profile within the initial dune footprint, with dune volume (D_v) calculated as the area under the dune portion of the profile (area above red line in Figure 4). Refitting a Gaussian curve to the post-storm profile can introduce fitting errors unrelated to profile response, therefore pre- and post-storm fore-dune width (D_W) was identified as the width of the initial dune footprint pre-storm and the width of the remaining dune feature within the initial dune footprint post-storm; if no portion of the profile was present within the initial dune footprint post-storm D_W was 0. The volume of overwash deposits was calculated as the amount of sediment above the pre-storm profile landward of the dune footprint position. For the 2D simulations, post-storm island characteristics were determined for each alongshore profile within the model domain but excluded the 2.5 km buffer areas on the lateral boundaries. The mean and standard deviation across the analysis area were calculated and compared to the pre-storm profile characteristics for each idealized profile. For the 1D simulations, the post-storm characteristics were derived from the post-storm profile directly for comparison.

3. Results

3.1. Morphologic Response

The pre- and post-storm profiles subjected to the varying storm conditions are shown in Figure 6 (1D) and Figure 7 (2D). The quantitative analysis of morphologic change for each idealized profile under the varying storm conditions for both dimensional setups are shown in Figures 8–11, presented as percent change relative to the pre-storm values. These qualitative and quantitative results were compared to one another for the corresponding storm scenarios and the predicted storm impact regime (Table 2). Additionally, modified idealized profile response was compared to the baseline idealized profile response for each storm scenario to illustrate and identify uncertainty related to antecedent morphology.

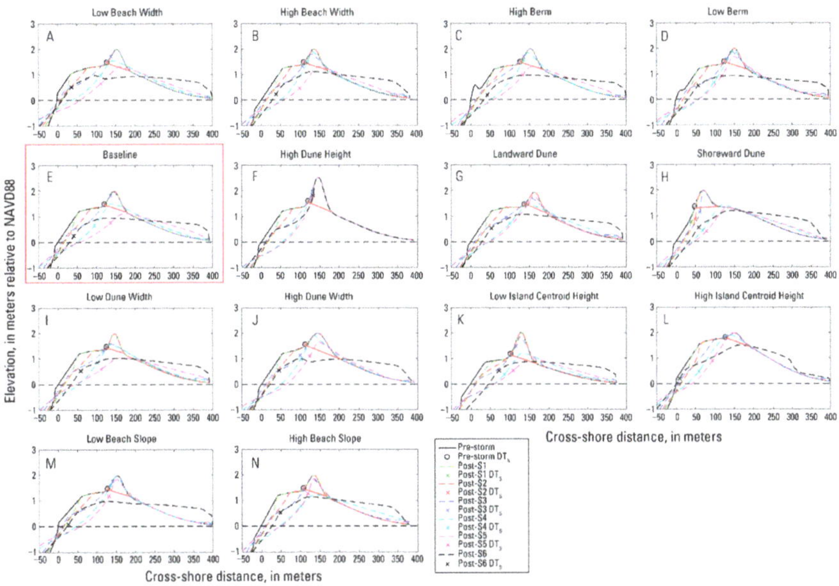

Figure 6. 1D pre- and post-simulation profiles for all 6 storm scenarios with shoreward dune toe position (DT_s) identified. (**A**) low beach width case, (**B**) high beach width case, (**C**) high berm height case, (**D**) low berm height case, (**E**) baseline case, (**F**) high dune height case, (**G**) landward positioned dune case, (**H**) shoreward positioned dune case, (**I**) low dune width case, (**J**) high dune width case, (**K**) low island centroid height case, (**L**) high island centroid height case, (**M**) low beach slope case, and (**N**) high beach slope case. Horizontal dashed line denotes the 0 m elevation line for reference. Red line indicates dune footprint area examined in post-storm analyses for dune volume and width changes.

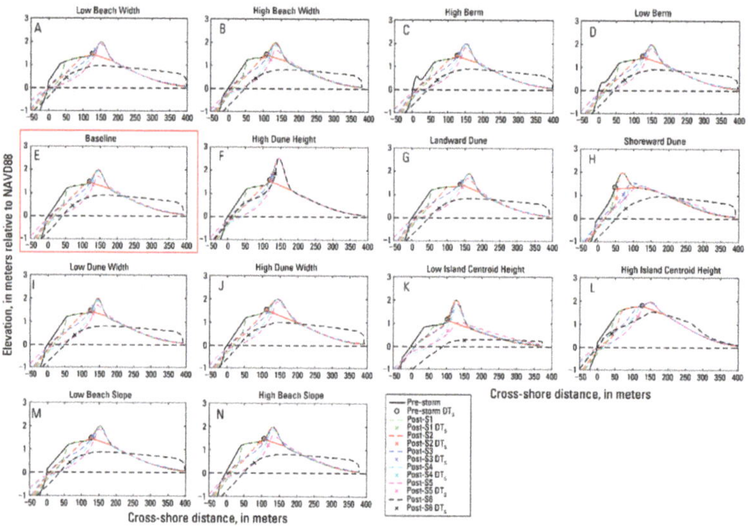

Figure 7. 2D pre- and averaged post-simulation profiles for all 6 storm scenarios with alongshore averaged shoreward dune toe position (DT_S) identified.(**A**) low beach width case, (**B**) high beach width case, (**C**) high berm height case, (**D**) low berm height case, (**E**) baseline case, (**F**) high dune height case, (**G**) landward positioned dune case, (**H**) shoreward positioned dune case, (**I**) low dune width case, (**J**) high dune width case, (**K**) low island centroid height case, (**L**) high island centroid height case, (**M**) low beach slope case, and (**N**) high beach slope case. Horizontal dashed line denotes the 0 m elevation line for reference. Red line indicates dune footprint area examined in post-storm analyses for dune volume and width changes.

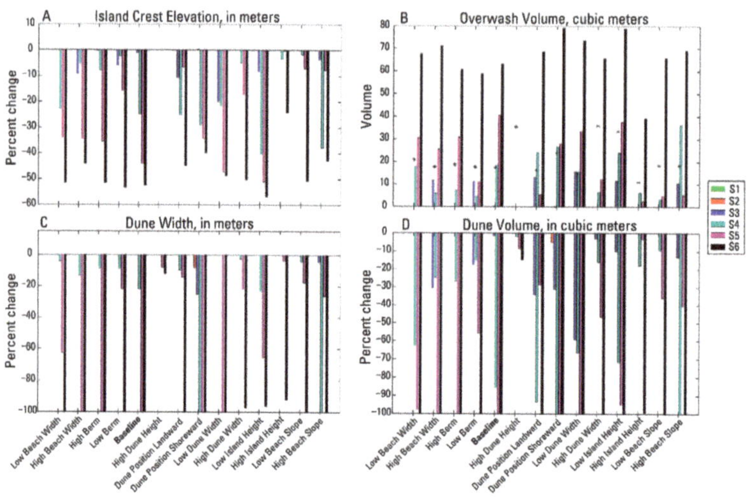

Figure 8. 1D post-storm percent change to the maximum island crest elevation (**A**), dune width (**C**), and dune volume (**D**) for storms 1 through 6 (S1–S6). 1D post-storm calculated overwash volume (**B**); note that asterisks indicate the pre-storm dune volume. Metrics are provided in the panel titles to indicate those features that were analyzed, not the values presented in the graphs. Note: y-axis scales differ among panels.

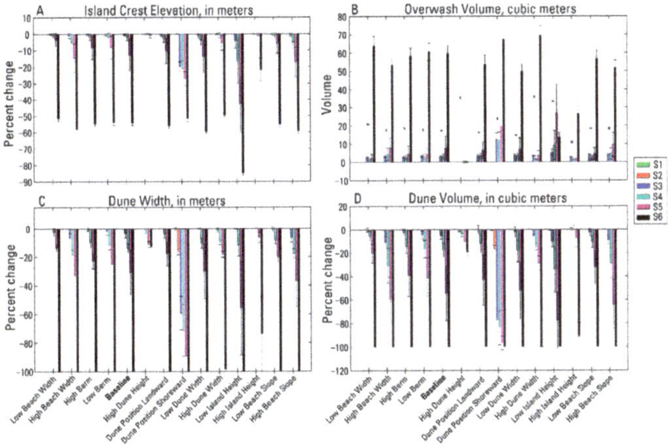

Figure 9. 2D post-storm percent changes to the alongshore averaged (with error bars of standard deviation) maximum island crest elevation (**A**), dune width (**C**), and dune volume (**D**) for storms 1 through 6 (S1–S6). 1D post-storm calculated overwash volume (**B**); note that asterisks indicate the pre-storm dune volume. Metrics are provided in the panel titles to indicate those features that were analyzed, not the values presented in the graphs. Note: y-axis scales differ among panels.

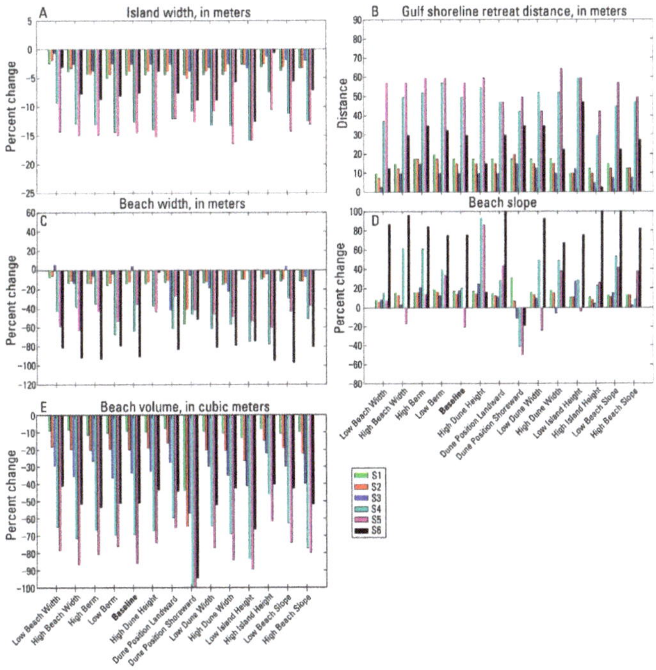

Figure 10. 1D post-storm percent change to island width (**A**), beach width (**C**), beach slope (**D**), and beach volume (**E**) for storms 1 through 6 (S1–S6). 1D post-storm calculated gulf shoreline retreat distance (**B**). Metrics are provided in the panel titles to indicate those features that were analyzed, not the values presented in the graphs. Note: y-axis scales differ among panels.

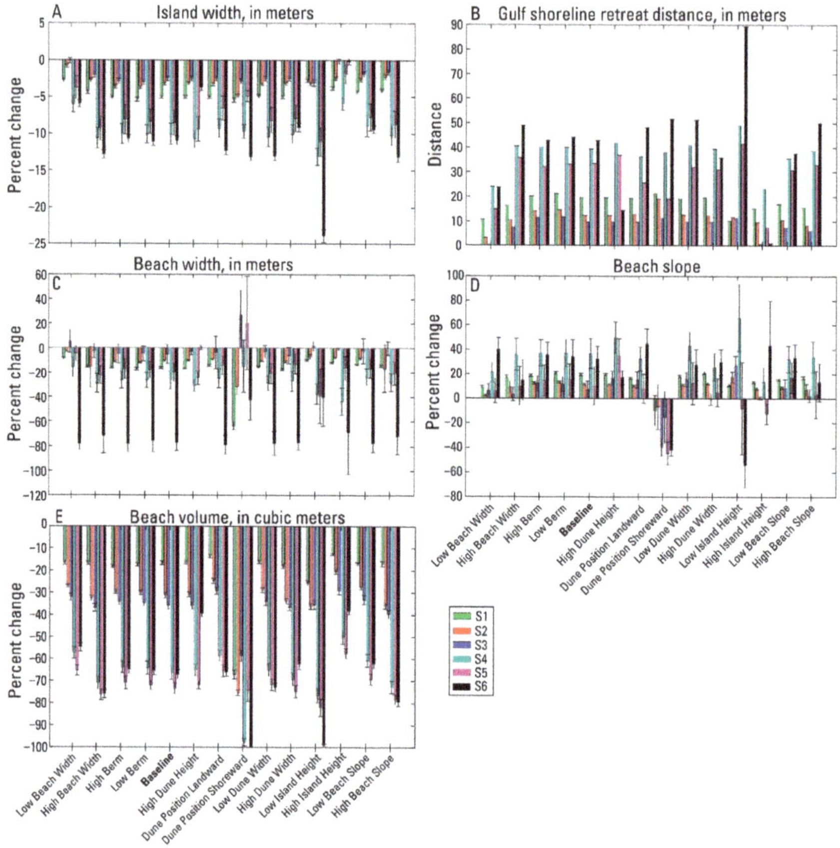

Figure 11. 2D post-storm percent changes to the alongshore averaged (with error bars of standard deviation) island width (**A**), beach width (**C**), beach slope (**D**), and beach volume (**E**) for storms 1 through 6 (S1–S6). 2D post-storm alongshore averaged gulf shoreline retreat distance (**B**). Metrics are provided in the panel titles to indicate those features that were analyzed, not the values presented in the graphs. Note: y-axis scales differ among panels.

The swash regime, in which no wave-dune interactions occur, was predicted for S1 and S2 in most of the profile simulations, except for collision predicted for the shoreward positioned dune and low island height profiles (Table 2). For these two storm scenarios, the beach area was more impacted than the dune feature in both dimensional setups of all profile cases except for the shoreward positioned dune profile (Figures 10 and 11). This profile has a small beach area and a high slope transition between beach and dune features so that even with low water levels dune avalanching can result through erosion of the upper beach and undermining of the dune face.

As storm scenario magnitude and duration increased, the 1D and 2D results began to diverge compared to their predicted impact regime (Table 2). The results for S3 through S5 show that there is a clear dimensional setup difference in morphologic response. The morphologic response of the dune for S3 in the 1D setup was slightly greater than that observed for the 2D setup in a few cases. The landward migration of the dune-toe for S3 shows that dune collision was minor but prevalent among the varying simulations (Figures 6 and 7), which follows the predicted impact regimes. For the profiles predicted to experience overwash during S3 (Table 2), none was observed in the 1D setup and only the shoreward positioned dune profile experienced overwash in the 2D setup. For S4 and S5, the profile and morphological change for the 1D setups were much greater in most of the simulations compared to the 2D setup. The S4 simulations in 1D all had some dune volume loss either through dune avalanching or overwash, the former being more likely for cases when overwash volumes were less than 10 m^3. Conversely, the 2D simulation response for S4 was more indicative of the collision regime based on the low dune volume and island crest elevation changes (Figure 9A,D). A similar pattern of profile response was observed for S5 although to a greater magnitude in both setups with more instances of overwash than in S4. Both dimensional setups for S6 were observed to have high instances of overwash based on morphological response and post-storm profile configuration (Figures 6 and 7) except for the high dune height profile.

3.2. Baseline Deviations and Uncertainty

The relative difference between baseline and modified profile response of various characteristics was analyzed by comparing the ratio of percent change to profile characteristics of all modified profiles to that of the baseline. These ratios are shown in Figure 12 (1D) and Figure 13 (2D) on a logarithmic scale to visualize the magnitude of differences (values equal to 0 were set to 10^{-5} for inclusion in Figures 12 and 13). In some cases, one of the relative difference values was less than 0. In Figures 12 and 13, a black square indicates that the pre- to post-storm difference in the baseline simulation was 0 or positive while the modified profile difference was negative. A black circle indicates the pre- to post-storm difference in the baseline simulation was negative while the modified profile difference was positive. These represent cases where modifications to profile characteristics lead to an opposite effect compared to the baseline profile. For dune volume (DV), island crest elevation (IC), and overwash volume (OV), there were no changes in S1 in either dimensional setup. For S2, there were only changes to the DV for the shoreward positioned dune profile in both setups (note the negative values here are based on no change to baseline profile DV). For DV, there was marked variation in the differences in response of modified profiles compared to the baseline, specifically for the 1D setup in S3, S4, and S5 (Figure 12). The variation in the 1D response difference for S3 of the modified profiles was either 0 or ranged from 1 to 40 times the dune volume change of the baseline profile, while S4–S5 had magnitude differences in some cases slightly greater than 1 but more generally less than 1, indicating the change to the baseline profile was greater than changes to the modified profile for these storm scenarios. A less defined pattern was observed for the 2D setup (Figure 13) where the response difference of S3 was not as pronounced and S4–S5 responses shifted closer to or greater than 1 in more cases compared to the 1D setup. The IC response differences followed a similar pattern as DV with S3 having the largest range of differences in the 1D setup (Figure 12C); the 2D setup showed minimal changes of IC in S3 for a few modified profiles (note the negative values here are based on no change to baseline profile IC), but S4 and S5 had a shift of response differences to greater than 1 (Figure 13C) in more profiles compared to the 1D setup of these simulations. The 1D OV differences, again, had a much higher magnitude range for S3 compared to S4–S5 (Figure 12D), although for the 2D setup the ranges between these three storms were similar and smaller with a shift to a difference in magnitude closer to or greater than 1 (Figure 13D). The response difference to the dune width (DW) was the most diverse among the storm scenarios for dune characteristics, in that no single storm scenario had a patterned response difference compared to others aside from S6 which was close to 1 for most other characteristics as well. S4 and S5 had the greatest difference in response compared to the baseline unlike the other three characteristics previously described for the 1D cases, whereas S3 had the highest range in response differences for the 2D setup (note the negative values in both 1D and 2D are based on no change to baseline profile DW). A total of 18 profile scenarios were evaluated showing change in the dune characteristics, including OV, for the 1D setup and a total of 19 profile scenarios were evaluated showing change in the 2D setup; cases with no difference in baseline and modified profile dune changes were excluded. The profiles with modified dune position and island base height had larger changes to these characteristics compared to the baseline among all the modified profiles evaluated. The shoreward positioned dune profile had the greatest amount of change in 6 of the 18 evaluated cases in 1D (S2: DV, DW; S3: DW, S4: DV, DW; S6: OV) and 12 of the 19 evaluated cases in 2D (S2: DV, DW; S3 and S4: DV, DW, IC, and OV; S5: DV, DW), while the low island height profile had the second highest amount of change in 4 of the 18 evaluated cases in 1D (S4: IC; S5: IC, OV; S6: IC) and 3 of the 19 evaluated cases in 2D (S5: IC, OV; S6: IC). Conversely, the high dune height profile had the lowest amount of change (0 which was set to 10^{-5} on the logarithmic y-axis) compared to the baseline profile for 8 of the 18 cases evaluated in the 1D setup (S4: DV, IC, OV; S5: OV; S6: DV, DW, IC, OV), while the high island height had the lowest amount of change for 2 of the 18 cases evaluated (S5 DV, DW). It is important to note, however, that for the remaining 8 cases, the response difference of 0 was the lowest observed for both these modified profiles. For the 2D setup, this pattern continued but with the high dune height cases having the lowest amount of change compared to the baseline for 6 of the 19 cases evaluated (S3: OV; S4: OV; S5: OV; S6: DV, DW, IC, OV), and 4 cases for the high island height profile (S4: DV, DW; S5: DV, DW), again for the remaining 8 cases, the response difference of 0 was the lowest observed, which was the case for both these modified profiles.

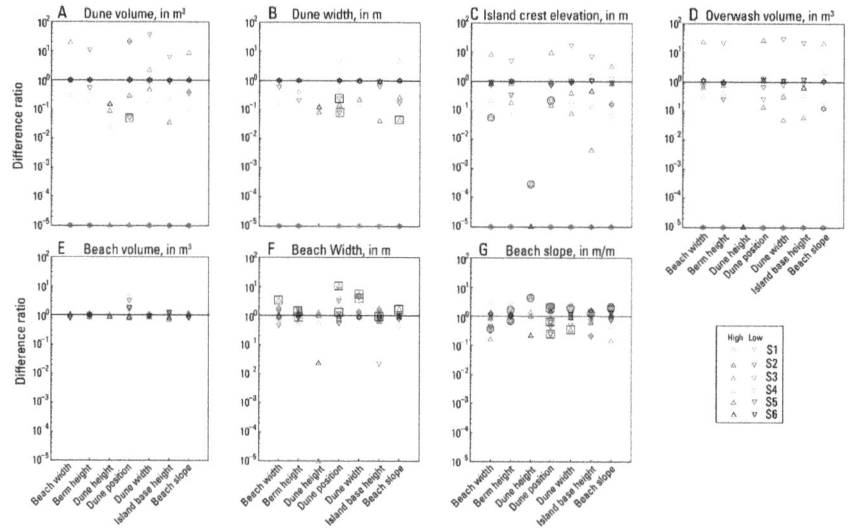

Figure 12. 1D difference ratio of percent change of dune volume (**A**), dune width (**B**), island crest elevation (**C**), overwash volume (**D**), beach volume (**E**), beach width (**F**), and beach slope (**G**) of the baseline and modified profiles; note circled markers around storms markers (S1–S6) indicate a negative relative response difference. Black squares indicate the pre- to post-storm difference in the baseline simulation was 0 or positive and black circles indicate the pre- to post-storm difference in the baseline simulation was negative. Horizontal black line indicates a difference ratio of 1; undefined or 0 ratio values were set to 10^{-5} for inclusion in the figures.

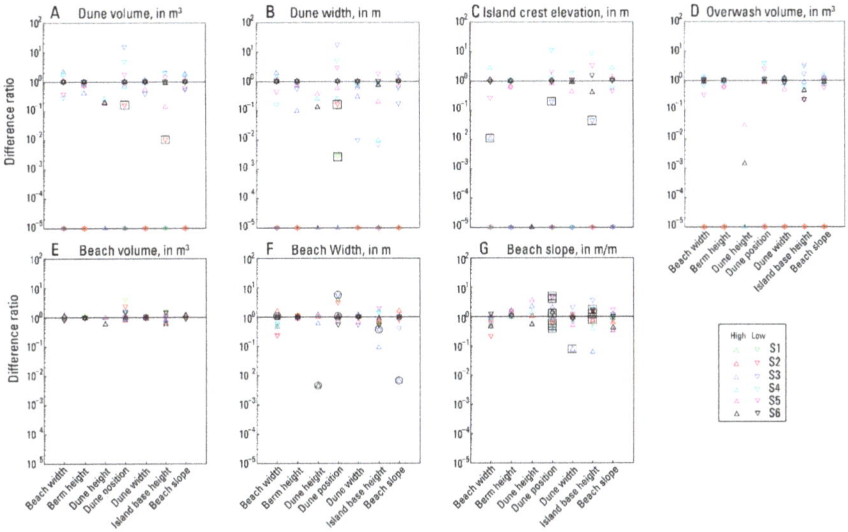

Figure 13. 2D difference ratio of percent change of alongshore averaged dune volume (**A**), dune width (**B**), island crest elevation (**C**), overwash volume (**D**), beach volume (**E**), beach width (**F**), and beach slope (**G**) of the baseline and modified profiles; note circled markers around storm markers (S1–S6) indicate a negative relative response difference. Black squares indicate the pre- to post-storm difference in the baseline simulation was 0 or positive and black circles indicate the pre- to post-storm difference in the baseline simulation was negative. Horizontal black line indicates a difference ratio of 1; undefined or 0 ratio values were set to 10^{-5} for inclusion in the figures.

The response difference of beach characteristics was less variable than those illustrated for dune characteristics. Response differences in beach volume (BV) were minor for all scenarios in both the 1D and 2D setups, except for the shoreward positioned dune profile which ranged up to 4.5 times the baseline for the 1D, and 4 times the baseline for the 2D (Figures 12E and 13E). For beach width (BW), S3 again had the highest range and magnitude of difference in both the 1D and 2D setups (Figures 12F and 13F). Note that S3 had difference magnitudes that ranged from −10 to less than 1 times the baseline for the 1D and −6 to less than 1 times the baseline for the 2D, while other scenarios had ranges slightly less than this for both dimensional setups. Interestingly, the negative values in the 1D and 2D BW comparisons are opposite, with baseline changes being positive in the 1D and negative in the 2D. This, again, suggested that the dimensional setup has some influence on the evolution of the varied profile response. The beach slope (BSL) response difference among scenarios was highly variable with no discernable pattern, S4 had the highest range of values for the 1D cases (4.6), while S3 had a larger range of response difference for the 2D cases (−5.2; Figures 12G and 13G). A total of 18 profile scenarios were evaluated showing change in the beach characteristics for the 1D and 2D setups; cases with no difference in baseline and modified profile beach changes were excluded. The shoreward positioned dune profile dominated the maximum response difference for the beach characteristic with 10 of the 18 evaluated cases in 1D (S1: BV, BW, BSL; S2: BV, BW; S3: BV, BW; S4: BV; S5: BV; S6: BV) and 10 of the 18 evaluated cases in 2D (S1 and S2: BV, BW; S3: BV, BW, BSL; S4: BV; S5: BSL; S6: BV). The high base island height profile was observed to have the least amount of change compared to the baseline for both dimensional setups, with 5 of the 18 cases in 1D (S2–S6: BV) and 8 of the 18 cases in 2D (S1: BV; S2: BV; S3 and S4: BV, BSL; S5 and S6: BV).

4. Discussion

The results of these simulations illustrate how different antecedent profile characteristics and their varying morphologic thresholds can influence the evolution of morphological characteristics, the effect of which scales with storm magnitude and duration for most cases. These relationships are easily identified in the 1D and 2D setups even though there was a marked difference in the amount of erosion in the medium level storm scenarios between these two setups (Figures 8 and 9). Interestingly, the similar response and magnitude of erosion for the beach and dune features for almost all the modified profiles for S1 and S2 in both dimensional setups suggest that XBeach tends to evolve different initial barrier island configurations toward a similar post-storm shape for less energetic storms. Additionally, like low magnitude storm conditions, XBeach simulations of the highest magnitude storm condition (S6) tended to evolve varying initial barrier island profiles to a similar post-storm shape. As for the medium level storm scenarios (S3–S5), dimensional setup differences have been shown in other studies, as well as the sensitivity testing, where the 1D setup of XBeach simulations tend to overpredict wave run up compared to a 2D setup of the same conditions which can directly correlate to beach and dune erosion [37,40]. This is evident by the response to the medium level storm scenarios (S3–S5) in the 2D case which were almost exclusively characterized by dune collision (Figure 7) while in the 1D setup the response was more varied with instances of overwash and extensive dune collapse (Figure 6). These differences in the mid-level storm magnitudes illustrate where non-static morphological thresholds, such as dune crest elevation decreases due to collapse or overwash, become very important. Accounting for the time-varying morphologic change, such as dune avalanching, throughout a simulation is critical to predicting regimes beyond using water level predictions alone. Through extensive modeling efforts with a larger range of storm magnitudes and durations, as well as expansion of the morphological dataset, the quantification of time-varying morphologic change can be captured and explored further for inclusion in coastal change hazard predictions.

For all modified profiles, their comparison to the baseline case for each scenario has shown that uncertainty in profile evolution can be attributed to antecedent morphology, morphologic evolution throughout a storm event, storm magnitude, or a combination of these factors. For dune features, the comparisons shown in Section 3.2 illustrate that while variation exists for all scenarios, the storm scenarios that represent medium magnitude storms (S3–S5) exhibit the greatest differences and therefore lead to more uncertainty. Specifically, the 1D response differences compared to the baseline for DV, IC, and OV in S3 (Figure 12) show that storm magnitude can play a major role in the variation in response for specific cases. The response difference for most of these cases when compared to the baseline is high even among highly varied initial profile morphologies. However, the lack of this pattern in the 2D setup suggests that the results from the 1D setup of the S3 cases could be highly dependent on the dimensional response difference instead of the storm magnitude. Putting dimensional setup aside and focusing on the response deviation of the modified profiles from the baseline shows that varied dune position and dune width are consistently highest for S3 in the 2D and 1D setup, respectively. The shoreward positioned dune has the greatest relative change for DV, DW, IC, and OV compared to the baseline in the 2D setup, while in the 1D setup the low dune width profile had the greatest relative change for DV, IC, and OV. For the latter case, the time-varying evolution of the antecedent morphology through the storm event causes beach slope to increase thereby increasing wave run-up and erosion of the dune feature. The high response difference for the shoreward positioned dune is directly related to the initial profile setup more so than the time varying evolved profile. In this case, the initial morphology of the dune feature was closer to the shoreline which caused earlier interaction between waves and the dune leading to more eroded features compared to the baseline.

For S4 and S5, the varied response in both dimensional setups is more attributable to the antecedent morphology and the time-varying evolution of that morphology throughout the storm scenarios than to the storm magnitude. Specifically, for the 1D cases there was a tendency for response differences in dune characteristics of the modified profiles to be less than or equal to the baseline magnitude (Figure 12). For the modified profiles that had response differences less than 1 in S4, in most cases the time-varying beach slope was greater at the peak of the storm for the baseline profile which likely led to greater wave run-up and dune erosion compared to the modified profiles. Alternatively, in a few of those cases, specifically the high dune height, high dune width, and high island base height, the antecedent morphology played a larger role in reducing the amount of erosion that occurred. These three profiles increased the pre-storm morphologic thresholds of dune crest elevation, dune volume, or dune toe elevation, respectively, which reduced the vulnerability of the dune feature to wave impacts compared to the baseline profile. For S5, the antecedent morphology played a much larger role in reducing the response difference compared to the baseline. Along with the three profiles just mentioned, the initial morphology of the landward positioned dune profile reduced the wave-dune interactions by essentially increasing the beach width and therefore lowering the slope for wave run-up. The other modified profiles that had response differences less than 1, followed the pattern as in S4 with time-varying slope reduction at or before the peak of the storm which likely led to reduced wave run-up compared to the baseline profile. For the 2D simulations, there was less distinction in the difference response for most of the dune characteristics evaluated, however for S4 and S5 there was an observable shift in differences toward the baseline change magnitude. This could be related to the overall difference in the dimensional setups for these two storm scenarios which had a reduction in the amount of erosion of the dune feature (Figures 8 and 9), as well as a reduction in the variability in erosion among the modified profiles.

The comparisons of beach characteristic changes between the baseline and modified profiles are less varied than those associated with dune characteristics. Specifically, the response difference in BV was minimal for almost all modified profiles and all scenarios in both the 1D and 2D setups, the main exceptions being the shoreward positioned dune and low island height profiles. The initial morphology of these two modified profiles is likely the root for these differences due to the volume loss of beach sediment of the initial morphology compared to the baseline. While these analyses compare relative differences based on normalized volumes, the reduction in overall beach volume of these two profiles led to high losses when impacted by the varying storm conditions (Figures 10 and 11). Additionally, for both dimensional setups the scenarios with the greatest differences in BV loss were S1 and S2 which suggests that lower magnitude storms can have a larger effect on the beach area compared to higher magnitude storms that exhibit wave-dune interactions that can transport sediment from the dune to the beach. The changes and response differences for BW and BSL give no clear indication of whether antecedent morphology or storm magnitude play a larger role, however for BW in the 1D simulations there is an observable pattern of S4 having response differences less than 1 in most cases while S5 response differences are mostly greater than 1. This is likely due to the difference in the duration of elevated TWL levels (Figure 5), where S5 had an earlier ramp up of storm conditions which likely increased the beach erosion in most cases compared to the later ramp up time and shorter duration of S4. Conversely, for the 2D setup the response differences were highly variable with increased uncertainty as to which factor plays a more significant role in profile evolution.

The time-varying evolution of beach slope is highly important when accounting for the post-storm changes between modified profiles and storm magnitudes. For example, in two cases for the 1D setup (landward dune and high beach slope), S4 resulted in greater dune erosion than S5 even though S5 has a much longer duration of elevated TWL. Investigation of morphologic evolution through time for these two simulations shows that the long period of elevated TWL before the time of peak TWL during S5 resulted in a shallow beach slope forming before peak TWL, and therefore low wave runup at this time. Conversely, the shorter duration of elevated TWL for S4 did not allow a shallow beach slope to form which caused higher wave run-up comparatively at the time of peak TWL and thus more dune erosion for both the landward dune and high beach slope simulations (Figure 6G,N). This reinforces the importance of accounting for time-varying morphologic characteristics, specifically beach slope, and their effect on predicting barrier island evolution.

5. Conclusions and Future Work

The suite of idealized profiles, synthetic storms, and morphologic responses presented in this paper has provided insight into how the process-based model XBeach evolves different barrier island profiles under the same hydrodynamic conditions. This study has shown that morphologic evolution in XBeach is sensitive to morphologic thresholds related to elevation and slope, both static and time-varying, especially for the medium storm magnitudes outlined in this study. Additionally, the setup of XBeach simulations in 1D has been observed to have larger morphologic changes to most of the profile setups compared to changes from the 2D simulations. Conversely, the model results indicate that XBeach simulation of low and high magnitude storm conditions usually evolve idealized profiles of a low elevation barrier island to a similar post-storm shape regardless of the pre-storm configuration or dimensional setup. However, there is a need for more sensitivity testing as it relates to determining what model parameters need modification in order to reproduce comparable results of 1D and 2D simulations, which are necessary for increasing the computational efficiency for a larger operational framework. The sensitivity to beach slope, whether modified by manipulating antecedent morphology or time-varying throughout a storm event, and its effect on morphologic change was highlighted in a way that magnifies its importance for coastal scientists studying or predicting barrier island morphologic response to storm impacts. This study has shown how varied antecedent morphology can influence differences in storm impact regardless of dimensional setup in XBeach with increases to dune volume (i.e., height or width) or island base height decreasing vulnerability to dune erosion, whereas dunes positioned closer to the shore (or by proxy a narrow beach width) and a reduction in island base height increased dune vulnerability. Building upon and expanding the scope of this work by incorporating larger and multiple dune features, wider beaches, vegetated dunes, nearshore sandbars, presence of a tidal bay or marsh flats, wider range of hydrodynamic conditions, and greater sensitivity analysis of dimensional setup differences could broaden the understanding of storm induced morphological response in XBeach and its use in an operational framework for predicting coastal change hazards.

Future work to expand the range of morphological characteristics, derived from multiple sites, could provide a greater variety of coastal environment representation. In addition, constraining observed data sets to specific island states that represent pre-storm, post-storm, and recovered states in order to provide morphological characteristic values for each state could enhance the representativeness of the idealized profiles to these specific island states and account for the correlation expected between individual profile characteristics. This would inform determination of coastal resiliency, restoration, and identification of vulnerabilities by providing specific profile configurations for the island states and determining how each respond to varying storm impacts. By applying these methods to different and more diverse coastal environments, a database of morphological characteristics, along with their morphological response to various storms could be harnessed to build on the work of others [15–17] to form an operational model for predicting magnitudes of coastal change, as well as a research tool for understanding how diverse coastal systems react to coastal change hazards.

Supplementary Materials: The following are available online at https://doi.org/10.5066/P9VD60JC, 1-dimensional coordinates (UTM Easting and UTM Northing) and topographic/bathymetric elevations for all initial profiles in Figure 4.

Author Contributions: Conceptualization, R.C.M. and P.S.D.; methodology, R.C.M. and P.S.D.; formal analysis, R.C.M., P.S.D., and R.M.; investigation, R.C.M.; writing—Original draft preparation, R.C.M.; writing—Review and editing, P.S.D., R.M., and D.P.; visualization, R.C.M., P.S.D., and R.M.; supervision, P.S.D. and D.P.; project administration, P.S.D. and D.P.; funding acquisition, P.S.D. and D.P. All authors have read and agreed to the published version of the manuscript.

Funding: This research was funded by the U.S. Geological Survey's Coastal and Marine Hazards and Resource Program.

Acknowledgments: R.C.M. would like to thank Ap van Dongeren and Ellen Quataert of Deltares for early help in the review of the methodology. The authors would also like to thank Justin Birchler and the two anonymous reviewers for providing input on this study. Any use of trade, firm, or product names is for descriptive purposes only and does not imply endorsement by the U.S. Government.

Conflicts of Interest: The authors declare no conflict of interest.

References

1. Cañizares, R.; Irish, J.L. Simulation of storm-induced barrier island morphodynamics and flooding. *Coast. Eng.* **2008**, *55*, 1089–1101. [CrossRef]
2. Sallenger, A.H. Storm impact scale for barrier islands. *J. Coast. Res.* **2000**, *16*, 890–895.
3. Morton, R.A. Factors Controlling Storm Impacts on Coastal Barriers and Beaches: A Preliminary Basis for near Real-Time Forecasting. *J. Coast. Res.* **2002**, *18*, 486–501.
4. Larson, M.; Kraus, N.C. *SBEACH: Numerical Model for Simulating Storm-Induced Beach Change Report 1: Empirical Foundation and Model Development*; Technical Report CERC-89-9; Coastal Engineering Research Center, Water-ways Experiment Station: Vicksburg, MS, USA, 1989.
5. Stockdon, H.F.; Holman, R.A.; Howd, P.A.; Sallenger, A.H. Empirical parameterization of setup, swash, and runup. *Coast. Eng.* **2006**, *53*, 573–588. [CrossRef]
6. Roelvink, D.; Reniers, A.; van Dongeren, A.; van Thiel de Vries, J.; McCall, R.; Lescinski, J. Modelling storm impacts on beaches, dunes and barrier islands. *Coast. Eng.* **2009**, *56*, 1133–1152. [CrossRef]
7. Plant, N.G.; Stockdon, H.F. Probabilistic prediction of barrier-island response to hurricanes. *J. Geophys. Res. Earth Surf.* **2012**, *117*, F03015. [CrossRef]
8. Stockdon, H.F.; Sallenger, A.H.; Holman, R.A.; Howd, P. A simple model for the spatially-variable coastal response to hurricanes. *Mar. Geol.* **2007**, *238*, 1–20. [CrossRef]
9. Lindemer, C.A.; Plant, N.G.; Puleo, J.A.; Thompson, D.M.; Wamsley, T.V. Numerical simulation of a low-lying barrier island's morphological response to Hurricane Katrina. *Coast. Eng.* **2010**, *57*, 985–995. [CrossRef]
10. Vousdoukas, M.I.; Ferreira, O.; Almeida, L.P.; Pacheco, A. Toward reliable storm-hazard forecasts: XBeach calibration and its potential application in an operational early-warning system. *Ocean Dyn.* **2012**, *62*, 1001–1015. [CrossRef]
11. Phillips, B.T.; Brown, J.M.; Bidlot, J.; Plater, A.J. Role of Beach Morphology in Wave Overtopping Hazard Assessment. *J. Mar. Sci. Eng.* **2017**, *5*, 1. [CrossRef]
12. Houser, C.; Hamilton, S. Sensitivity of post-hurricane beach and dune recovery to event frequency. *Earth Surf. Process. Landf.* **2009**, *34*, 613–628. [CrossRef]
13. Houser, C.; Wernette, P.; Rentschlar, E.; Jones, H.; Hammond, B.; Trimble, S. Post-storm beach and dune recovery: Implications for barrier island resilience. *Geomorphology* **2015**, *234*, 54–63. [CrossRef]
14. Thompson, D.M.; Dalyander, P.S.; Long, J.W.; Plant, N.G. *Correction of Elevation Offsets in Multiple Co-Located LiDAR Datasets*; Open File Report 2017-1031; U.S. Geological Survey: Reston, VA, USA, 2017; 10p. [CrossRef]
15. Doran, K.S.; Stockdon, H.F.; Sopkin, K.L.; Thompson, D.M.; Plant, N.G. *National Assessment of Hurricane-Induced Coastal Erosion Hazards: Mid-Atlantic Coast*; Open-File Report 2013-1131; U.S. Geological Survey: Reston, VA, USA, 2013; 28p. Available online: http://pubs.usgs.gov/of/2013/1131 (accessed on 24 March 2018).
16. Pearson, S.G.; Storlazzi, C.D.; van Dongeren, A.R.; Tissier, M.F.S.; Reniers, A.J.H.M. A Bayesian based system to assess wave-driven flooding hazards on coral reef-lined coasts. *J. Geophys. Res. Ocean.* **2017**, *122*, 10099–10117. [CrossRef]
17. Sanuy, M.; Jimenez, J.A.; Plant, N. A Bayesian Network methodology for coastal hazard assessments on a regional scale: The BN-CRAF. *Coast. Eng.* **2020**, *157*. [CrossRef]
18. De Groot, A.V.; De Vries, S.; Keijsers, J.G.S.; Riksen, M.J.P.M.; Ye, Q.; Poortinga, A.; Arens, S.M.; Van der Bochev-Burgh, L.M.; Wijnberg, K.M.; Schretlen, J.L.; et al. Measuring and modeling coastal dune development in The Netherlands. In *NCKDays, Crossing Borders in Coastal Research*; Kranenburg, W.M., Horstman, E.M., Wijnberg, K.M., Eds.; NCK: Enschede, The Netherlands, 2012; pp. 105–110.
19. Pender, D.; Karunarathna, H. A statistical-process based approach for modelling beach profile variability. *Coast. Eng.* **2013**, *81*, 19–29. [CrossRef]
20. Roelvink, D.; Costas, S. Coupling nearshore and aeolian processes: XBeach and duna process-based models. *Environ. Model. Softw.* **2019**, *115*, 98–112. [CrossRef]

21. Mickey, R.C.; Long, J.W.; Dalyander, P.S.; Jenkins, R.L., III; Thompson, D.M.; Passeri, D.L.; Plant, N.G. *Development of a Modeling Framework for Predicting Decadal Barrier Island Evolution*; Open-File Report 2019-1139; U.S. Geological Survey: Reston, VA, USA, 2020; 61p. [CrossRef]
22. McCall, R.T.; Van Thiel de Vries, J.S.M.; Plant, N.G.; Van Dongeren, A.R.; Roelvink, J.A.; Thompson, D.M.; Reniers, A.J.H.M. Two-dimensional time dependent hurricane overwash and erosion modeling at Santa Rosa Island. *Coast. Eng.* **2010**, *57*, 668–683. [CrossRef]
23. Sherwood, C.R.; Long, J.W.; Dickhudt, P.J.; Dalyander, P.S.; Thompson, D.M.; Plant, N.G. Inundation of a barrier island (Chandeleur Islands, Louisiana, USA) during a hurricane: Observed water-level gradients and modeled seaward sand transport. *J. Geophys. Res. Earth Surf.* **2014**, *119*, 18. [CrossRef]
24. Passeri, D.L.; Long, J.W.; Plant, N.G.; Bilskie, M.V.; Hagen, S.C. *XBeach Bottom Friction Scenarios: Model Inputs and Results*; U.S. Geological Survey Data Release: Reston, VA, USA, 2017. [CrossRef]
25. van der Lugt, M.A.; Quataert, E.; van Dongeren, A.; van Ormondt, M.; Sherwood, C.R. Morphodynamic modeling of the response of two barrier islands to Atlantic hurricane forcing, Estuarine. *Coast. Shelf Sci.* **2019**, *229*. [CrossRef]
26. Watkins, A.D. *A Synthesis of Alabama Beach States and Nourishment Histories*; University of Alabama: Tuscaloosa, AL, USA, 2011; Available online: http://acumen.lib.ua.edu/u0015/0000001/0000723/u0015_0000001_0000723.pdf (accessed on 11 February 2018).
27. Froede, C.R. Constructed Sand Dunes on the Developed Barrier-Spit Portion of Dauphin Island, Alabama (U.S.A). *J. Coast. Res.* **2010**, *26*, 699–703. [CrossRef]
28. Otvos, E.G. Coastal Barriers, Gulf of Mexico: Holocene Evolution and Chronology. *J. Coast. Res.* **2005**, *42*, 141–163.
29. Morton, R.A. Historical Changes in the Mississippi-Alabama Barrier-Island Chain and the Roles of Extreme Storms, Sea Level, and Human Activities. *J. Coast. Res.* **2008**, *24*, 1587–1600. [CrossRef]
30. Doran, K.S.; Long, J.W.; Birchler, J.J.; Brenner, O.T.; Hardy, M.W.; Morgan, K.L.M.; Stockdon, H.F.; Torres, M.L. *LiDAR-Derived Beach Morphology (Dune Crest, Dune Toe, and Shoreline) for U.S. Sandy Coastlines*; U.S. Geological Survey Data Release: Reston, VA, USA, 2017. [CrossRef]
31. Dalyander, P.S.; Mickey, R.C.; Passeri, D.L.; Plant, N.G. Development of an Empirical Dune Growth Model and Use in Evaluating Barrier Island Recovery from Storms. *Coast. Eng.* under review.
32. Stockdon, H.F.; Doran, K.J.; Thompson, D.M.; Sopkin, K.L.; Plant, N.G.; Sallenger, A.H. *National Assessment of Hurricane-INDUCED Coastal Erosion Hazards: Gulf of Mexico*; Open-File Report 2012-1084; U.S. Geological Survey: Reston, VA, USA, 2012. Available online: http://pubs.usgs.gov/of/2012/1084 (accessed on 24 March 2018).
33. Doran, K.S.; Long, J.W.; Overbeck, J.R. *A Method for Determining Average Beach Slope and Beach Slope Variability for U.S. Sandy Coastlines*; Open-File Report 2015-1053; U.S. Geological Survey: Reston, VA, USA, 2015. [CrossRef]
34. Mickey, R.C.; Long, J.W.; Plant, N.G.; Thompson, D.M.; Dalyander, P.S. *A Methodology for Modeling Barrier Island Storm-Impact Scenarios*; Open-File Report 2017-1009; U.S. Geological Survey: Reston, VA, USA, 2017. [CrossRef]
35. Federal Emergency Management Agency [FEMA]. Comparisons of FEMA Coastal Model Data with Data from Other Coastal Models and Long-Term Tidal Gauges: FEMA. 2014. Available online: http://www.southeastcoastalmaps.com/PublicDocs/Fact%20Sheet_Comparisons%20of%20Coastal%20Models%20and%20Gage%20Analyses_Info%20for%20Engineers_Aug%202014.pdf (accessed on 18 August 2018).
36. Faraci, C.; Scandura, P.; Foti, E. Bottom profile evolution of a perched nourished beach. *J. Waterw. Portcoastaland Ocean Eng.* **2014**, *140*, 04014021. [CrossRef]
37. Stockdon, H.F.; Thompson, D.M.; Plant, N.G.; Long, J.W. Evaluation of wave runup predictions from numerical and parametric models. *Coast. Eng.* **2014**, *92*, 1–11. [CrossRef]
38. Winter, R.C.; Ruessink, B.G. Sensitivity analysis of climate change impacts on dune erosion: Case study for the Dutch Holland coast. *Clim. Chang.* **2017**, *141*, 685–701. [CrossRef]

39. Mickey, R.C.; Long, J.W.; Dalyander, P.S.; Plant, N.G.; Thompson, D.M. A framework for modeling scenario-based barrier island storm impacts. *Coast. Eng.* **2018**, *138*, 98–112. [CrossRef]
40. Roelvink, D.; McCall, R.; Seyedabdolhossein, M.; Nederhoff, K.; Dastgheib, A. Improving predictions of swash dynamics in XBeach: The role of groupiness and incident-band runup. *Coast. Eng.* **2018**, *134*, 103–123. [CrossRef]

Publisher's Note: MDPI stays neutral with regard to jurisdictional claims in published maps and institutional affiliations.

Journal of
Marine Science and Engineering

MDPI

Article

Predicting Morphodynamics for Beach Intertidal Systems in the North Sea: A Space-Time Stochastic Approach

Patrick Bogaert [1,*], Anne-Lise Montreuil [2] and Margaret Chen [2]

1 Department of Environmental Sciences, Earth & Life Institute, Université Catholique de Louvain, 1348 Louvain-la-Neuve, Belgium

2 Hydrology and Hydraulic Engineering, Vrije Universiteit Brussel, Pleinlaan 2, 1050 Elsene, Brussels, Belgium; anne-Lise.Montreuil@vub.be (A.-L.M.); margaret.Chen@vub.ac.be (M.C.)

* Correspondence: patrick.bogaert@uclouvain.be

Received: 1 October 2020; Accepted: 6 November 2020; Published: 11 November 2020

Abstract: The ability to accurately predict beach morphodynamics is of primary interest for coastal scientists and managers. With this goal in mind, a stochastic model of a sandy macrotidal barred beach is developed that is based on cross-shore elevation profiles. Intertidal elevation was monitored from monthly to annually for 19 years through Real Time Kinematics-GPS (RTK-GPS) and LiDAR surveys, and monthly during two years with an RTK-GPS. In addition, during two campaigns of about two weeks, intensive surveys on a daily basis were performed with an RTK-GPS on a different set of profiles. Based on the measurements, space and time variograms are constructed in order to assess the spatial and temporal dependencies of these elevations. A separable space-time covariance model is then built from them in order to generate a large number of plausible future profiles at arbitrary time instants $t + \tau$, starting from observed profiles at time instants t. For each simulation, the total displaced sand volume is computed and a distribution is obtained. The mean of this distribution is in good agreement with the total displaced sand volume measured on the profiles, provided that they are lower than 45 m^3/m. The time variogram also shows that 90% of maximum variability is reached for a time interval τ of three years. These results demonstrate how the temporal evolution of an integrated property, like the total displaced sand volume, can be estimated over time. This suggests that a similar stochastic approach could be useful for estimating other properties as long as one is able to capture the stochastic space-time variability of the underlying processes.

Keywords: time and space variograms; intertidal barred beach morphology; stochastic modeling; space-time covariance model; data-based modeling

---

1. Introduction

Beaches act as a natural buffer to protect hinterland against significant waves and coastal erosion as well as flooding. Understanding and predicting coastal morphodynamics are critical for the safety of the existing ecosystems, infrastructures, and for mitigating expected damages from future changes. Morphodynamics of beach can also affect the biodiversity and biological component, like, e.g., nesting ecology of sea turtles, burrowing of bivalves and presence of sand crabs [1–3]. However, prediction of beach morphology remains a challenge due to the complex processes between forcing factors and sediment transport occurring on a broad range of temporal and spatial scales. These processes can involve regular seasonal events (tidal cycles, current patterns) or stochastic events, like storms. Inherent interactions between evolving morphological features, such as intertidal sand bars and external forcings, produce both linear and nonlinear feedback mechanisms [4]. The beach profile response may be related to the existing coastal state and to the forcings operating on varied spatial and temporal scales [5–7]. On short-time scale, ranging from hours to months, wave storms hitting the coast may be the dominant processes that impact the beach changes and sediment transport processes, while sea level change and sediment fluxes play a more important role on longer-time scale. In developing models for predicting beach morphology, it is thus crucial to distinguish between behaviour and evolution.

Over the last decades, complex process-based morphological models have been developed to simulate and predict beach morphological changes (e.g., [8–10]). Although they have performed well for short-term predictions, their computational costs, modelled instability, and misspecifications of the physics conditions prevent them from reproducing beach evolution. Therefore, morphological tendencies that are observed in practice are often difficult to model reliably [11]. This has encouraged the development of data-driven models that are statistical methods whose entire structure and associated parameters are estimated directly from the objective analysis of observational data. General characteristics of the natural system, such as beach width, volume, and slope, are deduced only based on data analysis without considering other sources of information about the system [12]. The key advantage is that no prior knowledge of the system dynamics is required which is appropriate when the dynamics are not fully understood, as well as their long-term simulation stability and robustness [13]. Stochastic approaches for assessing beach changes have been modeled early while using Markov-type processes [14,15]. Geostatistical estimation of beach properties that are based on variogram method [16], linear [4], and nonlinear [17–19] data-based modelling techniques have been applied in the past in order to investigate beach morphology, mostly in a purely spatial mapping framework. Additionally, the beach equilibrium model is another parametric method which has been widely accepted and often used as a proxy for shoreline (e.g., [20–22]) and more recently to study the variation of beach profile position [23]. The success of data-based modelling depends on the total size of the observational data set, characterized by its specific spatial coverage, its spatial and temporal resolution and its overall length in time. A blind testing comparison of various models (ranging from established models to machine learning algorithms) making use of cross-shore profiles or shorelines is provided by Montaño et al. [24].

As emphasized by [25], two approaches have been generally adopted to better study beach morphodynamics: long-term approach that is based on seasonal to decadal data, and recovery from dramatic events. This study focuses on the morphodynamics of sandy beaches characterized by an intertidal bars-trough system commonly found along meso- and macro-tidal environments by taking advantage of high-quality data sets (with high spatial resolution and a combination of low to high temporal resolutions) over a time span of 19 years along the Belgian coast. The aim is to investigate the benefit of using a space-time stochastic modeling approach to capture both the spatial variability and temporal evolution of the beach elevation along profiles. Proxies of the beach evolution, such as dune foot position, limit of wet and dry sand, shoreline position, or isocontours (e.g., [26–28]) are out of reach, as our dataset is restricted to one dimensional measurements repeated along the same few cross-shore profiles over time. The developed space-time model is thus used to predict the integrated sand volume change over time along these profiles. Relationships between the morphological development of the multi-barred beaches and controlling factors as meteo-marine conditions will be subject of future studies, as the main goal of the paper is to assess the ability of such a space-time model to predict sand volume change from data that where mostly sampled every few months, thus precluding the direct use of the direct use of meteo-marine factors at a much finer time scale, as done in [25]. As noted by [16], accurate sand volume estimation is required to monitor rates of beach erosion and accration, so sand volume change assessment is particularly relevant for beach preservation and management.

2. Study Site

The North Sea is an arm of the Atlantic Ocean, with a majority of ocean current coming from the northwest opening, while a lesser portion of warm current comes from the English Channel. The main flow pattern of water is an anti-clockwise rotation along its edges. The fetch length is limited so that wave period is shorter, and amplitudes are higher than in deep ocean water, as the North Sea is located on the continental shelf. The Belgian coast is boarding the east part of North Sea and is about 65 km long, mainly oriented from South-West (SW) to North-East (NE). The typical wind speed is about 8 m/s, with winds most frequently coming from SW. The time averaged significant wave height is about 100 cm, with typical wave period of about 6 s. Most of the time waves and winds come from the same direction. Waves that propogate in the shallow Belgian coastal zone are transformed due to the interaction with the sea bottom, so that wave climate varies from place to place. Frequent storms occur during the winter months, where most severe events are characterized by wave heights exceeding 4 m and storm duration of more than 30 h. For several centuries, anthropogenic activities along the Belgian coastline have attempted to reshape the coast to maintain the coastline position, or even to claim land from the sea, so that the natural interaction between the beach and coastal dunes has been disrupted. The protection of the Belgian coastline is a combination of natural and artificial defence [29]. Nowadays, 60% of the coast has a hard coastal protection with seawalls, revetments and groynes [30]. Belgian beaches mainly consist of fine to medium sand and mainly characterized by quartz enriched by calcareous particles, with an average median grain size of 200–220 μm and a cross-shore variation in median grain size of 160–380 μm. Sediments tend to become coarser from the west to the east [31].

Our study site is located at Groenendijk beach (sections 48–50, as defined by the Coastal Division of the Flemish Authority) situated in the western part of the Belgian coast (Figure 1). This part of the coast is oriented SW-NE and is characterized by sandy dissipative beaches (surf scaling parameter >20) with a gentle slope of less than 1% and a width >300 m. There is a northeastward longshore sediment drift in this zone, with a total longshore SW-NE transport evaluated at 145,000 m^3/year along the Belgian coast [32]. The beach is backed by well-developed coastal dunes of 10 m TAW (Tweede Algemeene Waterpassing, the Belgian datum tied to the lowest spring tide in Ostend). Several large shallow sand banks (known as the Flemish banks) can be found offshore on the shallow continental shelf (−3 m TAW). They dissipate part of the wave energy and, thus, contribute to lower wave height reaching the beaches [33]. The coast experiences a macro-tidal regime with a tidal range from 3 to 5 m [34]. During energetic events, storm surge is over 1.0 m. The onshore wave buoy at Trapegeer, located 7 km offshore and at −10 m TAW, has a mean significant wave height around 0.6 m with a period of 3–4 s. Wave direction measured at Kwintebank wave buoy, which is 15 km offshore and at −3.5 m TAW, is characterized by two dominant wave directions coming primarily southwest and to a lesser extent northeast.

This beach is characterized by a multiple intertidal bar-trough system, also called a ridge and runnel beach, as classified by [35]. In general, the system consists of three to five parallel bars between the high and low water lines (Figure 2). In general too, bars and troughs are continuous in alongshore direction, except when intertidal drainage channels are present. These features are separated by an average distance of 90 m cross-shore and an average bar-trough amplitude of 60 cm. Though no data are available for the submerged area in this study, it is worth noting that the bar system generally extend beyond the intertidal zone, with subtidal bars that play an important role in dissipating wave energy. The intertidal beach is composed of fine to medium sand with a median diameter of 200 μm. The beach has prograded at a rate of about 21 $m^3/m{\cdot}y$ (cubic meters per meter width and per year) over the last decades [36].

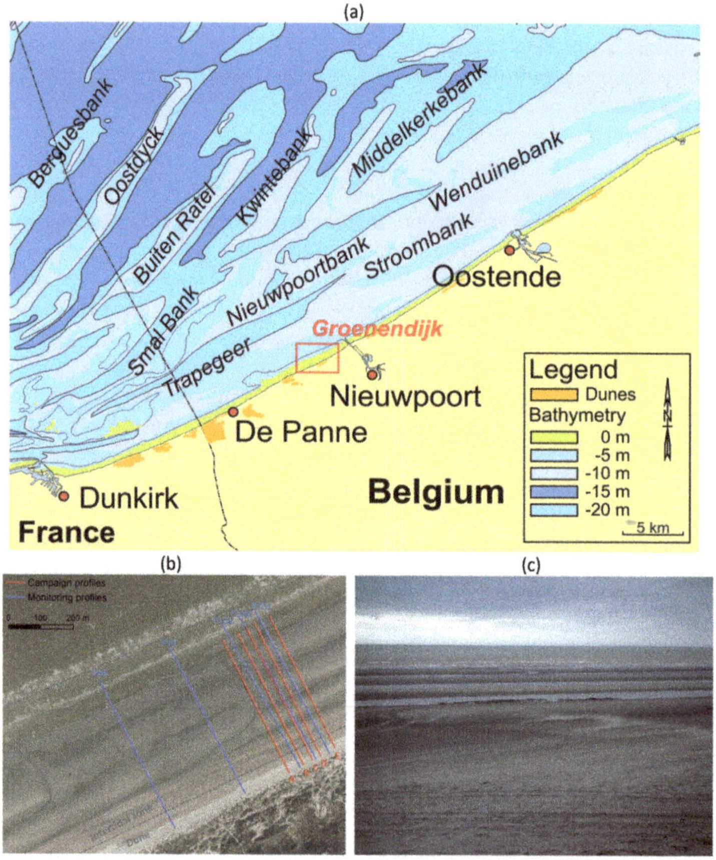

Figure 1. Location of study site and cross-shore profiles. (**a**) Study site position on the Belgian Coast. (**b**) Continuous monitoring profiles (in blue), campaign profiles (in red), intertidal zone (between dashed lines). (**c**) Photo of the Groenendijk beach showing the intertidal bars.

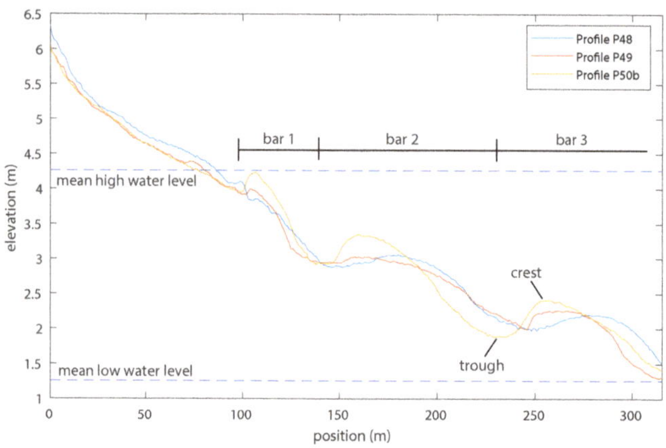

Figure 2. An example of three cross-shore Coastal Division profiles (out of the five) sampled at the same date, emphasizing the presence of sand bars over all profiles as well as the variability between profiles.

3. Material and Methods

3.1. Monitoring Profiles

Elevation measurements that were obtained along profiles P48 to P50c were surveyed by the Coastal Division of the Flemish Authority in order to monitor changes in beach volume over time. Surveys were carried out on a monthly basis between February 2017 and May 2019 while using a Real Time Kinematics (RTK) GPS system, except in November 2017 and in summers. No survey was completed in summer season (i.e., in July–August due to the touristic season) and in November 2017. The distance between profiles range from 50 m to 240 m (see Figure 1). The location of these profiles depends on the monitoring sections, as delimited by the Coastal Division. All of the profiles start at a fixed location near the dune toe and are covering the first three bars of the intertidal zone. No measurements were made for the submerged zone. The Root Mean Square Error (RMSE) was about 2.5 cm in vertical and horizontal directions. In general, distances between successive measurements varied from 5 to 20 m, with an average distance of about 15 m. Profiles were rebuilt at 1 m spacing resolution with unmodified Akima spline interpolation function [37]. This function was compared to other splines and selected as it had the best fit and no overshooting effect. An example of these profiles is given in Figure 2.

In addition, airborne Light Detection And Ranging (LiDAR) surveys were commissioned by Coastal Division on a yearly basis between 2000 and 2012, and carried out twice per year between 2013 and 2019. LiDAR surveys were conducted during low spring tides in order to cover a large part of the beach. A timeline of these surveys is presented in Figure 3 and the exact dates are given in Table 1. Each survey provided a three-dimensional (3D) (horizontal position and vertical elevation) cloud of points over the emerged and intertidal zone of the beach, with a density ranging from one to five points per square meter. RMSE of LiDAR is reported to be equal to 2 cm under test conditions. The point clouds were used to produce Triangulated Irregular Networks (TIN) that were converted afterwards to raster Digital Terrain Models (DTMs) at a resolution of 25 cm. The elevations were extracted from the DTMs along the P48 to P50c profiles at the same 1 m spacing than for the RTK GPS data. Thus, combining these LiDAR and RTK GPS data provides a set of five profiles where elevation is measured at a 1 m spacing at 54 distinct dates from 2000 to 2019, with data from years 2000 to 2016 that are only coming from LiDAR measurements, while data for years 2017 to 2019 are coming both from LiDAR and RTK GPS measurements.

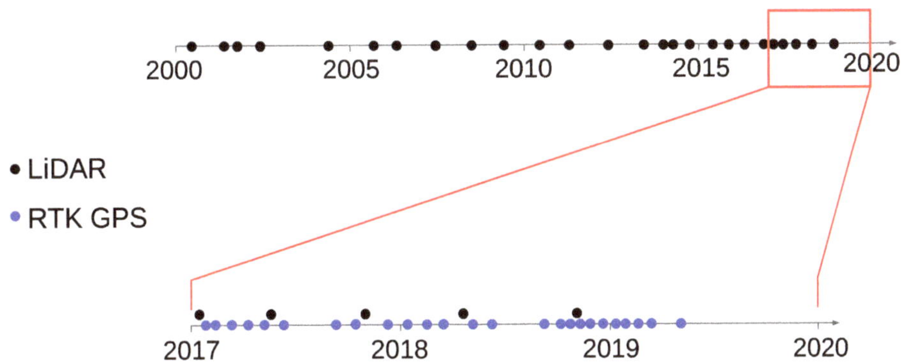

Figure 3. Timeline of the monitoring surveys used in this study. Top: LiDAR surveys since 2000. Bottom: detail of the timeline with RTK GPS (blue) and LiDAR (black) surveys since 2017.

Table 1. Timeline of the monitoring surveys used in this study. Italicized dates correspond to the surveys for the CREST campaigns.

System	Dates (day/month/year)
LiDAR	16/06/2000, 10/05/2001, 28/09/2001, 12/05/2002, 17/04/2004, 23/08/2005, 18/04/2006, 03/06/2007, 19/06/2008, 24/05/2009, 28/05/2010, 17/04/2011, 07/05/2012, 29/04/2013, 10/12/2013, 15/04/2014, 06/11/2014, 17/05/2015, 27/10/2015, 10/04/2016, 14/12/2016, 17/01/2017, 26/05/2017, 06/11/2017, 17/04/2018, 06/11/2018, 20/04/2019
Monitoring RTK-GPS profiles	01/02/2017, 16/02/2017, 15/03/2017, 11/04/2017, 11/05/2017, 14/06/2017, 20/09/2017, 25/01/2017, 19/12/2017, 23/01/2018, 22/02/2018, 21/03/2018, 16/05/2018, 13/06/2018, 26/09/2018, 23/10/2018, 04/11/2018, 12/11/2018, 26/11/2018, 17/12/2018, 15/01/2019, 24/01/2019, 21/02/2019, 12/03/2019, 19/03/2019, 18/04/2019, 20/05/2019
Campaign RTK-GPS profiles	24/01/2018, 25/01/2018, 27/01/2018, 28/01/2018, 29/01/2018, 30/01/2018, 31/01/2018, 01/02/2018, 01/11/2018, 02/11/2018, 03/11/2018, 04/11/2018, 05/11/2018, 06/11/2018, 07/11/2018, 08/11/2018, 09/11/2018, 10/11/2018, 11/11/2018, 12/11/2018, 13/11/2018

3.2. Campaign Profiles

Extra elevation profiles were obtained in the framework of the Climate Resilient Coast project between 2015 and 2019 (http://www.crestproject.be). Two campaigns were performed at the study site. The profiles were surveyed every day for eight and ten days, respectively, in January and November 2018. Cross-shore elevation of the beach was measured along five profiles in section 50 of the beach (Figure 1b) using an RTK GPS in walking mode to automatically measure elevation with 1 m measurement interval. Benchmarks that were near the dune toe were used to ensure that sampled positions along the profiles were the same for all dates. Profiles were not interpolated due to the high density of measurements. Similarly, measurements are only covering the emerged and intertidal zones.

3.3. Average Profiles and Progradation

In order to study the space-time dynamics of the profiles over time, we will first investigate the long-term evolution of profiles over time. Consider that $Z(x,t)$ is the measured elevation at position x along a given profile at a time instant t. Thus, we can characterize the long term evolution at any location x by the mean function $\mu(x,t) = E[Z(x,t)]$, where $E[.]$ refers to the expectation. We will consider here that $\mu(x,t) = \mu(x) + [a(x) + b(x)t]$, where $\mu(x)$ is a time-invariant mean profile while the second term quantifies the linear progradation over time, with parameters $a(x)$ and $b(x)$ that depend on the position x along the profile. From the data at hand, it is easy to estimate $a(x)$ and $b(x)$ by looking at the values $Z(x,t)$ along the time axis for every position x. Substracting this progradation effect allows for us to obtain $\mu(x)$. These time-invariant mean profiles $\mu(x)$ can then easily be compared and progradation rates $b(x)$ can be integrated to obtain the beach progradation rate ΔV, with

$$\Delta V = \int b(x)dx$$

where ΔV is usually expressed in $m^3/m{\cdot}y$. It is worth noting that we only consider here a linear progradation over time, but more elaborate temporal evolution could be used if relevant, of course.

3.4. Spatial/Temporal Variograms

Assessing the way that elevation changes over time and along these profiles is done by relying on the spatial and temporal variograms. To the opposite of previous studies related to the use of these tools in beach studies (see e.g., Swales [16], Dolan et al. [38], Harley et al. [39]) that were restricted to separate spatial and temporal analyses in a mapping framework, we will consider both space and time together. Consider that $Z(x,t)$ is the measured elevation at position x along a given profile at a time instant t. It is characterized by a mean elevation $\mu(x)$ (after correcting for progradation) that depends on the position x along the profile. The variation of elevation around this mean profile is denoted $Y(x,t) = Z(x,t) - \mu(x)$. The spatial and temporal variograms are then, respectively, defined as

$$\gamma(h) = \frac{1}{2}Var[Y(x+h,t) - Y(x,t)] \quad ; \quad \gamma(\tau) = \frac{1}{2}Var[Y(x,t+\tau) - Y(x,t)]$$

where h is the distance separating two locations and τ is the time lag separating two measurements. These variograms measure the dissimilitude between measurements, either as a function of the distance h or the time lag τ that separates them. For measurements that can be considered to be zero-mean and second-order stationary over space and time, they are linked to the traditional covariance functions $C(h)$ and $C(\tau)$, which are defined as

$$C(h) = E[Y(x+h,t)Y(x,t)] \quad ; \quad C(\tau) = E[Y(x,t+\tau)Y(x,t)]$$

so that $\gamma(h) = \sigma^2 - C(h)$ and $\gamma(\tau) = \sigma^2 - C(\tau)$, where σ^2 is the variance of the measurements. These covariance functions are also related to the correlation functions $\rho(h)$ and $\rho(\tau)$, with $C(h) = \sigma^2\rho(h)$ and $C(\tau) = \sigma^2\rho(\tau)$. More details can be found in [40,41] about the properties and the estimation of these various functions. The main advantage of the variogram is to allow for the user to interpret the extent of the spatial/temporal dependence based on the shape of the function, as illustrated in Figure 4. The distance/time lag that is needed to reach 95% of the variance σ^2 is called the range r of the variogram. It corresponds to the distance/time lag beyond which no significant correlation exists. Measurements separated by at least this range can be considered statistically unrelated for practical purposes. The vertical jump at the origin of the function is called the nugget effect σ_0^2 and corresponds here to the variance of the measurement errors (i.e., the errors that are associated with the RTK GPS and LiDAR systems). If a periodic spatial/temporal behaviour is present in the measurements, it will be translated as a periodic behaviour as well in the corresponding spatial/temporal variograms and covariance functions. There is a direct connection with spectral theory, as the power spectral density is the Fourier transform of the covariance function (see e.g., Jenkins and Watts [42]), which is itself linearly related to the variogram. When it comes to relating beach profiles to various scales, alternate approaches, like wavelet analysis, could be relevant (see e.g., Li et al. [43]).

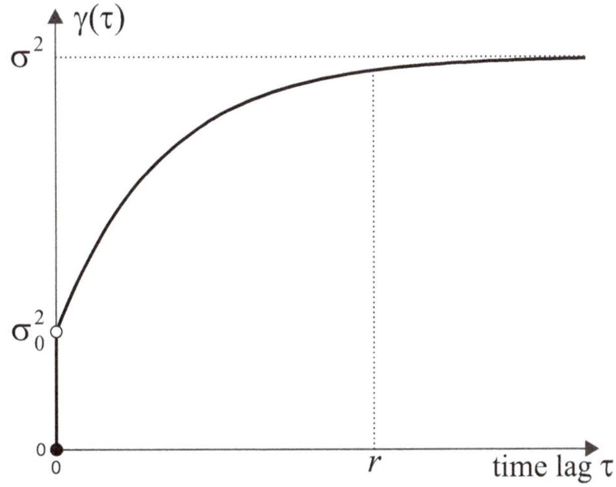

Figure 4. Generic features for a temporal variogram $\gamma(\tau)$, showing the variance σ^2, the nugget effect σ_0^2 and the range r. The same generic features are relevant for a spatial variogram $\gamma(h)$ in general.

3.5. Space-Time Model and Simulations

Elevation change $Y(x,t)$ needs to be statistically assessed in space and time together. This requires a space-time (ST) covariance model combining the spatial and temporal correlation functions. Such a ST model is given by the separable covariance model

$$C(h,\tau) = E[Y(x+h,t+\tau)Y(x,t)] = \sigma^2\rho(h)\rho(\tau) \tag{1}$$

that combines in a simple way the correlation functions $\rho(h)$ and $\rho(\tau)$ that were defined separately over space and over time, respectively. This model provides the covariance matrix for any vector of positions and time instants. Instead of using this space-time model for doing spatial interpolation or temporal prediction (i.e., using classical kriging algorithms as usually done), we will use it instead in a conditional simulation framework for assessing the volume changes over time along a cross-shore profile.

Given two time instants t and $t+\tau$, the vector $\mathbf{Z}_t$ is defined as the elevations that are measured along a given profile at the time instant t, with $\mathbf{Z}_t = (Z(x_1,t),\ldots,Z(x_n,t))'$. Similarly, the vector $\mathbf{Z}_{t+\tau}$ contains the elevations measured at the same set of n locations but at the time instant $t+\tau$. Clearly, the $n \times n$ spatial covariance matrix $\mathbf{\Sigma}_s$ for both $\mathbf{Z}_t$ and $\mathbf{Z}_{t+\tau}$ is the same and it only depends on the spatial covariance function $C(h) = \sigma^2\rho(h)$. Moreover, from Equation (1), the $n \times n$ covariance matrix between $\mathbf{Z}_t$ and $\mathbf{Z}_{t+\tau}$ is given by $\rho(\tau)\mathbf{\Sigma}_s$. Consequently, the $2n \times 2n$ covariance matrix $\mathbf{\Sigma}$ for the stacked vector $\mathbf{Z} = (\mathbf{Z}_t, \mathbf{Z}_{t+\tau})$ that involves the elevation at two time instants that are separated by a time lag τ is then given by

$$\mathbf{\Sigma} = \begin{pmatrix} \mathbf{\Sigma}_s & \rho(\tau)\mathbf{\Sigma}_s \\ \rho(\tau)\mathbf{\Sigma}_s & \mathbf{\Sigma}_s \end{pmatrix} \tag{2}$$

This ST covariance matrix is at the root of any simulation procedure that simultaneously involves the elevation measurements at the same positions along the profile at two time instants t and $t+\tau$. Starting from observed elevations at time t, it is thus possible to simulate elevations at time $t+\tau$ by accounting for the ST dependence between all measurements as given by $\mathbf{\Sigma}$. Repeating this several times produces a distribution of simulated profiles at $t+\tau$. If $\mathbf{Z}_t$ and $\mathbf{Z}_{t+\tau}$ are assumed to be jointly Gaussian random vectors, simulations are easy to implement. Indeed, what is sought for is the conditional distribution $\mathbf{Z}_{t+\tau}|\mathbf{z}_t$ of possible random vectors $\mathbf{Z}_{t+\tau}$ at time $t+\tau$, given that $\mathbf{z}_t$ has been observed at time t.

In statistical theory, if one assumes, in general, a Gaussian random vector $\mathbf{Z} = (\mathbf{Z}_1, \mathbf{Z}_2) \sim N(\boldsymbol{\mu}, \mathbf{\Sigma})$ where $\mathbf{Z}_1 \sim N(\boldsymbol{\mu}_1, \mathbf{\Sigma}_{11})$ and $\mathbf{Z}_2 \sim N(\boldsymbol{\mu}_2, \mathbf{\Sigma}_{22})$, it then comes that

$$\mathbf{Z}_2|\mathbf{z}_1 \sim N(\boldsymbol{\mu}_2 + \mathbf{\Sigma}_{21}\mathbf{\Sigma}_{11}^{-1}(\mathbf{z}_1 - \boldsymbol{\mu}_1), \mathbf{\Sigma}_{22} - \mathbf{\Sigma}_{21}\mathbf{\Sigma}_{11}^{-1}\mathbf{\Sigma}_{12})$$

Setting now $\mathbf{Z}_2 = \mathbf{Z}_{t+\tau}$ and $\mathbf{Z}_1 = \mathbf{Z}_t$ with $\boldsymbol{\mu}_1 = \boldsymbol{\mu}_2 = \boldsymbol{\mu}$ and knowing from Equations (1) and (2) that $\mathbf{\Sigma}_{11} = \mathbf{\Sigma}_{22} = \mathbf{\Sigma}_s$, while $\mathbf{\Sigma}_{12} = \mathbf{\Sigma}_{21} = \rho(\tau)\mathbf{\Sigma}_s$, this gives

$$\mathbf{Z}_{t+\tau}|\mathbf{z}_t \sim N(\boldsymbol{\mu} + \rho(\tau)(\mathbf{z}_t - \boldsymbol{\mu}), (1-\rho^2(\tau))\mathbf{\Sigma}_s) \tag{3}$$

Using now a Choleski decomposition of the $\mathbf{\Sigma}_s$ matrix, such that $\mathbf{\Sigma}_s = \mathbf{L}\mathbf{L}'$ where $\mathbf{L}$ is a lower triangular matrix, it is easy to simulate a random vector having the properties specified in Equation (3), with

$$\mathbf{Z}_{t+\tau} = \sqrt{1-\rho^2(\tau)}\,\mathbf{L}\boldsymbol{\varepsilon} + \boldsymbol{\mu} + \rho(\tau)(\mathbf{z}_t - \boldsymbol{\mu}) \tag{4}$$

where $\boldsymbol{\varepsilon}$ is a $n \times 1$ vector of independent zero-mean unit variance Gaussian distributed values, which can obtained from any random number generator, and where $\boldsymbol{\mu} = (\mu(x_1),\ldots,\mu(x_n))'$. Using Equation (4) repeatedly allows for simulating an arbitrary large number of independent $\mathbf{Z}_{t+\tau}$ profiles, which are all conditioned on the same initial observed profile $\mathbf{z}_t$ and obey the properties given in

Equation (3). In other words, all of the simulated $\mathbf{Z}_{t+\tau}$'s are plausible elevation profiles at time instant $t+\tau$. They are all consistent both with the observed values $\mathbf{z}_t$ at time t and with the covariance model specified in Equation (1).

4. Results

Most of the results thatare presented here are based on analyses conducted jointly on the five monitoring profiles P48 to P50c, covering a time span of 19 years. The 1 m spacing elevations as interpolated from the Akima splines are used in all subsequent analyses. The campaign profiles, which are covering a time span of only few days, are only used for aspects related to the estimation of the temporal variogram.

Depending on the date and the time of measurements, the length of the profiles built by combining LiDAR and RTK GPS measurements may vary. In order to process comparable measurements for all profiles, only the first 316 meters have been considered for the subsequent computations. They all covered the emerged and major part of the intertidal zone of the beach, which can extend close to 400 m at some dates.

4.1. Average Profiles and Progradation

The mean elevations of the five profiles surveyed at the same 54 dates were first computed in order to investigate the dynamic of the bars over time and along the profiles. The sampling frequencies ranged from a few days to two years (with a yearly average sampling frequency from 2000 to 2016 and a monthly average sampling frequency from 2017 to 2019). Accordingly, for a given position along the profile, the elevation at each date was weighted by the average time that separates it from the previous and next sampling dates (i.e., the closer two dates are, the lower is their weight). The corresponding weighted mean of the elevation over time was computed every meter along each of the 5 profiles, thus yielding five temporally averaged profiles $\widehat{\mu}(x)$ that are presented in Figure 5.

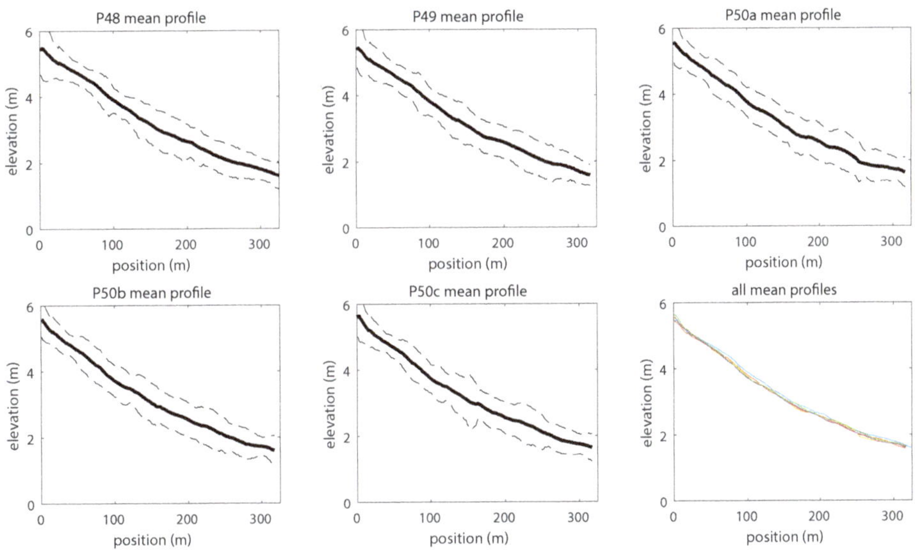

Figure 5. Average Coastal Division profiles as computed from a weighted mean of the elevation measured at 54 dates (plain line), along with the standard deviation of the measurements (dashed line). The last figure shows all average profiles as superimposed for comparison purpose.

Although bar systems appear clearly on any single profile at any date, these features do not appear anymore on the average profiles. The average profiles also do not differ from each other.

The magnitude of the variations around the average profile (as measured by the variance σ^2 of the elevations around the mean profile) is similar for all positions along the profile, with an estimated value $\widehat{\sigma}^2$ = 0.0510 m^2. This emphasizes the fact that, even if the bars and troughs are ubiquitous features, their position is subject to major changes over a time span of several years (see Section 4.2). Interpolating all of the profiles at the same location for all dates makes possible to look at the time series of elevations at any chosen position, as presented in Figure 6. The results indicate that a progradation mainly occurs close to the dunes, where elevation increases over time. The amount of beach progradation can be assessed from the slope of the regression line fitted on these time series, at each position and for the five profiles (Figure 7).

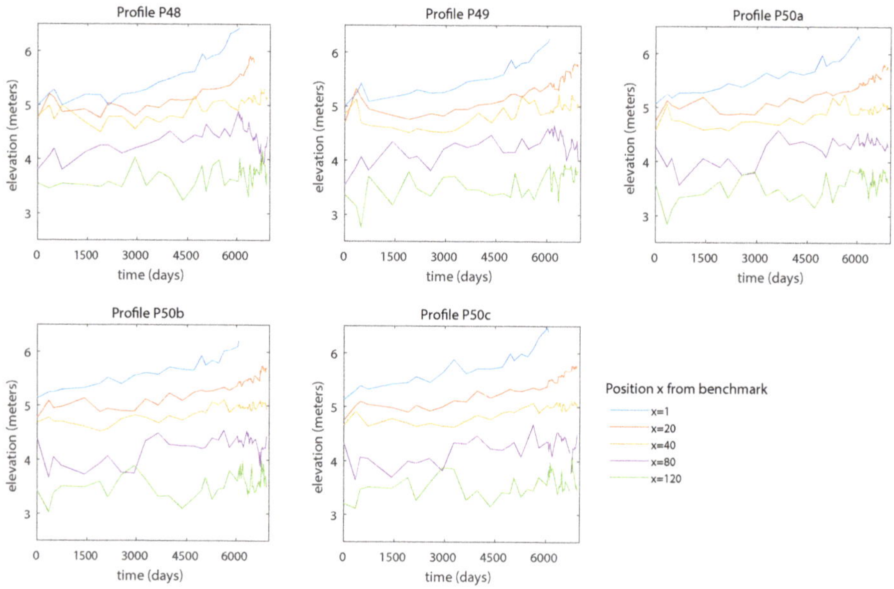

Figure 6. Time series of elevation at a set of chosen positions x along the Coastal Division cross-shore profiles (where $x = 0$ corresponds to the benchmark at the dune toe), showing that progradation (as quantified by the slope of the time series) mainly occurs close on the upper beach part.

The progradation rate is maximum at the origin of the profiles and decreases when moving away from the dunes. Consequently, mean slope decreases when moving away from the dune. Sediment transport increases from the intertidal zone to the dune, that propagated seaward over the last decades. This long-term accretion, observed for a long time at the study site [36], is probably the combination of sand supply from the sand banks, the littoral drift and the adjacent nourished beaches [44]. Integrating this progradation rate along the profile yields an average vertical elevation increase of 0.019 m/y. This corresponds to an accumulated sand volume of about 5.7 m^3/m·y along the profile. The local elevation increase is however close to 0 beyond the first 80 m and varies greatly within these 80 first meters, as seen from Figure 7.

For the subsequent analyses, these average profiles $\widehat{\mu}(x)$ are substracted from the raw elevation measurements $Z(x,t)$ in order to obtain zero-mean residual elevations $Y(x,t) = Z(x,t) - \widehat{\mu}(x)$. These are used for assessing the spatial and temporal variability using the spatial and temporal variograms. Beach progradation is accounted for in all subsequent computations that involve distinct time instants, based on the curve of the average progradation rates, as given in Figure 7. Elevations were corrected over time by considering at each location the slope estimated in Figure 7 and by multiplying it by the time that separates sampled profiles. Therefore, the average profiles $\widehat{\mu}(x)$ are corrected in order to account for the progradation that occurs between time instants t and $t + \tau$. It is worth noting that all of the subsequent computations were also done by omitting this progradation correction for comparison purposes (results not shown here) and the results were qualitatively identical.

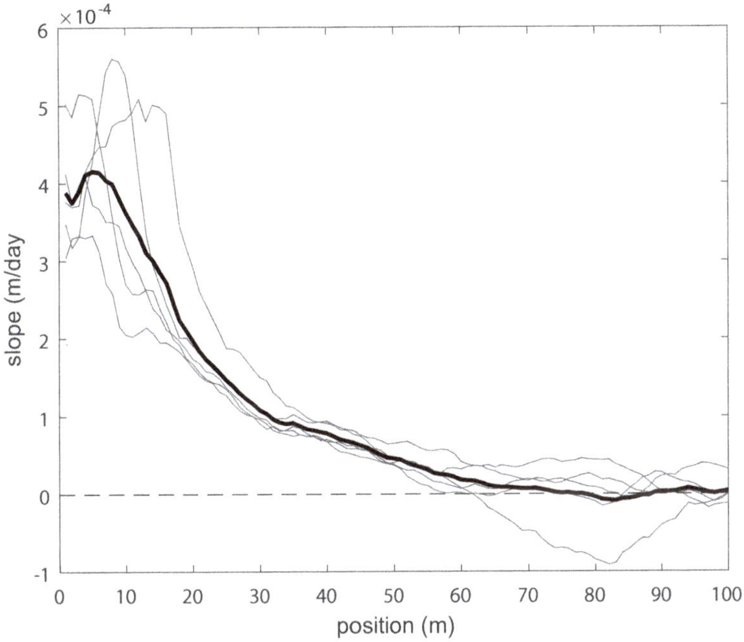

Figure 7. Estimation of the progradation rate as a function of the position along the Coastal Division profiles. Gray lines correspond to the slopes estimated separately for the five profiles, while thick black line gives the corresponding average slopes.

4.2. Spatial and Temporal Variograms

Based on the zero-mean $Y(x,t)$ values, the temporal variability of the elevations is determined by estimating the temporal variogram $\gamma(\tau)$ from the five monitoring profiles at the 54 sampling dates. The normalized temporal variogram (i.e., the temporal variogram divided by the total variance σ^2) is presented in Figure 8. The vertical axis represents the fraction of the total variance σ^2, which is reached as a function of the time lag τ between measurements. It can be seen that about 90% of the total variance is reached for $\tau = 1095$ days (i.e., 3 years). Thus, elevations measured at the same position on a profile can be considered as uncorrelated for time lag greater than a few years. This fraction of the total variance is about 25% and 65% for measurements separated by a time lag of one month ($\tau = 31$) and one year ($\tau = 365$), respectively. This fast increase of the dissimilitude between elevations over time is consistent with the fact that the bars and troughs cannot be observed on the average profiles computed over a time span of 19 years. It emphasizes the stochastic nature of the processes driving bar dynamics through time.

It is not possible to reliably estimate the temporal variogram for any time lag lower than a month due to the fact that monitoring profiles were typically sampled on a monthly basis over the last few years. This estimation can however be done using the campaign profiles sampled daily over short time spans, yielding estimates that match well the general shape of the fitted model (Figure 8). They can also be used to estimate the nugget effect associated with the measurement errors, yielding an estimated value $\widehat{\sigma}_0^2$ = 7.5 10^{-4} m^2. The corresponding estimated standard deviation $\widehat{\sigma}_0 = 0.027$ m is in close agreement with the reported RMSE of RTK GPS and LiDAR systems, as given in Section 3.1.

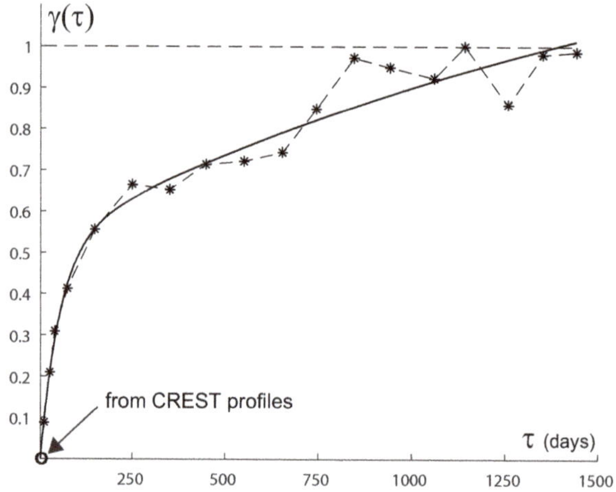

Figure 8. Estimation of the (normalized) temporal variogram from the Coastal Division profiles (plain circles and dashed line), along with the estimation of the short-range variability obtained from the CREST profiles (white circles). The fitted model corresponds to the thick plain line.

While using each monitoring profile for the 54 distinct dates, the spatial variability of the $Y(x,t)$ values around the mean profile is assessed by estimating the spatial variogram $\gamma(h)$. As seen in Figure 9, the presence of the bars and troughs is obvious from the periodic behaviour of the variogram. The average distance separating a bar from its neighbouring trough is about 45 m, as measured on this estimated variogram. This means that even if the presence of the bars and troughs cannot be detected on the temporally averaged profiles (Figure 5), these bar systems are, however, present along the profiles. The spatial variogram is not strictly periodic, which emphasizes that the distance between these bars and their amplitude are not constant along the profiles and are also subject to noticeable major fluctuations over time. This is confirmed by the temporal variogram, which shows that elevations are not correlated anymore for time lags longer than few years.

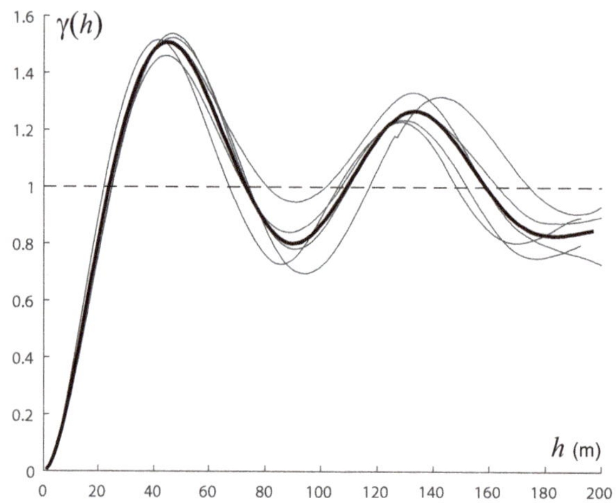

Figure 9. Estimation of the (normalized) spatial variogram from the five Coastal Division profiles (gray lines), along with the average of these variograms (thick black line). The periodic behaviour of the variogram is associated with the alternance between bars and troughs along the profiles, with an average bar-trough distance of about 45 m.

4.3. Space-Time Model and Volume Changes

From the estimated variance $\widehat{\sigma}^2$ and variogram models $\widehat{\gamma}(h)$ and $\widehat{\gamma}(\tau)$, a ST model as given in Equation (1) is obtained. This ST model can be used in a simulation framework. The beach volume change over time is assessed in order to illustrate the benefits of this space-time modeling approach for our study. The goal is to estimate the displaced volume $V(\tau)$ between two time instant t and $t+\tau$, as defined by

$$\begin{aligned} V(\tau) &= \int |Z(x,t+\tau) - Z(x,t)|dx \\ &\simeq \Delta_x \sum_{i=1}^{n} |Z(x_i,t+\tau) - Z(x_i,t)| \end{aligned} \tag{5}$$

where $|\cdot|$ denotes the absolute value and $\Delta_x = 1$ m in our case. The true displaced volumes $V(\tau)$ between two successive sampling dates is first computed for the five monitoring profiles from the measured elevations at the 54 sampling dates. They are compared to the displaced volumes computed while using simulated profiles obtained from Equation (4), to which the prograded volume over the corresponding time span is added. For the five monitoring profiles and the 53 first sampling dates (thus excluding the last date), a large numbers of simulated vectors $\mathbf{Z}_{t+\tau}$ of elevations at the next sampling time instant $t+\tau$ are generated conditionally to the observed $\mathbf{z}_t$ elevations at time instant t. The predictor of $V(\tau)$ is the average of the volume changes as computed from the various simulated $\mathbf{Z}_{t+\tau}$'s, while using of course the same initial observed elevations $\mathbf{z}_t$ in Equation (5). The procedure is illustrated in Figure 10 for the 200 first meters of profile P48 and for a time lag τ of 34 days. The corresponding volume change values are then computed for each simulation and a distribution of volume change is obtained, from which the average volume change and confidence limits are estimated, as illustrated in Figure 11. The same operation can then be repeated for all profiles and time instants t. The dissimilitude between simulated profiles at time $t+\tau$ and initial profile at time t increases with τ, so that the mean volume change and confidence limits increase as well with τ (Figure 12).

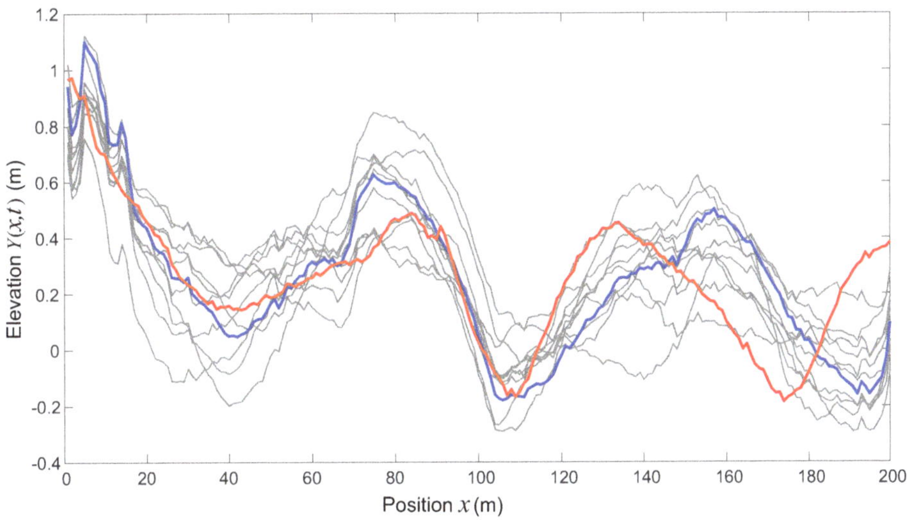

Figure 10. Illustration of the simulation procedure for profile P48 and a time lag $\tau = 34$ days, where the 10 (out of 10,000) simulated profiles at time instant $t+\tau$ (gray lines) are all conditioned on the same initial observed values at time instant t (blue line) and where simulated profiles need to be compared to observed values at time $t+\tau$ (red line). For the sake of readability, only the centered values $Y(x,t) = Z(x,t) - \widehat{\mu}(x)$ around the average profile $\widehat{\mu}(x)$ are represented here.

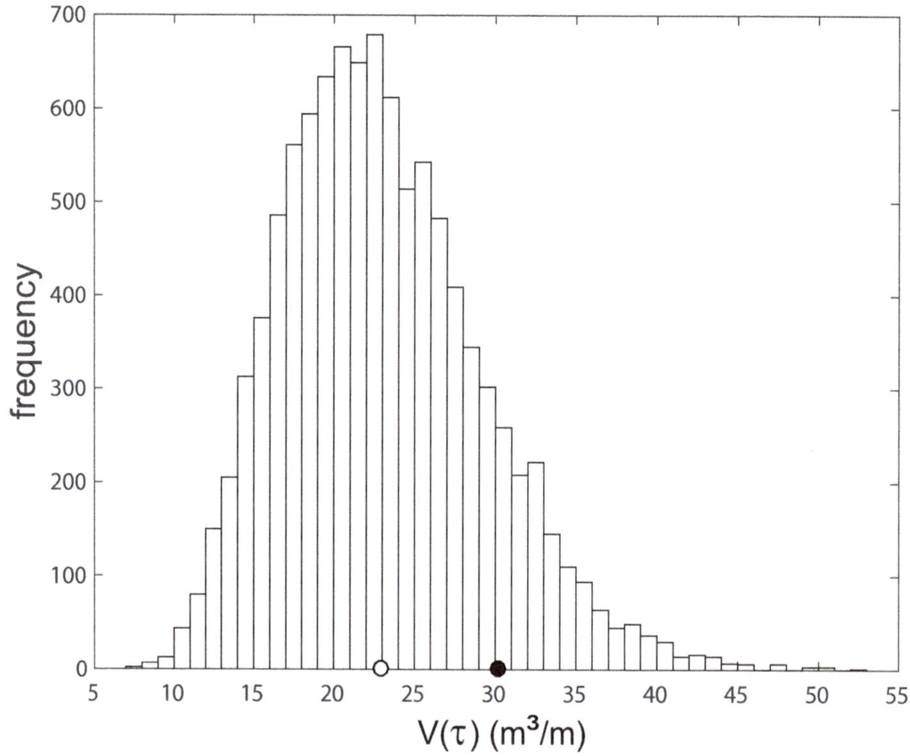

Figure 11. Distribution of the volume change built from a set of 10,000 simulations, based on the profiles at time instant t and $t+\tau$, as given in Figure 10. The open disk is the mean simulated volume change and the black disk is the observed volume change between time instants t and $t+\tau$.

The distribution of volume change that was obtained in Figure 11 for $\tau = 34$ days shows that this volume change ranges from 10 m^3/m to 45 m^3/m. It is not expected that a single value, like the average volume change (i.e., the mean of this distribution), will closely relate to the true volume change. However, Figure 13 indicates that the true volume changes of each date are in good agreement with those that were obtained from the simulations, except when the true volume changes are higher than 45 m^3/m (which is typically observed for $\tau > 6$ months). The underestimation of the predicted volume change when the observed volume changes are very high shows here the limitations of the modeling approach. Indeed, Figure 12 shows that the predicted volume change (i.e., mean of simulated volume change distribution) is bounded from above and is not expected to exceed 45 m^3/m. This is also seen in Figure 13, where predicted volume change only reaches this upper bound when $\tau \geq 2$ years.

Predicting a volume change for such high time lags is challenging. Firstly, the corresponding profiles at time t and $t+\tau$ are expected to be poorly correlated (see Figure 8). Secondly, this study is conducted without accounting for the prevailing meteorological and marine conditions that are likely to impact the evolution of the beach between two time instants t and $t+\tau$. When τ increases, so does the probability of observing energetic events, such as storms, which would naturally lead to a higher volume change compared to what would be expected under typical conditions (e.g., [29]). Unfortunately, the effect of meteorological and marine conditions on this study site are still not completely understood. The effective impact of energetic events on volume change is not obvious either, as it depends on various factors (such as wind speed and direction, wave height and energy, etc.) interacting in a nontrivial way.

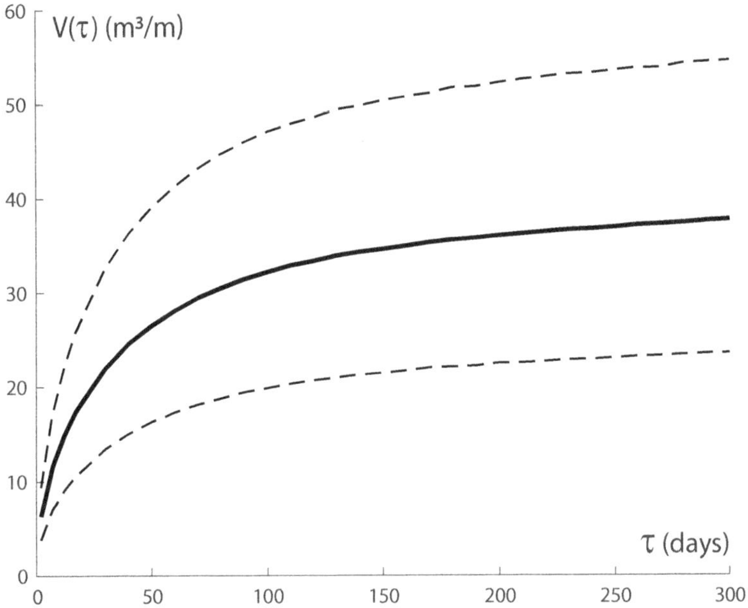

Figure 12. Mean volume change (plain line) and bounds of the 90% confidence interval of the volume changes (dashed lines) as a function of τ, as computed here for the 200 first meters of the profile.

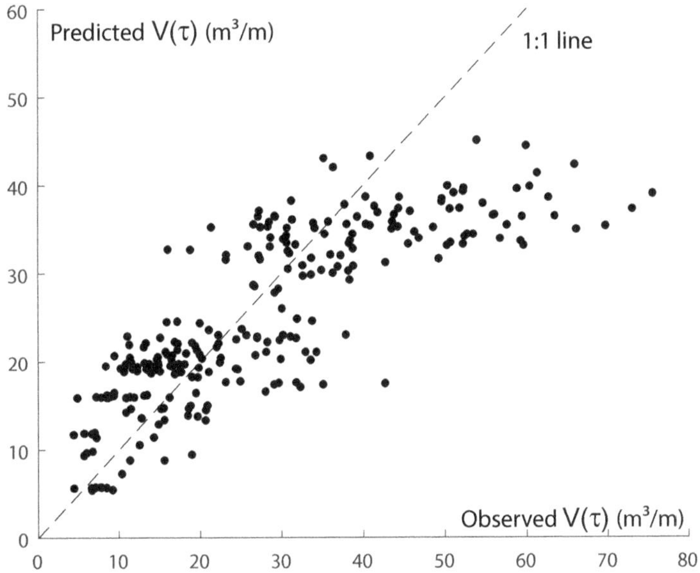

Figure 13. Comparison of predicted volume changes between time instants t and $t + \tau$ (i.e., average of the simulated distributions) versus observed volume changes, based on the five Coastal Division profiles at 53 distinct time instants t.

5. Discussion

5.1. Advantages of this Innovative Model

This study presents an innovative data-driven method for modeling morphodynamics of multi-barred beaches while using high-quality measurements of cross-shore profiles based on a stochastic approach. This is undertaken by first estimating and modeling the spatial and temporal evolution of beach elevations. Subsequently, a separable space-time model is built in order to evaluate the temporal evolution of spatially integrated morphological characteristics. Based on simulations, the temporal evolution of the total displaced sand volume can then be estimated for arbitrary time lags.

A similar stochastic approach could be useful for estimating other beach feature characteristics as long as one is able to capture the stochastic space-time variability of the underlying process. As previously reported [12], the strengths of such stochastic approach are to provide evidence for distinguishing generic characteristics of the data and also that no prior assumptions on the controlling processes are required. The combination of new statistical techniques, the improvement of computer power, as well as increasing length and quality of beach morphology measurements favor the application of a space-time stochastic approach. The findings, specifically the matching of observed and predicted volume changes, also suggest that beach elevation could benefit from being modeled as a stochastic process. Although elevation acts in a non fully predictable way, it can be assessed over both space and time while using a space-time covariance model.

When compared to other approaches, the stochastic method applied in this article provides reliable predictions of barred beach morphology, along with confidence bounds. Other approaches are not usually suited to process intertidal bars. Equilibrium profile models can rarely model intertidal bars. Process-based models still lack the complete understanding of intertidal bar behaviour to accurately process intertidal barred beach morphology (e.g., [45]). In contrast, the method employed to build the model described in this article can be applied to any morphological data set with sufficient measurements or any study site where such measurements can be carried out. Additionally, the model's poor prediction of long-term large volume change challenges the significance of studies using low measurement frequency (e.g., a survey per year) [46].

5.2. Restrictions of the Model

Regarding the prediction of volume change over time, several improvements need to be considered for further studies. The first one is to include information regarding the prevailing meteorological and marine conditions over a given time interval $[t,t+\tau]$. These conditions are known to drive morphological change on the beach (e.g., [6,7,47,48]). Energetic events, such as storms, are likely to have a major impact on a beach profile (e.g. [49]). This would thus affect the observed volume change in a way that cannot be predicted by our simple space-time covariance model. Thus, it expected that these volume change predictions would be improved if the effect of these meteorological and marine conditions are accounted for in the simulation procedure. However, in the framework of long-term studies, it is quite unclear how these forcing factors can be easily incorporated in the analysis. Indeed, beach evolution is typically driven by few energetic (erosive) events occurring over few days under stormy conditions, intertwined with long periods of low energetic (recovery) conditions. It is thus unlikely that volume changes as computed from cross-shore profile measurements separated by several months could be related to the largely changing meteo-marine conditions that were acting during that period of time. Additionally, it is worth noting that accounting for meteo-marine conditions this is not even necessarily enough for predicting the evolution, even over short time scales (see, e.g., Biausque and Senechal [25]. Indeed, field observations suggest that alongshore sand displacement driven by tide- and wave- currents might play a role in the intertidal bar morphology. It is already known that sediment transport at this study site has a strong alongshore component [50]. Previous studies found important longshore morphological changes of the bar systems with almost no cross-shore change, under both energetic and calm conditions at another multi-barred macrotidal beach (e.g., in Merlimont,

France, [51,52]). The intertidal troughs themselves are also a driver of longshore sediment transport (e.g., [53]). Thus it would be appropriate to undertake a two-dimensional (2D) analysis on both cross-shore and alongshore dimensions, and this analysis should be conducted at a much shorter time scale than for our data sets in order to unambiguously capture the effect of the forcing factors as close as possible to the time of the measurements. Finally, it is worth remembering here that our study focused on the emerged and intertidal zones of the beach, without including the contribution of the submerged area to the volume change assessment. Subtidal bars are likely participate to the dynamics of the intertidal zone. Thus, volume changes that are indicated in this study are subject to revision provided that additional measurements are made available.

5.3. Findings in Bar Morphodynamics

Morphodynamics of beaches with multiple intertidal bar-trough systems have been well documented over the past (e.g., [35,46,47,54,55]). They mainly occur on restricted coasts with a surplus of sand, exposed fetch-limited waves, and large tidal range above 3 m [56]. Previous studies on morphological dynamics of multiple intertidal have reported that the bars tend to be permanent and often appearing at preferential locations [55]. Although intertidal bar are ubiquitous features along the Belgian coast, the findings here indicate that they are subject to major changes over a time span of several years. The temporal and spatial variograms (Figures 7 and 8) suggest no correlation of the position and amplitude of the bars and troughs after three years, with 90% of the total variance reached for $\tau = 1095$ days. Thus, a resetting of the intertidal bar profile occurs on annual basis. This is agreement with [57] who found little resemblance of the bars along the Lincolnshire coast (east of England) tracked over time using annual aerial photographs. The morphodynamics of intertidal bar system are principally governed by the tidal range, the incident wave energy level and variability [47]. Bars tend to migrate onshore under calm wave conditions, while they tend to flatten and migrate offshore during storms, with migration rates of up to 10 m per month [58]. However, other factors, such as wave angle and water depth at the bar crest, may force the effect of particular wave conditions on bars development. Additionally, alongshore sediment transport can play an important role on the dynamics of the system, especially in the troughs due to longshore currents over the tidal cycle [7]. The morphological response of the intertidal bars can be affected by relation time effects and morphological feedbacks. The complex and stochastic nature of the processes driving the position and behaviour of the bar-through system are highlighted in the space-time covariance model. Improved knowledge on the development of intertidal bar systems will be beneficial to coastal managers, for whom the evolution of bars is a key component in preventing coastal erosion.

6. Conclusions

To the best of our knowledge, this study is the first to investigate the morphodynamics of sandy multibarred beach that was based on a space-time stochastic approach by relying on a space-time covariance model and a stochastic simulation procedure. It presents a space-time analysis and modeling of cross-shore profiles of a macrotidal beach on the Belgian coast, characterized by intertidal bar systems exhibiting significant changes over time. Often traditional morphodynamic studies characterize the main spatial properties and the temporal evolution of the bars and troughs. However, this article suggests that a stochastic approach is relevant when modeling the temporal evolution of integrated properties along such cross-shore profiles.

The spatial and temporal evolution of beach elevation is estimated and modeled for a set of high-quality measurement profiles over 19 years in order to build up a space-time model. Spatial and a temporal variograms are used in order to quantify the extent of the spatial and temporal dependence of elevation. The findings provide evidence of counter-intuitive facts, such as the absence of bars and troughs on the temporally averaged profiles despite their ubiquitous presence. Based on a space-time separable covariance model, a spatially integrated beach property, like the volume change occurring between two time instants t and $t + \tau$, can be assessed through a simulation approach for providing a predicted volume change as a function of τ, along with corresponding confidence bounds. These findings also suggest that beach elevation could benefit from being modeled as a stochastic process. Besides volume change, other integrated beach morphological characteristics could be predicted in a similar stochastic framework as long as the spatial and temporal evolution of the corresponding processes can be estimated and modeled from the data at hand.

Author Contributions: Conceptualization, P.B.; writing—original draft preparation, P.B.; data analysis, P.B.; data curation, A.-L.M. and M.C.; writing—review and editing, A.-L.M. and M.C. All authors have read and agreed to the published version of the manuscript.

Funding: This research was funded by the Belgian Science Policy Office in the framework of the thematic RS4MODY project (STEREO III project SR/00/360).

Acknowledgments: The authors thank the Coastal Division for providing the RTK-GPS and LiDAR data and Alexis Esquerre for his help in preparing an earlier version of this manuscript.

Conflicts of Interest: The authors declare no conflict of interests.

Data Availability Statement: Research data are not shared.

References

1. Cheng, I.; Huang, C.; Hung, P.; Ke, B.Z.; Kuo, C.W.; Fong, C.L. Ten Years of Monitoring the Nesting Ecology of the Green Turtle, Chelonia mydas, on Lanyu (Orchid Island), Taiwan. *Zool. Stud.* **2009**, *48*, 83–94.
2. Sönmez, B. Morphological variations in the green turtle (Chelonia mydas): A field study on an eastern Mediterranean nesting population. *Zool. Stud.* **2019**, *58*. [CrossRef]
3. Boyco, C. New records of sand crabs (Crustacea: Decapoda: Albuneidae and Blepharipodidae) from the western Pacific with description of two new species of Paralbunea Serène, 1977. *Zool. Stud.* **2019**, *59*. [CrossRef]
4. Larson, M.; Capobianco, M.; Jensen, H.; Rozynski, G.; Southgate, H.; Stive, M.; Wijnberg, K.; Hulscher, S. Analysis and modelling of field data on coastal morphological evolution over yearly and decadal time scales. Part 1: Background and linear techniques. *J. Coast. Res.* **2003**, *19*, 760–775.
5. Morton, R.; Paine, J.; Gibeaut, J. Stages and durations of post-storm beach recovery, southeastern Texas coast. *J. Coast. Res.* **1994**, *10*, 884–908.
6. Anthony, E.; Levoy, F.; Monfort, O. Morphodynamics of intertidal bars on a megatidal beach, Merlimont, Northern France. *Mar. Geol.* **2004**, *208*, 73–100. [CrossRef]
7. Masselink, G.; Kroon, A.; Davidson-Arnott, R. Morphodynamics of intertidal bars in wave-dominated coastal settings—A review. *Geomorphology* **2006**, *73*, 33–49. [CrossRef]
8. De Vriend, H.; Capobianco, M.; Chesher, T.; de Swart, H.; Latteux, B.; Stive, M. Approaches to long-term modelling of coastal morphology: A review. *Coast. Eng.* **1993**, *21*, 225–269. [CrossRef]
9. Roelvink, D.; Reniers, A.; Van Dongeren, A.; de Vries, J.; McCall, R.; Lescinski, J. Modelling storm impacts on beaches, dunes and barrier islands. *Coast. Eng.* **2009**, *56*, 1133–1152. [CrossRef]
10. Roelvink, D.; Costas, S. Coupling nearshore and aeolian processes: XBeach and dune process-based models. *Environ. Model. Softw.* **2019**, *115*, 98–112. [CrossRef]
11. Horrillo-Caraballo, J.; Karunarathna, H.; Pan, S.; Reeve, D. Performance of a data-driven technique applied to changes in wave height and its effect on beach response. *Water Sci. Eng.* **2016**, *9*, 42–51. [CrossRef]
12. Southgate, H.; Wijnberg, K.; Larson, M.; Capobianco, M.; Jansen, H. Analysis of field data of coastal morphological evolution over yearly and decadal timescales: Part 2. Non-linear techniques. *J. Coast. Res.* **2003**, *19*, 776–789.
13. Reeve, D.; Horrillo-Caraballo, J.; Magar, V. Statistical analysis and forecasts of long-term sandbank evolution at Great Yarmouth, UK. *Estuarine Coast. Shelf Sci.* **2008**, *79*, 387–399. [CrossRef]
14. Sonu, C.; James, W. A Markov Model for beach profile changes. *J. Geophys. Res.* **1973**, *78*, 1462–1471. [CrossRef]
15. Mason, S.; Hansom, J. A Markov model for beach changes on the holderness coast of England. *Earth Surf. Process. Landf.* **1989**, *14*, 731–743. [CrossRef]
16. Swales, A. Geostatistical Estimation of Short-Term Changes in Beach Morphology and Sand Budget. *J. Coast. Res.* **2002**, *18*, 338–351.

17. Ruessink, B.; Van Enckevort, I.; Kuriyama, Y. Nonlinear principal component analysis of nearshore bathymetry. *Mar. Geol.* **2004**, *203*, 185–197. [CrossRef]
18. Gunawardena, Y.; Ilic, S.; Pinkerton, H.; Romanowicz, R. Nonlinear transfer function modeling of beach morphology at Duck, North Carolina. *Coast. Eng.* **2008**, *56*, 46–58. [CrossRef]
19. Pape, L.; Ruessink, B.; Wiering, M.; Turner, I. Neural network modelling of nearshore sandbar behaviour. In Proceedings of the International Joint Conference on Neural Networks '06, Vancouver, BC, Canada, 16–21 July 2006; pp. 8743–8750.
20. Yates, M.; Guza, R.; O'Reilly, W. Equilibrium shoreline response: Observations and modeling. *J. Geophys. Res.* **2009**, *114*. [CrossRef]
21. Davidson, M.; Splinter, K.; Turner, I. A simple equilibrium model for predicting shoreline change. *Coast. Eng.* **2013**, *73*, 191–202. [CrossRef]
22. Castelle, B.; Marieu, V.; Bujan, S.; Ferreira, S.; Parisot, J.P.; Capo, S.; Sénéchal, N.; Chouzenoux, T. Equilibrium shoreline modelling of a high-energy meso-macrotidal multiple-barred beach. *Mar. Geol.* **2014**, *347*, 85–94. [CrossRef]
23. Lemos, C.; Floc'h, F.; Yates, M.; Le Dantec, N.; Marieu, V.; Hamon, K.; Delacourt, C. Equilibrium modeling of the beach profile on a macrotidal embayed low tide terrace beach. *Ocean Dyn.* **2018**, *68*, 1207–1220. [CrossRef]
24. Montaño, J.; Coco, G.; Antolínez, J.; Beuzen, T.; Bryan, K.R.; Cagigal, L.; Idier, D. Blind testing of shoreline evolution models. *Nat. Sci. Rep.* **2020**, *10*. [CrossRef]
25. Biausque, M.; Senechal, N. Seasonal morphological response of an open sandy beach to winter wave conditions: The example of Biscarrosse beach, SW France. *Geomorphology* **2019**, *332*, 157–169. [CrossRef]
26. Battiau-Queney, Y.; Billet, J.F.; Chaverot, S.; Lanoy-Ratel, P. Recent shoreline mobility and geomorphologic evolution of macrotidal sandy beaches in the north of France. *Mar. Geol.* **2003**, *194*, 31–45. [CrossRef]
27. Van de Lageweg, W.I.; Bryan, K.; Coco, G.; Ruessink, G. Observations of shoreline—Sandbar coupling on an embayed beach. *Mar. Geol.* **2013**, *344*, 101–114. [CrossRef]
28. Angnuureng, D.; Almar, R.; Senechal, N.; Castelle, B.; Addo, K.A.; Marieu, V.; Ranasinghe, R. Two and three-dimensional shoreline behaviour at a MESO-MACROTIDAL barred beach. *J. Coast. Conserv.* **2017**, *21*, 381–392. [CrossRef]
29. Haerens, P.; Bolle, A.; Trouw, K.; Houthuys, R. Definition of storm thresholds for significant morphological change of the sandy beaches along the Belgian coastline. *Geomorphology* **2012**, *143*, 104–117. [CrossRef]
30. Lebbe, L.; Van Meir, N.; Viaene, P. Potential implications of sea-level rise for Belgium. *J. Coast. Res.* **2008**, *24*, 358–366. [CrossRef]
31. Speybroeck, J.; Bonte, D.; Courtens, W.; Gheskiere, T.; Grootaert, P.; Maelfait, J.P.; Van Landuyt, W. The Belgian sandy beach ecosystem: A review. *Mar. Ecol.* **2008**, *29*, 171–185. [CrossRef]
32. Verwaest, T.; Delgado, R.; Janssens, J.; Reyns, J. Longshore sediment transport along the Belgium coast. In Proceedings of the Coastal Sediments 2011, Miami, FL, USA, 2–6 May 2011; pp. 1528–1538.
33. MacDonald, N.; O'Connor, B. Influence of Offshore banks on the adjacent coast. In Proceedings of the Coastal Engineering 1994, Kobe, Japan, 23–28 October 1994; pp. 2311–2324.
34. Afdeling Kust. Getijtafels voor Nieuwpoort, Oostende, Blankenberge, Zeebrugge, Vlissingen, Prosperpolder, Antwerpen en Wintam (Technical Report). In *Agentschap Maritieme Dienstverlening en Kust*; Vlaamse Overheid: Brussels, Belgium, 2019. Available online: https://www.vlaanderen.be/publicaties (accessed on 25 September 2020).
35. King, C.; Williams, W. The formation and movement of sand bars by wave action. *Geogr. J.* **1949**, *113*, 70–85. [CrossRef]
36. Houthuys, R. *Morfologische Trend van de Vlaamse Kust in 2011*; Technical Report; Agentschap Maritieme dienstverlening en Kust, Afdeling Kust: Oostende, Belgium, 2012; p. 150. (In Dutch)
37. Akima, H. A new method of interpolation and smooth curve fitting based on local procedures. *J. ACM* **1970**, *17*, 589–602. [CrossRef]
38. Dolan, R.; Fenster, M.; Holme, S. Spatial Analysis of Shoreline Recession and Accretion. *J. Coast. Res.* **1992**, *8*, 263–285.
39. Harley, M.; Turner, I.; Short, A.; Ranasinghe, R. Assessment and integration of conventional, RTK-GPS and image-derived beach survey methods for daily to decadal coastal monitoring. *Coast. Eng.* **2011**, *58*, 194–205. [CrossRef]

40. Cressie, N.; Wikle, C.K. *Statistics for Spatio-Temporal Data*; Wiley: New-York, NY, USA, 2011; p. 624.
41. Chun, Y.; Griffith, D.A. *Spatial Statistics and Geostatistics: Theory and Applications for Geographic Information Science and Technology*; Kwan, M.-P., Ed.; SAGE Advances in Geographic Information Science and Technology Series; University of California: Berkeley, CA, USA, 1993; p. 200.
42. Jenkins, G.; Watts, D. *Spectral Analysis and Its Applications*; Emerson-Adams Press: Boca Raton, FL, USA, 2000; p. 525.
43. Li, Y.; Lark, M.; Reeve, D. Multi-scale variability of beach profiles at Duck: A wavelet analysis. *Coast. Eng.* **2005**, *52*, 1133–1153. [CrossRef]
44. Brand, E. Intertidal Beach Morphodynamics of a Macro-Tidal Sandy Coast (Belgium). Ph.D. Thesis, Vrije Universeit Brussel, Brussel, Belgium, 2019.
45. Cohn, N.; Hoonhout, B.; Goldstein, E.; De Vries, S.; Moore, L.; Durán Vinent, O.; Ruggiero, P. Exploring Marine and Aeolian Controls on Coastal Foredune Growth Using a Coupled Numerical Model. *J. Mar. Sci. Eng.* **2019**, *7*, 13. [CrossRef]
46. Miles, A.; Ilic, S.; Whyatt, D.; James, M. Characterizing beach intertidal bar systems using multi-annual LiDAR data. *Earth Surf. Process. Land.* **2019**, *44*, 1572–1583. [CrossRef]
47. Kroon, A.; Masselink, G. Morphodynamics of intertidal bar morphology on a macrotidal beach under low-energy wave conditions, North Lincolnshire, England. *Mar. Geol.* **2002**, *190*, 591–608. [CrossRef]
48. Cartier, A.; Héquette, A. The influence of intertidal bar-trough morphology on sediment transport on macrotidal beaches, northern France. *Z. Geomorphol.* **2013**, *57*, 325–347. [CrossRef]
49. Karunarathna, H.; Brown, J.; Chatzirodou, A.; Dissanayake, P.; Wisse, P. Multi-timescale morphological modelling of a dune-fronted sandy beach. *Coast. Eng.* **2018**, *136*, 161–171. [CrossRef]
50. Voulgaris, G.; Simmonds, D.; Michel, D.; Howa, H.; Collins, M.; Huntley, D.A. Measuring and Modelling Sediment Transport on a Macrotidal Ridge and Runnel Beach: An Intercomparison. *J. Coast. Res.* **1998**, *14*, 315–330.
51. Anthony, E.J.; Levoy, F.; Monfort, O.; Degryse-Kulkarni, C. Short-term intertidal bar mobility on a ridge-and-runnel beach, Merlimont, northern France. *Earth Surf. Process. Land.* **2005**, *30*, 81–93. [CrossRef]
52. Sedrati, M.; Anthony, E. Sediment dynamics and morphological change on the upper beach of a multi-barred macrotidal foreshore, and implications for mesoscale shoreline retreat: Wissant Bay, northern France. *Z. Geomorphol. Suppl. Issues* **2008**, *52*, 91–106. [CrossRef]
53. Pedreros, R.; Howa, H.; Michel, D. Application of grain size trend analysis for the determination of sediment transport pathways in intertidal areas. *Mar. Geol.* **1996**, *135*, 35–49. [CrossRef]
54. King, C. *Beaches and Coasts*; St Martin Press: New York, NY, USA, 1972; p. 570.
55. Masselink, G.; Anthony, E. Location and height of intertidal bars on macrotidal ridge and runnel beaches. *Earth Surf. Process. Land.* **2004**, *26*, 759–774. [CrossRef]
56. Orford, J.; Wright, P. What's in a name? Descriptive or genetic implications of ridge and runnel topography. *Mar. Geol.* **1978**, *28*, M1–M8. [CrossRef]
57. Van Houwelingen, S.; Masselink, G.; Bullard, J. Characteristics and dynamics of multiple intertidal bars, north Lincolnshire, England. *Earth Surf. Process. Land.* **2006**, *31*, 428–443. [CrossRef]
58. Van Houwelingen, S.; Masselink, G.; Bullard, J. Dynamics of multiple intertidal bars over semidiurnal and lunar tidal cycles, North Lincolnshire, England. *Earth Surf. Process. Land.* **2008**, *33*, 1473–1490. [CrossRef]

Publisher's Note: MDPI stays neutral with regard to jurisdictional claims in published maps and institutional affiliations.

Journal of
Marine Science and Engineering

MDPI

Article

Development and Application of an Empirical Dune Growth Model for Evaluating Barrier Island Recovery from Storms

P. Soupy Dalyander [1,*], Rangley C. Mickey [2], Davina L. Passeri [2] and Nathaniel G. Plant [2]

[1] The Water Institute of the Gulf, Baton Rouge, LA 70802, USA
[2] U.S. Geological Survey, St. Petersburg Coastal and Marine Science Center, St. Petersburg, FL 33701, USA; rmickey@usgs.gov (R.C.M.); dpasseri@usgs.gov (D.L.P.); nplant@usgs.gov (N.G.P.)
* Correspondence: sdalyander@thewaterinstitute.org

Received: 13 November 2020; Accepted: 26 November 2020; Published: 1 December 2020

Abstract: Coastal zone managers require models that predict barrier island change on decadal time scales to estimate coastal vulnerability, and plan habitat restoration and coastal protection projects. To meet these needs, methods must be available for predicting dune recovery as well as dune erosion. In the present study, an empirical dune growth model (EDGR) was developed to predict the evolution of the primary foredune of a barrier island. Within EDGR, an island is represented as a sum of Gaussian shape functions representing dunes, berms, and the underlying island form. The model evolves the foredune based on estimated terminal dune height and location inputs. EDGR was assessed against observed dune evolution along the western end of Dauphin Island, Alabama over the 10 years following Hurricane Katrina (2005). The root mean square error with EDGR (ranging from 0.18 to 0.74 m over the model domain) was reduced compared to an alternate no-change model (0.69–0.96 m). Hindcasting with EDGR also supports the study of dune evolution processes. At Dauphin Island, results suggest that a low-lying portion of the site was dominated by overwash for ~5 years after Katrina, before approaching their terminal height and becoming growth-limited after 2010. EDGR's computational efficiency allows dune evolution to be rapidly predicted and enables ensemble predictions to constrain the uncertainty that may result if terminal dune characteristics are unknown. In addition, EDGR can be coupled with an external model for estimating dune erosion and/or the long-term evolution of other subaerial features to allow decadal-scale prediction of barrier island evolution.

Keywords: dune growth; storm recovery; barrier island; Dauphin Island

---

1. Introduction

Numerous approaches exist to estimate the response of dunes to extreme events, including storm-impact scales [1–3], relatively simple 1D models [4,5], probabilistic approaches [6–9], and complex 2D numerical models that can resolve longshore variability in dune erosion, overwash, and breaching [10]. Approaches to describe the recovery and rebuilding of dunes following major disturbances have also been developed [11,12]. The timescale of dune recovery, which is long compared to storm-response timescale, is particularly relevant to decisions regarding restoration and management of ecologically and economically valuable coastal systems [13]. Techniques are required for predicting dune formation and growth that can be used in conjunction with storm-response models to understand the evolution of the beach–dune system on decadal timescales.

Coastal dunes build when Aeolian transport carries sediment from the subaerial beach landward or seaward until a sufficient perturbation (ripples, vegetation, sand fencing, etc.) disturbs the flow. A positive feedback develops, resulting in sediment deposition and the formation of dunes that may vary spatially and temporally in height, width, and form [14–18]. The morphology of dunes is a key factor in the response of barrier islands to storms and, ultimately, their long-term evolution [1,19,20]. Several formulations have been established to relate sediment flux via Aeolian transport to wind speed [21,22]. Intermittent transport begins when a critical threshold of wind speed is reached, then increases with wind speed until continuous transport (but variable mass flux) occurs [23]. Sediment flux in beach environments rarely reaches the maximum potential transport predicted by these formulations [24–26]. For example, flux is reduced when the fetch, i.e., the length over which the wind blows, is below a critical distance [24,25,27]. In the coastal zone, fetch length is controlled by the width of the beach and the wind angle [14,28]. Fetch length at a given site varies with time due to variation in beach width and shoreline orientation driven by short- and long-term processes of coastal erosion and accretion (tides, storms, sea-level rise, etc.) [29–32]. Aeolian mass flux also depends on sediment characteristics, such as particle size and density, and bed characteristics such as sorting or the presence of surface crusts that inhibit suspension [25]. In addition, topographic features (berms, dunes, etc.) alter the local wind field and sediment entrainment [14,25,33]. Flux is also reduced below theoretical maximum levels if the sediment is moist, which increases the threshold wind speed necessary to induce transport [27,34]. Many of the factors controlling Aeolian transport are highly variable in space and time, are difficult to measure in nearshore environments, particularly on regional scales (e.g., wetting and drying, sediment grain size, and wind gustiness), or both [23,25]. For this reason, probabilistic approaches to estimate Aeolian sediment flux have been considered as an alternative to strictly deterministic methods [23,35]. Integrating models of Aeolian sediment flux up to the formation and building of coastal dunes on the landform scale remains a challenge [36], and attempts to do so on the landscape scale have occurred only recently and with limited testing in real-world environments [37,38].

The complexity of controlling factors has led to conceptual models being developed to describe the formation and evolution of coastal dunes without explicitly resolving the underlying process of Aeolian transport. Coastal dune formation following destruction by storms has been categorized into four states: (1) rapid sediment deposition at the foreshore, (2) accretion of the back beach, (3) incipient (also known as embryonic) dune formation, and (4) dune and vegetation establishment [39,40]. One of the earliest frameworks used to evaluate the terminal state of dune systems is the surf-beach-dune model, which extended the Wright and Short beach state model [41,42] to the evolution and form of coastal dunes [18,43]. Dissipative beaches, characterized by wide, shallow-sloping beaches resulting from low-energy wave climates, are associated with higher and more persistent dune features. In contrast, reflective beaches are associated with higher energy environments, narrower beaches, and lower/less persistent coastal dunes. Recent testing with computational fluid dynamics (CFD) modelling has supported this conceptual model [33]. However, there are many challenges to extending these conceptual models into numerical frameworks that are broadly applicable at the landscape scale [12]. For example, the rebuilding of dunes following a storm event relies first on the recovery of the lower and upper beach [44], and robust and accurate quantitative models for capturing this process are limited [45].

Multiple additional factors have been identified as driving the location and development of dunes [46], which also confounds the development of predictive models. A strong link exists between dune morphodynamics and vegetation, the evolution of which is complex [14,37,47]. Vegetation type and density vary considerably in space and time, driven by prevailing environmental conditions (rainfall, temperature, etc.) and feedback with dune evolution [48,49]. Sediment supply in part determines dominant vegetation species and thus affects bio-physical feedbacks [50]. Increased frictional roughness over vegetation results in sediment deposition and incipient dune formation, which can also occur along wrack and other perturbations on the beach [37,51]. The sediment trapping characteristics of vegetation control the rate of sediment deposition, while the burial and saltwater sensitivity of individual species controls vegetation survival. As a result of these feedbacks, some plant species are associated with higher, narrow dunes, while others are associated with shorter and wider landforms [37,49,52]. Because of the link between dunes and vegetation and the influence of sediment moisture on Aeolian transport, dune growth may exhibit seasonal variability [53].

As an alternative to process-based methods of predicting dune growth, Houser et al. [11] proposed an empirical approach for estimating coastal dune crest elevation increase based on a formulation developed for modeling dune building in interior (non-coastal) plains [54]. In this approach, the maximum elevation of the foredune is predicted using a sigmoid growth curve following growth rate patterns observed for dune-building vegetation. Houser et al. found that parameterized sigmoid growth curves were capable of capturing dune elevation change for two barrier islands in the northern Gulf of Mexico (Santa Rosa Island, Florida and Galveston Island, Texas) [11].

An empirical dune growth model (EDGR) has been developed to predict the evolution of barrier islands foredunes based on the sigmoid growth tendency of dunes. The model was calibrated using light detection and ranging (lidar) data from Dauphin Island, Alabama, a barrier island in the northern Gulf of Mexico that has seen extensive storm damage and recovery in previous decades (Figure 1). In the previous work, EDGR was coupled with the XBeach storm impact model [10] and the Delft3D hydrodynamic and morphodynamic model [55] to hindcast [56] and forecast [57,58] the decadal scale evolution of the island. The coupled model framework was shown to have good skill in evolving characteristics of the island; however, EDGR was not run or assessed as a standalone model in this prior work. In addition, the relative influence of different processes driving dune recovery were not explored.

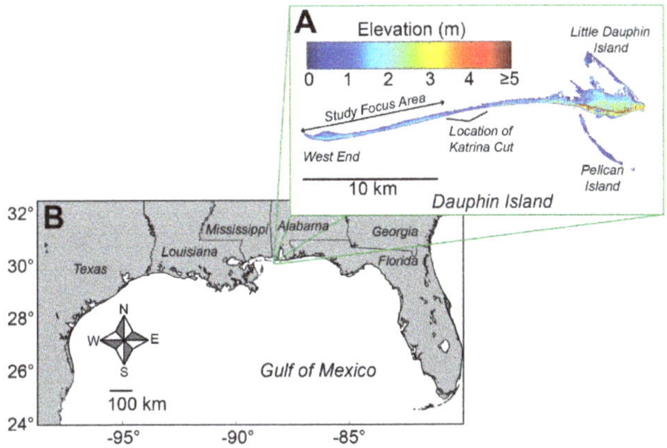

Figure 1. (**A**) Location of Dauphin Island within (**B**) the northern Gulf of Mexico.

In the current work, the EDGR model is evaluated under multiple model configurations and used to explore processes driving dune evolution at Dauphin Island. The mechanics of the model are included in the next section, along with a description of the configurations used in the current work. The results section assesses model skill under these varying configurations through comparison to observational data collected between 2005 and 2015. The discussion section illustrates how the model can be used to investigate spatial and temporal variability in dune growth, including evaluating the relative contribution of erosion and accretion processes. This section also considers the scope of model applicability and discusses its underlying assumptions.

2. Materials and Methods

2.1. Empirical Dune Growth (EDGR) Model

The EDGR dune model is briefly reviewed here with a complete description available in Table A2 of the work in [56]. The foundation of EDGR is a novel mechanism for decomposing beach or barrier island elevation profiles into the sum of a set of shape functions representing landform features, referred to hereafter as the parameterized island Gaussian fit (PIGF) method. Each cross-profile ($Z_E(x,y,t)$) is represented as the sum of Gaussian curves ($Z_{G,n}(x,y,t_1)$), where "n" enumerates each curve) representing the underlying island platform and, if present, one or more berm and/or dune features as (Figure 2; Table 1)

$$Z_E(x,y,t_1) = Z_G(x,y,t_1) + \varepsilon(x,y) = \sum_{n=1}^{n=n_{max}} Z_{G,n}(x,y,t_1) + \varepsilon(x,y). \tag{1}$$

The fit deviation (error) is captured in $\varepsilon(x,y)$. Each Gaussian is defined by height ($G_{high,n}(x,t_1)$), full width at half maximum ($G_{wide,n}(x,t_1)$), and location ($G_{Y,n}(x,t_1)$) as

$$Z_{G,n}(x,y,t_1) = G_{high,n}(x,t_1) \times \exp\left(-\frac{4\cdot \ln(2) \times (y - G_Y,n(x,t_1))^2}{G_{wide,n}(x,t_1)^2}\right). \tag{2}$$

Dune height is predicted following an equation presented in Houser et al. [11]. The total maximum elevation of the foredune ($D_{high}(x,t)$), equivalent to the height of the foredune Gaussian curve ($G_{high}(x,t)$) plus the height of the underlying island platform, is calculated from the longshore-variable initial dune elevation ($D_{high,0}$), terminal dune elevation ($D_{high,F}$), growth rate (r), and time of incipient dune formation (t_0) as

$$D_{high}(x,t) = \frac{D_{high,F}(x) \times D_{high,0}(x)}{\left(D_{high,F}(x) - D_{high,0}(x)\right)\cdot \exp(-r \times (t - t_0(x))) + D_{high,0}(x)}. \tag{3}$$

A parameterized linear model is used to calculate the foredune Gaussian width ($G_{wide}(x,t)$) from the foredune Gaussian height ($G_{high}(x,t)$) using prescribed values for slope (m_{dWdH}) and y-intercept (b_{dWdH}) as (Figure 2)

$$G_{wide}(x,t) = m_{dWdH} \times G_{high}(x,t) + b_{dWdH}. \tag{4}$$

After the foredune is evolved, EDGR reconstructs the cross-shore profiles by replacing the foredune in the island profile from the prior time step with the predicted foredune curve (Figure 2).

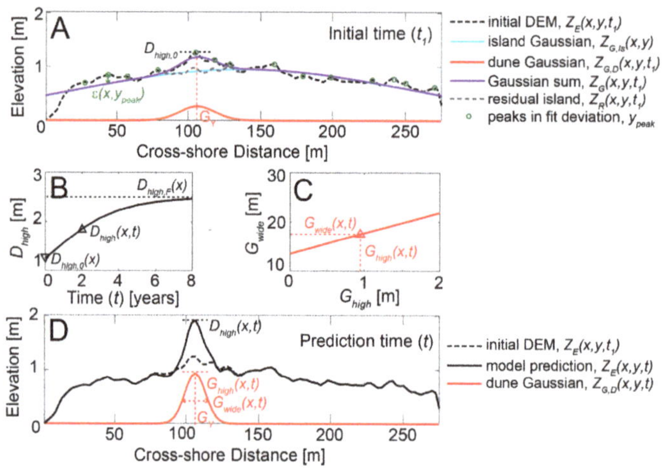

Figure 2. (**A**) Diagram illustrating the Gaussian decomposition method and empirical dune growth model (EDGR) formulations for growth in (**B**) dune height and (**C**) width. (**D**) The evolving foredune is incorporated into the rest of the cross-shore profile as it evolves. Model formulations may be found in Equations (1)–(5) with descriptions of model parameters provided in Table 1. Figure from the work in [56].

Table 1. Description of EDGR model parameters. Model formulations may be found in Equations (1)–(5) with a visual representation of model application given in Figure 2.

Parameter	Description
x	Longshore position
y	Cross-shore position
$Z_E(x,y,t_1)$	Measured cross-shore elevation profile at initial time, t_1
$Z_G(x,y,t_1)$	Gaussian decomposition of cross-shore profile at initial time, t_1
$e(x,y)$	Fit deviation (error) between $Z_E(x,y,t_1)$ and $Z_G(x,y,t_1)$
$Z_{G,n}(x,y,t)$	Gaussian curves representing the underlying island platform ($Z_{G,Is}(x,y,t)$) and one or more dunes ($Z_{G,D}(x,y,t)$), if present
$Z_R(x,y,t_1)$	Residual elevation of the cross-shore profile excluding the foredune ($Z_R(x,y,t_1) = Z_E(x,y,t_1)–Z_{G,D}(x,y,t_1)$).
$Z_E(x,y,t)$	EDGR-predicted profile at time t, derived by adding the predicted foredune elevation to the residual profile ($Z_E(x,y,t) = Z_{G,D}(x,y,t) + Z_R(x,y,t_1)$).
$D_{high}(x,t)$	Total elevation of the foredune at time t, equivalent to the height of the Gaussian curve representing the foredune ($G_{high}(x,t)$) plus the height of the residual island at the foredune location ($Z_R(x,y_D,t_1)$)
$D_{high,0}(x)$	Initial height of the foredune
$D_{high,F}(x)$	Terminal height of the foredune
r	Foredune growth rate
$G_{high}(x,t)$	Height of the Gaussian curve representing the foredune at time t
$G_{wide}(x,t)$	Width of the Gaussian curve representing the foredune at time t
m_{dWdH}	Slope of the linear model used to predict $G_{wide}(x,t)$ from $G_{high}(x,t)$
b_{dWdH}	Y-intercept of the linear model used to predict $G_{wide}(x,t)$ from $G_{high}(x,t)$

In addition to allowing foredune growth to be modeled, the PIGF method allows characteristics of individual dunes to be rapidly identified and quantified [59]. EDGR includes formulations for incipient dune formation, with dunes forming either at a prescribed location ($G_Y(x)$) for use in hindcasting or at a location based on the distance ($D_{SL,0}(s)$) to a post-storm shoreline ($SL_Y(x)$) to allow for barrier island retreat when forecasting:

$$G_Y(x) = SL_Y(x) + \Delta_{SL,0}(x) \tag{5}$$

The model can be run standalone or in conjunction with an external model that predicts storm impacts or other changes to the beach and dune not associated with foredune growth.

Model parameters ($D_{high,F}$, G_Y, r, m_{dWdH}, b_{dWdH}) can be derived in multiple ways depending on the application of the model and available data. In cases where EDGR is used to fill gaps between elevation surveys, Gaussian basis set decomposition of the initial and final elevation data can be used. Otherwise, template (proxy) data from other time periods or sites can be used under the assumption that growth will be similar for the modelled dunes. When template data are used and the final configuration of the dunes is unknown, the longshore distribution of $D_{high,F}$ for the template data is decomposed via fast Fourier transform (FFT) [60], the phases of the wavenumbers are randomized, and an inverse FFT is applied to reconstruct a new longshore distribution of $D_{high,F}$. This approach enables sensitivity testing of model predictions by allowing multiple scenarios of longshore varying dune height to be modeled, each of which retains the statistical characteristics (mean, standard deviation, and spatial scales of variability) of the template data.

2.2. Study Site and Model Configuration

The focus of the current study was on the western half of Dauphin Island, an uninhabited region that is separated from the community of Dauphin Island by Katrina Cut. This breach originally formed during Hurricanes Ivan (2004) and Katrina (2005). Although it was subsequently closed with a rock wall and has begun to infill with sediment, it provides a natural barrier between areas to the west of the Cut that evolve predominantly under natural processes and areas to the east that are heavily driven by anthropogenic influence. The island has been frequently inundated or overwashed during storm events over the last decades, resulting in multiple cycles of dune erosion and loss followed by recovery and growth. Additional information on the geophysical characteristics of Dauphin Island can be found in [61–65].

The island was divided into four analysis regions based on observed variability in dune evolution over the study period: R1, the wide western terminus of the island; R2, a narrower portion of the island associated with overwash fans in 2005 that approached or reached the bay shoreline; R3, a narrow portion of the island that incurred numerous small breaches during Katrina; and R4, the immediately adjacent to the region of Katrina Cut (Figure 3). Model parameters for Dauphin Island were previously derived from two lidar elevation surveys taken in 2005 and 2015 for each region as well as the complete model domain (Table 2 [56]). In the current study, the sensitivity of model predictions to these parameters is explored.

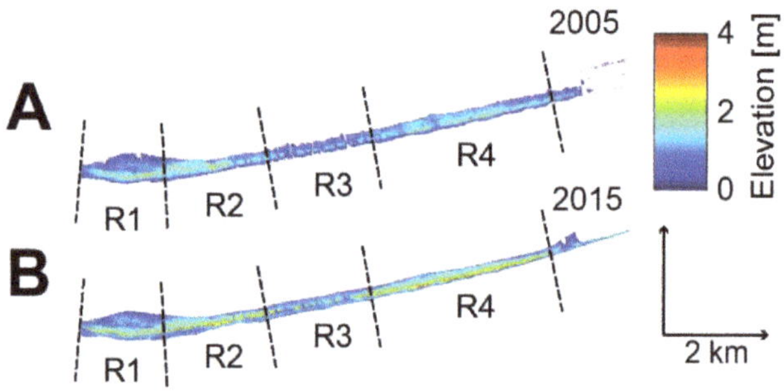

Figure 3. Elevation of Dauphin Island from light detection and ranging (lidar) surveys taken in (**A**) 2005 and (**B**) 2015; also shown are analysis regions used in the current study (dashed black lines), which were selected based on longshore variation in observed dune characteristics and behavior over the period of 2005–2015.

Table 2. EDGR model parameters calculated for each analysis region (Figure 3) based on lidar surveys in 2005 and 2015. Model parameters include initial elevation, $D_{high,0}$; terminal elevation, $D_{high,F}$; growth rate, r; distance to the 2005 shoreline, $\Delta_{SL,0}$; and m_{dWdH} and b_{dWdH}, the slope and y-intercept of the linear model used to estimate foredune Gaussian width from height.

	$D_{high,0}$ (m)	$D_{high,F}$ (m)	r (m/yr)	$D_{SL,0}$ (m)	m_{dWdH}	b_{dWdH} (m)
R1	1.46	2.51	0.58	83	3.54	18.18
R2	1.50	3.05	0.45	78	2.91	18.33
R3	0.63	2.25	0.55	71	12.66	12.23
R4	1.48	2.89	0.56	62	16.92	7.91
R1–R4	1.29	2.75	0.53	71	9.50	13.90

The model framework was applied to predict the recovery of the dunes at Dauphin Island following Hurricane Katrina over the period of August 2005 to January 2015. Data for model assessment were taken from 12 topographic lidar surveys collected over this period (Figure 4). Additional information on the processing of these data and citations for data access may be found in [56,66]. For the EDGR model application, data were interpolated to cross-shore transects with 5 m longshore resolution and 2.5 m cross-shore resolution. There were seven tropical storm events and hurricanes that passed within 200 km of the study area during this period (Figure 4; https://coast.noaa.gov/hurricanes/), resulting in dune erosion and occasional overwash and inundation of lower-lying areas of the island. Additional information on historical storms and their impacts at Dauphin Island can be found in [62,67–70].

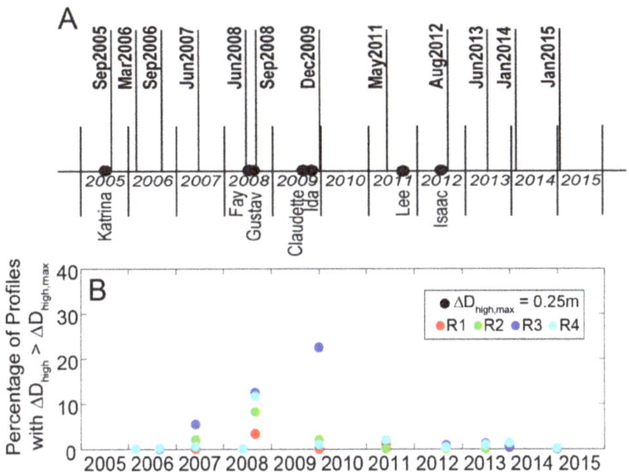

Figure 4. (**A**) Lidar surveys (black lines) taken over the study time period of August 2005 to January 2015, along with tropical storms or hurricanes that passed within 200 km of Dauphin Island (black dots). Survey data are not available for R1 and R2 for the March 2006 survey. (**B**) Percentage of cross-shore profiles for each analysis region for which the loss in elevation of dune height (ΔD_{high}) was great than 0.25 m compared to the previous survey. Taken from the work in [56].

EDGR was applied at Dauphin Island to predict dune growth from 2005 to 2015 using six model configurations (Table 3) that evaluated the model's skill and capacity for analyzing dune evolution, including testing the sensitivity of the model to the quantity of input data and applicability in predicting dune growth where local, historical dune growth data are unavailable. In the first configuration, denoted E1, EDGR operated standalone with $D_{high,F}$ and G_Y derived from 2015 data. This configuration benchmarks the mode skill in predicting dune width, cross-shore dune profile shape, and dune height between initial and terminal values. Three standalone experiments were run using FFT randomization to generate $D_{high,F}$ and $D_{SL,0}$ parameters from the 2015 lidar using different sections of the island as the terminal dune template: the entire site, denoted E2; data from region R3, denoted E3; and data from region R4, denoted E4. These configurations evaluated the skill of the model when parameterized with nonlocal data, such as in cases where historical data are unavailable for the site of interest. An additional experiment, denoted E5, was conducted using the observed dune locations and the FFT-randomized dune height values from configuration E2; this configuration allowed the error associated with a randomized terminal dune height to be isolated from errors in dune position. In the final configuration, denoted E6, EDGR was coupled to the lidar observations as a proxy erosion model to identify if elevation loss of the dune crest exceeded a defined threshold to restart dune growth. A threshold of 0.25 m was used, which was estimated to be greater than the error in the bias-corrected lidar [66]. In this case, $D_{high,0}$ was reset as the elevation of the lidar dune crest with no other data assimilation. Observed dune position and FFT-generated dune heights from experiment E5 were used so that comparison between E5 and E6 quantifies the model improvement if dune erosion is accounted for. This configuration evaluates and illustrates the use of EDGR in analyzing the relative contribution of erosion and accretion processes to dune evolution.

Table 3. Model configurations used to predict dune growth at Dauphin Island from 2005 to 2015. L2005 is a "no-change" model taken as the 2005 lidar, and the other six models are different configurations of EDGR. $D_{high,F}$ is terminal dune height and $D_{SL,0}$ is distance to shore. In all experiments, the growth rate ® is fixed at 0.53, and the slope and y-intercept of the linear model used to estimate foredune Gaussian width from height (m_{dWdH} and b_{dWdH}) are fixed at 9.5 and 13.9.

Experiment ID.	$D_{high,F}$ (m)	$D_{SL,0}$	Realizations	Coupled to Lidar?	Threshold for Reset of $D_{high,0}$
L2005	n/a: no change model using 2005 lidar				
E1	Prescribed from 2015 values	Prescribed from 2015 values	1	No	n/a
E2	FFT with Regions 1–4 in template	FFT with Regions 1–4 in template	100	No	n/a
E3	FFT with Region 3 in template	FFT with Region 3 in template	100	No	n/a
E4	FFT with Region 4 in template	FFT with Region 4 in template	100	No	n/a
E5	FFT with Regions 1–4 in template	Prescribed from 2015 values	100	No	n/a
E6	FFT with Regions 1–4 in template	Prescribed from 2015 values	100	Yes	0.25 m

The 0.5 m contour in the 2005 lidar survey was taken as the shoreline location in calculating $D_{SL,0}$ so that profiles with incomplete coverage of the intertidal zone could be included. For experiments using FFT randomization (E2–E6), 100 different sets of terminal dune characteristics were generated and run for each template configuration and model performance is presented as the mean and standard deviation of error statistics calculated individually for each of the 100 initializations. As a benchmark of EDGR's skill, an alternate "no-change" model (the 2005 lidar, hereafter referred to as L2005) was also evaluated. The L2005 model assessment represents the error that would result from using increasingly outdated dune topography. Dune height and position were extracted from L2005 as the point of maximum island elevation. EDGR time steps were chosen such that the model generated output at the times of the observed lidar survey (Figure 4, time step of several months to upwards of two years). The longshore and cross-shore resolution of the model matched the gridded lidar (2.5 m cross-shore resolution and 5 m longshore resolution).

3. Results

3.1. Prescribed Terminal Dunes (Experiment E1)

EDGR formulations for identification of $Z_R(x,y,t_1)$ and growth of the foredune as a Gaussian form were robust in application at Dauphin Island (Figure 5). Across R1–R4, the Gaussian fitting method for isolating the residual island profile in the 2005 initial DEM converged for 1912 out of 1918 of the cross-shore profiles (99.7%). A portion of these profiles were subaqueous in 2005 and/or a foredune was not identifiable in the 2015 lidar survey. The Gaussian foredune building method was successfully applied to all other profiles (100%, 94%, 62%, and 96% of profiles in R1–R4). The largest errors in the predicted DEM were seaward and shoreward of the EDGR-evolved foredune, i.e., in areas where the initial elevations were not updated by the model (Figure 5).

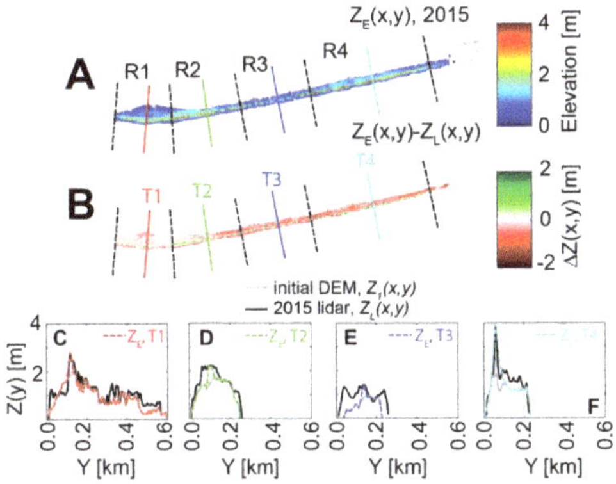

Figure 5. (**A**) EDGR 2015 model prediction for experiment E1 (Table 3) and (**B**) difference between the prediction and 2015 lidar. From within each region R1–R4 (demarcated by dashed black lines), a representative cross-shore transect has been selected (shown as colored lines in panels (**A**,**B**) and denoted as T1–T4). The initial (gray line) and 2015 lidar elevation profiles (black line) are shown for each of these transects (panels (**C**–**F**)), along with the EDGR prediction for 2015 (colored dashed line, where the color for each transect is the same as in panels (**A**,**B**).

$Z_E(x,y,t)$ was compared to the corresponding lidar ($Z_L(x,y,t)$) over the area ($Z_{dune}(x,y,t)$) that EDGR updated elevation ($|Z_E(x,y,t){-}Z_L(x,y,t_0)| > 0.05$ m) (Figure 6). EDGR's RMSE for Z_{dune} increased slightly over the first 1–4 years post-storm before leveling out (0.31–0.50 m in 2015; lowest RMSE in R1). Bias was relatively consistent over time (−0.35 to 0.14 m). R2 decreased somewhat over the first 1–4 years in R2–R4 before leveling out to 0.51–0.61, while remaining consistently high for R1 (0.66–0.77). These predictions were benchmarked against L2005 over the same cross-shore area; the no-change model prediction skill initially deteriorated before leveling off in 2012–2015.

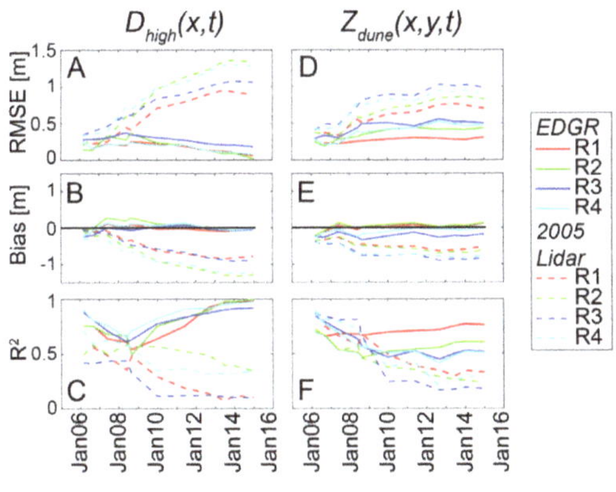

Figure 6. (**A**,**D**) Root mean square error (RMSE), (**B**,**E**) bias, and (**C**,**F**) coefficient of determination (R2) for experiment E1 (Table 3). D_{high} (left column) was taken as the maximum dune elevation for each cross-shore profile. The same statistics are shown for the 2005 lidar. Statistics for $Z_{dune}(x,y,t)$ (right column) are calculated at all cross- and longshore locations where EDGR updated the elevation for a given lidar survey.

3.2. Template Terminal Dunes (Experiments E2–E6)

Using R1–R4 as the EDGR template (experiment E2) improved predictions of $D_{SL,0}$ over L2005 in R1–R3 while performing similarly to L2005 in R4 (Figure 7; Tables A1 and A2). EDGR had a negative bias in $\Delta_{SL,0}$ for the western regions (−15 ± 10 m and −8 ± 8 m for R1 and R2) and a slight positive bias for the eastern regions (3 ± 8 m and 7 ± 7 m for R3 and R4). RMSE in $\Delta_{SL,0}$ was similar throughout the study site, decreasing slightly from east to west (R1, 36 ± 6 m; R2, 27 ± 4 m, 27 ± 3 m, and 25 ± 3 m). R2 exhibited no spatial pattern and ranged from 0.06 to 0.18, with low values reflecting the high amount of scatter associated with randomizing terminal dune position. Spatial variation in the post-Katrina shoreline with a length scale greater than the 100 m Hanning filter is evident in the error in $\Delta_{SL,0}$ (Figure 8). Both EDGR and L2015 have spikes in error at locations where there are small discontinuities in the foredune, such as where overwash or breaching resulted in the dune reforming farther inland than the primary foredune at adjacent locations.

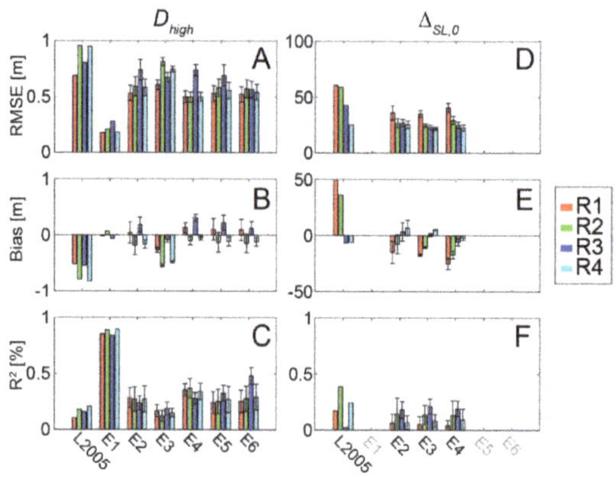

Figure 7. (**A**,**D**) Root mean square error (RMSE), (**B**,**E**) bias, and (**C**,**F**) coefficient of determination (R2) for model predictions using six experiment configurations of EDGR (E1–E6; Table 3) of D_{high} (**A**–**C**) and $\Delta_{SL,0}$ (**D**–**F**). Statistics are calculated over all lidar surveys from 2006 to 2015. For EDGR configurations using template dunes, the value shown is the mean over the 100 realizations, with the vertical bar indicating plus and minus one standard deviation. Error statistics for $\Delta_{SL,0}$ are not shown in cases (gray) where dune location was prescribed using historical data.

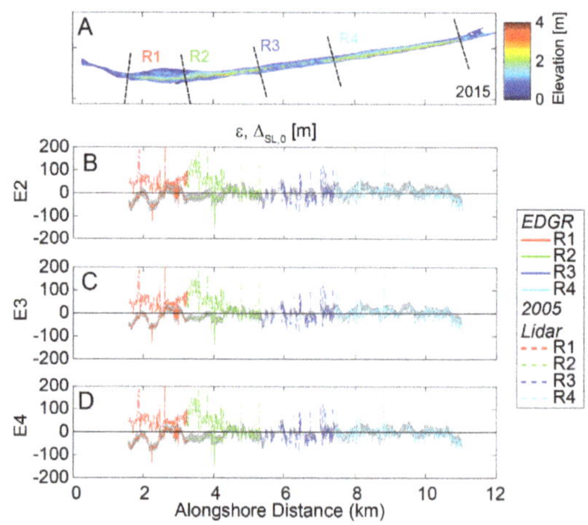

Figure 8. (**A**) Dauphin Island (2015 topography shown). Error in terminal dune position for three EDGR configurations (**B**–**D**; see Table 3) referenced as distance to the post-Katrina shoreline (ε,$\Delta_{SL,0}$) for 100 independently initialized FFT randomizations (colored lines represent the mean, gray represents the standard deviation). The same statistics are shown for the 2005 lidar, which was used as the initial elevation in EDGR.

Experiment E2 had similar bias to L2005 for D_{high} in the first 1–2 years following Hurricane Katrina, after which time EDGR began to outperform L2005 (Figure 9; Tables A1 and A2). EDGR bias was positive for D_{high} in R1 (0.16 ± 0.26 m in January 2015) and R3 (0.44 ± 0.23 m in January 2015) and negative in R2 (−0.30 ± 0.24 m in January 2015) and R4 (−0.18 ± 0.11 m in January 2015). RMSE was similar for regions R1, R2, and R4 (0.69 ± 0.10 m, 0.74 ± 0.12 m, and 0.74 ± 0.10 m in January 2015) and lower than for L2005 after the first 2–4 years post-Katrina. RMSE in EDGR was somewhat higher for R3 (0.93 ± 0.16 m in January 2015) and was of a similar magnitude to RMSE for L2005 until 7–8 years post-Katrina. R2 for D_{high} in EDGR was similar across all regions (mean of 0.03–0.05 for R1–R4 in 2015) and lower than for E1 and L2005.

Temporal patterns in model performance in experiment E3 (data from R3 as the dune template) and experiment E4 (data from R4 as the dune template) were similar to experiment E2 (full island used as the dune template), but use of a more spatially confined template with less longshore variability reduced the range of variability in predictions and error (Figures 7 and 9). Experiment E3 had similar error statistics to E2 in $\Delta_{SL\prime 0}$ (mean RMSE varied by less than 3 m; mean bias varied by less than 3 m; and mean R2 varied by less than 0.03). A larger effect was seen in experiment E4 (mean RMSE varied by up to 5 m and mean bias became more negative by up to 10 m). For D_{high}, use of R3 as the template (experiment E3) increased RMSE in R1, R2, and R4 by up to 0.22 m; created a more negative bias; and reduced R2. As expected, RMSE and magnitude of bias in R3 were reduced slightly (by 0.07 and 0.10 m). RMSE for configuration E4 decreased in R1, R2, and R4 and increased in R3, resulting in an RMSE for this region of ~0.25 m more than the rest of the study site. Bias became more positive compared to experiment E2 (up to 0.31 m in R3), while R2 increased somewhat for each region. The most notable effect of using prescribed dune locations (experiments E5 vs. E2) on D_{high} was that bias became slightly more positive in all regions (change of 4–6 cm).

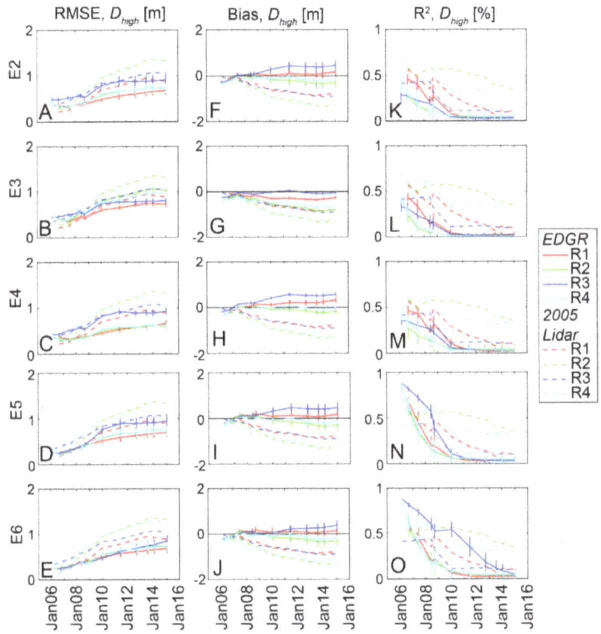

Figure 9. (**A–E**) Root mean square error (RMSE), (**F–J**) bias, and (**K–O**) coefficient of determination (R2) of D_{high} for EDGR applied to predict dune growth from 2005 to 2015. Each row represents a different configuration of EDGR (Table 3), with model predictions benchmarked against a no-change model represented by 2005 lidar (L2005).

The coupled configuration (E6) allowed evaluation of dune crest erosion through time. Elevation loss (ΔD_{high}) throughout the study domain occurred more frequently prior to 2010 (Figure 4). The most widespread losses occurred following Hurricane Gustav in 2008, with $\Delta D_{high} > 0.25$ m in 3–13% of profiles across R1–R4. The following year, $\Delta D_{high} > 0.25$ m occurred in 22% of profiles in R3, with minimal exceedance of this threshold (<2%) in R1, R2, and R4. After 2010, $\Delta D_{high} > 0.25$ m was minimal (<2% for all regions and surveys). Comparing experiment E6 to the identical standalone run (experiment E5), RMSE decreased in all regions with the largest improvement (0.17 m) in R3 (Figure 7, Figure 9). This region also saw a slight improvement in magnitude of model bias (0.05 m) and R2 (0.09), while error statistics in the other regions negligibly improved or degraded slightly (e.g., increase in magnitude of negative bias in R2).

4. Discussion

EDGR relies on several assumptions that define the scope of model applicability, including that (1) the cross-shore structure and alongshore variability of a barrier island can be characterized by spatial decomposition with a basis set of Gaussian curves, and (2) this spatial decomposition supports a temporal process decomposition, separating dune growth from other coastal changes. In the following sections, we discuss the validity of these assumptions and draw conclusions on the generalizability of the modeling approach for broader scientific, engineering, and coastal management applications.

4.1. Dune Characterization

The EDGR model may be applied when a scale separation allows the forms of the barrier island platform and foredune to be isolated from each other. In its present formulation, the model uses the PIGF method and assumes Gaussian forms for the platform and any dune or berm features. Additional testing at other sites is required to test the validity of this approach at other barrier islands and to determine if alternate approaches can be used at sites where features are not well approximated by a Gaussian form. For example, a linear approximation for the residual elevation could be applied in application to mainland coasts, or a half-waveform model could be used to represent the ocean-facing half of the domain [71]. The PIGF method can also be modified to use a piecewise function approach to capture different regions of the cross-shore profile, such as using a linear fit to approximate the beach [59].

A benefit of spatial decomposition is that it enables EDGR as a mechanism for characterizing dunes in addition to predicting growth. Applying a dune template and benchmarking against historical data allows evaluation of the similarity of the study site to the template and sensitivity testing to dune drivers and response. For example, the bias in $\Delta_{SL,0}$ became more positive moving from west to east in all experiments at Dauphin Island using FFT-randomization, reflecting formation of the dunes somewhat closer to the post-Katrina shoreline where the island narrows in width. D_{high} did not exhibit the same trend and was overpredicted in R1 and R3 when R1–R4 or R4 was used as the template, and underpredicted in the other two regions. D_{high} bias became more positive in all four regions when known terminal dune locations were prescribed in the model (experiment E5) rather than using template locations (experiment E2) despite using an identical distribution of $D_{high,F}$. This variation was found to have occurred because the 2005 elevation tended to be higher at the location of actual foredune formation than at the template location, resulting in larger values of $D_{high,I}$. However, the foredune did not typically form at the overall elevation maximum in each post-storm cross-shore profile (captured in the error in L2005 dune location). Incipient dunes are known to form at the site of flow perturbations, which include local elevation highs. The results at Dauphin Island suggest there may be a site-specific window of potential dune formation constrained by distance to shore.

Decomposition of lidar profiles into a Gaussian basis set for parameterization of m_{dWdH} and b_{dWdH} (Equation (4)) allowed rapid and automated analysis of the relationship between dune width (G_{wide}) and height above the residual island (G_{high}). There was considerable profile-to-profile variation in the relationship of G_{wide} to G_{high}, but a temporal pattern emerged wherein low dunes in the first 1–2 years after Katrina had a relatively steep relationship between G_{wide} and G_{high} (Figure 10). As the dunes grew over time, they also became narrower for a given height (decrease in m_{dWdH}). The relationship between G_{wide} to G_{high} followed this trend over the first 6–8 years following Katrina, after which the relationship stabilized in most of the analysis regions. This result is consistent with prior findings suggesting that the dune ramp—a relatively shallow sloping feature at the base of the dune that is incorporated in the Gaussian form—is a source of sediment for dune crest growth [72].

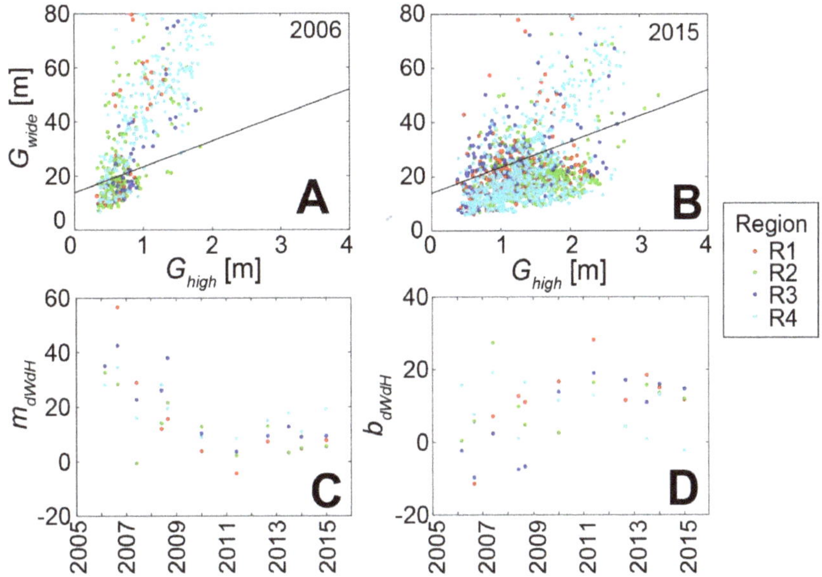

Figure 10. Relationship between the height (G_{high}) and width (G_{wide}) of the foredune identified via sum-of-Gaussians fitting of observed cross-shore profiles (Figure 2). Shown are (**A**) September 2006 (~1 year post-Katrina) and (**B**) January 2015 (~10 years post-Katrina). Boundaries of the analysis regions (R1–R4) are found in Figure 3. The relationship between G_{high} and G_{wide} was found to vary in time (linear model slope, m_{dWdH}, and intercept, b_{dWdH}, shown in panels (**C**,**D**). This temporal variability was beyond the scope of this initial study and a single relationship (black line in panels (**A**,**B**)) was used in all EDGR applications.

4.2. Dune Evolution

A requirement for EDGR application is the availability of sufficient data for model initialization, which can be a challenge for coastal models in general [73]. The skill of the model using R3 and R4 as the template (experiments E3 and E4) suggests that the minimum longshore spatial scale of data needed to parameterize the model is on the order of 2–4 km. Data on terminal dune state and the time scale of post-storm dune recovery are required, while testing of the influence of the growth rate parameter suggests that the model is relatively less sensitive to the availability of interim data points during post-storm recovery. In cases where local dune recovery data are unavailable, the limited impact on prediction error for a given region of the study site when using a different region as the template indicates that dune data may be transferred effectively between similar sites using the FFT method. In the application at Dauphin Island, EDGR outperformed the no-change model (L2005) within each region in terms of bias and RMSE even when R3, the region most dissimilar to the rest of the study site, was used as the template (experiment E3). Data from a prior time period are also likely to be an appropriate template for future growth at any given site under the assumption that recovery from different storms will be broadly similar. It may be possible to generate template values from a more distant location when dune recovery data at a target site are limited or unavailable, thus increasing the general applicability of the model to include sites with insufficient historical data for model parameterization. Gutierrez et al. [74] found that D_{high} is correlated to factors such as the shoreline change rate and beach width, suggesting that such information can be used to constrain the choice of template at sites where long-term dune evolution data are unavailable. Factors known to control dune building rate and form (vegetation type, dominant wind direction and speed, sediment grain size, etc.) should also be assessed and compared between the study area and potential EDGR template sites.

Where sufficient data are available, EDGR can be used to evaluate the relative contribution of processes of erosion and accretion to dune evolution. Using R3 as the template (experiment E3) increased the negative bias in D_{high} for R2 and R4 and introduced a negative bias in R1, while excluding R3 from the template (experiment E4) improved model performance in all the other regions. These results suggest that either R3 has the lowest terminal dune heights of the study site (i.e., $D_{high,F}$ is constrained by dune-building processes) or frequent overwash inhibits dunes from approaching their terminal state (i.e., D_{high} is constrained by erosion). The relative influence of these processes can be considered by comparing otherwise identical standalone (experiment E5) to lidar-coupled (experiment E6) model predictions. From 2005 to 2010, R3 had the highest percentage of profiles with dune elevation loss between lidar surveys exceeding 0.25 m (Figure 4). In particular, R3 exhibited widespread erosion in the 2008–2009 time period that was not observed in the other regions. Coupling to lidar reduced the RMSE and bias, and increased R2 for R3 over the period of 2005–2012 (Figure 9). From 2010 to 2015, few profiles in any of the regions exhibited loss of D_{high} greater than the 0.25 m, indicating limited overwash during this time. However, positive bias in R3 predictions began to grow in the coupled model after 2012 and exceeded the bias in the standalone run using R3 as the template (experiment E3). This result suggests that reduced capacity for dune building is also contributing to the relatively low dune heights found in R3. Saltwater inundation impacts vegetation type, resulting in a secondary effect of storms on dune evolution [75]. Therefore, one possible explanation may be that the higher frequency of overwash during the 2005–2010 period altered the local vegetation. Alternately, sediment supply and/or other factors controlling Aeolian flux may differ in this region. The coupled model is also informative as to what types of events are driving significant dune erosion at Dauphin Island. Two periods of widespread elevation loss (in 2008 and 2009) corresponded to passage of tropical storms within 200 km of Dauphin Island, while elevation loss in 2007 suggests that extratropical storms and/or far-field tropical storms have also resulted in overwash over portions of the island.

The application of EDGR to investigate dune growth also highlighted potential model modifications. In the current version, incremental changes to the dune profile from an external erosion model (those which do not reset D_y or $D_{high,0}$) are discarded in a coupled framework. Dune elevation change in EDGR could alternately be formulated based on differential dune growth [11] as

$$\frac{dD_{high}(x,t)}{dt} = r \times D_{high}(x,t) \times \left[1 - \frac{D_{high}(x,t)}{D_{high,F}(x)}\right]. \tag{6}$$

Positive elevation change (from EDGR) and negative elevation change (from an erosion model) could then be explicitly resolved in a coupled framework. Doing so would require iterative calibration of the growth parameter to describe uninhibited dune elevation gain, which cannot be derived from observational field data that includes times of overwash and inundation. Dune width change could similarly be reformulated and calibrated within EDGR as dD_{wide}/dt, which may resolve some of the challenges found in identifying a robust relationship between G_{high} and G_{wide}. Reformulation in this manner would further segregate the impacts of Aeolian and hydrodynamic sediment transport on dune evolution, thus facilitating explicit examination of these processes and the factors that control them in addition to potentially improving model performance.

The EDGR model could also be modified to replace the Houser dune growth formulation with a formulation that included time-varying environmental conditions that influence dune growth [14,21–28].

For example, growth rate (r) could be formulated as a function of predicted Aeolian sediment flux, which could either be derived from an existing process-based model (see, e.g., in [37,38]) or parameterized as a function of wind speed, fetch length, and sediment bed characteristics (grain size, moisture content). Vegetation growth and sediment trapping [14,37,47] could similarly be explicitly modeled or characteristics of dune form (width, height) parameterized as a function of dominant vegetation type [53]. Modification of the model formulation to include explicit linkage to environmental characteristics or processes correlated to dune growth would replace the need for historical dune growth data with a requirement for, e.g., site-specific wind and vegetation data to predict dune growth parameters.

4.3. Broader Model Application

When initial and terminal dune characteristics are known and prescribed (experiment E1), EDGR predicts the interim evolution of dune height and the cross-shore profile of the dune. In this case, the model has the highest accuracy in predicting spatially and temporally variable dune elevations. Even when the precise spatial distribution of terminal dune characteristics is unknown, EDGR captures spatially averaged characteristics in dune parameters and had lower region-wide bias than the no-change alternative model (Figure 7). Capturing island-scale dune statistics (experiments E2–E6) is of similar value to predicting the precise longshore distribution (experiment E1) in some cases. For example, dune habitat may be evaluated based on regional or island-wide acreage [65]. Similarly, estimates of the probability of island-scale overwash and inundation [2] may be improved by updating estimates of dune maximum elevation. In cases where the longshore distribution of dune state is required [8,76], EDGR's computational efficiency enables extensive sensitivity testing and ensemble prediction to constrain possible future scenarios. For reference, the 1918 cross-shore profiles initialized in under 30 s and the model completed simulations for 2005 to 2015 in less than 5 min while running on a Windows desktop computer with 2.4 GHz Intel Xeon CPU. This efficiency enables a wide range of dune outcomes and associated uncertainty to be generated and evaluated. Last, EDGR may be coupled with other models that evolve the storm impacts and other changes to the coast not associated with the foredune as part of a model framework for predicting decadal-scale coastal evolution [56–58]. Therefore, the EDGR method is readily applicable to coastal vulnerability and restoration studies and associated management applications.

5. Conclusions

In the current study, an empirical dune growth model (EDGR) was developed and used to predict the recovery of the primary foredune of a barrier island following erosion during storms. Cross-shore profiles are first approximated as the sum of a set of Gaussian curves representing dune and berm features atop an underlying island form using a novel parameterized island Gaussian fit (PIGF) method. The foredune Gaussian is evolved with maximum height following the Houser et al. [11] growth model and width calculated from a new linear formulation relating dune width to height. Terminal dune location and height can be prescribed if known, such as when EDGR is being used to fill temporal data gaps between post-storm and recovered lidar surveys. Alternatively, EDGR can create a longshore distribution of terminal dune parameters via fast Fourier transform randomization of available template data from a region that is assumed to be similar to the study site. Bulk characteristics (mean and standard deviation) and length-scales of spatial variability from the template are preserved, while EDGR's high computational efficiency allows multiple potential realizations to be rapidly generated and tested to constrain the range of possible outcomes. EDGR can operate as a standalone model or be coupled with an external model that predicts dune erosion and/or evolution of other characteristics of the barrier island, such as the shoreline.

EDGR was used to predict foredune evolution of the western portion of Dauphin Island over the 10-year period following widespread dune destruction by Hurricane Katrina (2005) and compared to 12 topographic lidar surveys. Model predictions were benchmarked against 2005 lidar, representing elevation values that would be used if updated lidar were unavailable. EDGR improved predictions of maximum dune elevation over the no-change model, with highest performance when known terminal dune characteristics were prescribed. This application also illustrated the potential of the model to be used in investigating dune growth processes. Use of the Gaussian basis set to extract dune width and height for model parameterization revealed that the relationship between dune height and width evolved as the dunes grew; over time, dunes became relatively narrower for a given height. In addition, comparison between a standalone model configuration and one wherein EDGR was coupled to lidar surveys as a proxy for a dune erosion model suggests that dune heights in a lower elevation portion of the study site were constrained by erosion in the first few years following Katrina. Relatively limited elevation gain compared to other portions of the study site in more recent years, when dune crest erosion was not identified in the lidar, suggests that the area has lower terminal heights, and is thus now growth-limited. This application illustrates that EDGR is a computationally efficient and landform-resolving method for predicting the decadal-scale evolution of barrier island foredunes and for evaluating dune evolution drivers and response.

Author Contributions: Conceptualization, P.S.D.; methodology, P.S.D., R.C.M., and N.G.P.; software, P.S.D., and R.C.M; validation, D.L.P. and R.C.M.; formal analysis, P.S.D. and R.C.M.; investigation, P.S.D. and N.G.P.; data curation, P.S.D. and R.C.M.; writing—original draft preparation, P.S.D.; writing—review and editing, D.G.P, N.G.P., and R.C.M.; visualization, P.S.D.; supervision, P.S.D.; project administration, P.S.D.; funding acquisition, P.S.D. and N.G.P. All authors have read and agreed to the published version of the manuscript.

Funding: EDGR was developed as part of the Alabama Barrier Island Restoration Assessment, funded by the National Fish and Wildlife Foundation (NFWF). Portions of the research presented here were also funded by the USGS.

Acknowledgments: P.S.D. would like to acknowledge Joseph Long (U.N.C. Wilmington), Robert Jenkins (USGS), and Dave Thompson (USGS) for input during the development of EDGR. The authors would also like to thank several anonymous reviewers who provided feedback on drafts of the manuscript. The Dauphin Island project for which EDGR was developed was a collaborative effort between the U.S. Geological Survey (USGS), U.S. Army Corps of Engineers (USACE), and the State of Alabama to investigate restoration options that protect and restore the natural resources of Dauphin Island, Alabama. The authors would like to acknowledge these organizations for their role in motivating and refining this research. Any use of trade, firm, or product names is for descriptive purposes only and does not imply endorsement by the U.S. Government.

Conflicts of Interest: The authors declare no conflict of interest.

Appendix A

Included below are skill metrics of EDGR predictions of dune position (Table A1) and height (Table A2) for six configurations of the model (Table 3) applied to Dauphin Island. Configurations E2–E6 are based on 100 realizations of randomized dune configurations based on a template of historical dune data; values given in this case are the mean and standard deviation over the set of realizations for each model configuration.

Table A1. Root mean square error (RMSE), bias, and coefficient of determination (R^2) of dune position referenced as distance to the 2005, post-storm shoreline ($\Delta_{SL,0}$) for six configurations of EDGR and an alternate, no-change model taken as the 2005 lidar (Table 3).

Region	R1			R2			R3			R4		
Statistic	RMSE (m)	Bias (m)	R^2 (%)	RMSE (m)	Bias (m)	R^2 (%)	RMSE (m)	Bias (m)	R^2 (%)	RMSE (m)	Bias (m)	R^2 (%)
L2005	61	50	0.17	59	36	0.39	43	−7	0.02	25	−6	0.24
E1	0	0	1	0	0	1	0	0	1	0	0	1
E2	36 ± 6	−15 ± 10	0.06 ± 0.08	27 ± 4	−8 ± 8	0.14 ± 0.14	27 ± 3	3 ± 8	0.18 ± 0.07	25 ± 3	7 ± 7	0.07 ± 0.06
E3	35 ± 3	−18 ± 1	0.05 ± 0.07	25 ± 2	−10 ± 1	0.14 ± 0.08	23 ± 2	1 ± 1	0.21 ± 0.06	22 ± 1	5 ± 0	0.08 ± 0.06
E4	41 ± 4	−25 ± 5	0.04 ± 0.05	30 ± 3	−17 ± 3	0.13 ± 0.13	25 ± 3	−6 ± 3	0.19 ± 0.07	23 ± 3	−2 ± 2	0.09 ± 0.10
E5	0 [1]	0 [1]	1 [1]	0 [1]	0 [1]	1 [1]	0 [1]	0 [1]	1 [1]	0 [1]	0 [1]	1 [1]
E6	0 [1]	0 [1]	1 [1]	0 [1]	0 [1]	1 [1]	0 [1]	0 [1]	1 [1]	0 [1]	0 [1]	1 [1]

[1] EDGR perfectly captures $\Delta_{SL,0}(x)$ in configurations where D_y is prescribed.

Table A2. Root mean square error (RMSE), bias, and coefficient of determination (R^2) of dune height (D_{high}) for six configurations of EDGR and an alternate, no-change model taken as the 2005 lidar (Table 3).

Region	R1			R2			R3			R4		
Statistic	RMSE (m)	Bias (m)	R^2 (%)	RMSE (m)	Bias (m)	R^2 (%)	RMSE (m)	Bias (m)	R^2 (%)	RMSE (m)	Bias (m)	R^2 (%)
L2005	0.69	−0.52	0.11	0.96	−0.79	0.18	0.81	−0.54	0.16	0.95	−0.82	0.21
E1	0.18	−0.02	0.85	0.21	0.07	0.89	0.28	−0.06	0.84	0.18	0	0.89
E2	0.53 ± 0.07	0.04 ± 0.19	0.28 ± 0.09	0.59 ± 0.09	−0.18 ± 0.17	0.27 ± 0.11	0.74 ± 0.09	0.18 ± 0.14	0.24 ± 0.06	0.58 ± 0.07	−0.16 ± 0.08	0.27 ± 0.12
E3	0.61 ± 0.04	−0.27 ± 0.04	0.17 ± 0.05	0.81 ± 0.04	−0.54 ± 0.03	0.12 ± 0.05	0.67 ± 0.04	−0.08 ± 0.04	0.19 ± 0.06	0.75 ± 0.03	−0.48 ± 0.02	0.15 ± 0.04
E4	0.50 ± 0.05	0.13 ± 0.09	0.35 ± 0.05	0.50 ± 0.05	−0.10 ± 0.07	0.37 ± 0.09	0.74 ± 0.05	0.31 ± 0.06	0.28 ± 0.05	0.50 ± 0.04	−0.05 ± 0.04	0.34 ± 0.07
E5	0.53 ± 0.07	0.10 ± 0.19	0.24 ± 0.10	0.58 ± 0.08	−0.13 ± 0.17	0.25 ± 0.11	0.69 ± 0.09	0.22 ± 0.14	0.33 ± 0.07	0.56 ± 0.07	−0.11 ± 0.08	0.27 ± 0.12
E6	0.52 ± 0.07	0.10 ± 0.19	0.25 ± 0.10	0.57 ± 0.08	−0.14 ± 0.17	0.28 ± 0.11	0.57 ± 0.07	0.12 ± 0.12	0.48 ± 0.08	0.54 ± 0.07	−0.12 ± 0.08	0.29 ± 0.11

References

1. Sallenger, J. Storm impact scale for barrier islands. *J. Coast. Res.* **2000**, *16*, 890–895.
2. Stockdon, H.F.; Sallenger, A.H.; Holman, R.A.; Howd, P.A. A simple model for the spatially-variable coastal response to hurricanes. *Mar. Geol.* **2007**, *238*, 1–20. [CrossRef]
3. Long, J.W.; De Bakker, A.T.M.; Plant, N.G. Scaling coastal dune elevation changes across storm-impact regimes. *Geophys. Res. Lett.* **2014**, *41*, 2899–2906. [CrossRef]
4. Larson, M.; Erikson, L.; Hanson, H. An analytical model to predict dune erosion due to wave impact. *Coast. Eng.* **2004**, *51*, 675–696. [CrossRef]
5. Palmsten, M.L.; Holman, R.A. Laboratory investigation of dune erosion using stereo video. *Coast. Eng. Proc.* **2012**, *60*, 123–135. [CrossRef]
6. Heijer, C.D.; Knipping, D.T.; Plant, N.G.; Vries, J.S.V.T.D.; Baart, F.; Van Gelder, P. Impact Assessment of Extreme Storm Events Using a Bayesian Network. *Coast. Eng. Proc.* **2012**, *1*. [CrossRef]
7. Lentz, E.E.; Hapke, C.J.; Stockdon, H.F.; Hehre, R.E. Improving understanding of near-term barrier island evolution through multi-decadal assessment of morphologic change. *Mar. Geol.* **2013**, *337*, 125–139. [CrossRef]
8. Plant, N.G.; Stockdon, H.F. Probabilistic prediction of barrier-island response to hurricanes. *J. Geophys. Res. Space Phys.* **2012**, *117*, 17. [CrossRef]
9. Wilson, K.E.; Adams, P.N.; Hapke, C.J.; Lentz, E.E.; Brenner, O.T. Application of Bayesian Networks to hindcast barrier island morphodynamics. *Coast. Eng.* **2015**, *102*, 30–43. [CrossRef]
10. Roelvink, J.A.; Reniers, A.; Van Dongeren, A.; Vries, J.V.T.D.; McCall, R.; Lescinski, J. Modelling storm impacts on beaches, dunes and barrier islands. *Coast. Eng.* **2009**, *56*, 1133–1152. [CrossRef]
11. Houser, C.; Wernette, P.; Rentschlar, E.; Jones, H.; Hammond, B.; Trimble, S. Post-storm beach and dune recovery: Implications for barrier island resilience. *Geomorphology* **2015**, *234*, 54–63. [CrossRef]
12. Keijsers, J.G.S.; Poortinga, A.; Riksen, M.J.P.M.; Maroulis, J. Spatio-Temporal Variability in Accretion and Erosion of Coastal Foredunes in the Netherlands: Regional Climate and Local Topography. *PLoS ONE* **2014**, *9*, e91115. [CrossRef]
13. Dalyander, P.S.; Meyers, M.; Mattsson, B.J.; Steyer, G.; Godsey, E.; McDonald, J.; Byrnes, M.R.; Ford, M. Use of structured decision-making to explicitly incorporate environmental process understanding in management of coastal restoration projects: Case study on barrier islands of the northern Gulf of Mexico. *J. Environ. Manag.* **2016**, *183*, 497–509. [CrossRef]
14. Luna, M.C.D.M.; Parteli, E.J.; Durán, O.; Herrmann, H.J. Model for the genesis of coastal dune fields with vegetation. *Geomorphology* **2011**, *129*, 215–224. [CrossRef]
15. Jackson, N.L.; Nordstrom, K.F. Aeolian sediment transport and landforms in managed coastal systems: A review. *Aeolian Res.* **2011**, *3*, 181–196. [CrossRef]
16. Psuty, N.P. Sediment budget and dune/beach interaction. *J. Coast. Res.* **1988**, *3*, 1–4.
17. Psuty, N.P. Spatial Variation in Coastal Foredune Development. In Proceedings of the 3rd European Dune Congress, Galway, Ireland, 17–21 June 1992; pp. 3–13.
18. Hesp, P.A. Surfzone, beach, and foredune interactions on the Australian SE coast. *J. Coast. Res.* **1988**, *3*, 15–25.
19. Larson, M.; Palalane, J.; Hallin, C.; Hanson, H. Simulating cross-shore material exchange at decadal scale. Theory and model component validation. *Coast. Eng.* **2016**, *116*, 57–66. [CrossRef]
20. Lorenzo-Trueba, J.; Ashton, A.D. Rollover, drowning, and discontinuous retreat: Distinct modes of barrier response to sea-level rise arising from a simple morphodynamic model. *J. Geophys. Res. Earth Surf.* **2014**, *119*, 779–801. [CrossRef]
21. Bagnold, R.A. The movement of desert sand. *Proc. R. Soc. Math Phys. Eng. Sci.* **1936**, *157*, 594–620. [CrossRef]
22. Sherman, D.J.; Jackson, D.W.; Namikas, S.L.; Wang, J. Wind-blown sand on beaches: An evaluation of models. *Geomorphology* **1998**, *22*, 113–133. [CrossRef]
23. Davidson-Arnott, R.; Bauer, B. Aeolian sediment transport on a beach: Thresholds, intermittency, and high frequency variability. *Geomorphology* **2009**, *105*, 117–126. [CrossRef]
24. Davidson-Arnott, R.G.; Law, M.N. Measurement and Prediction of Long-term Sediment Supply to Coastal Foredunes. *J. Coast. Res.* **1996**, *12*, 654–663.
25. Delgado-Fernandez, I. A review of the application of the fetch effect to modelling sand supply to coastal foredunes. *Aeolian Res.* **2010**, *2*, 61–70. [CrossRef]

26. Jackson, D.W.; Cooper, A. Beach fetch distance and aeolian sediment transport. *Sedimentology* **1999**, *46*, 517–522. [CrossRef]
27. Davidson-Arnott, R.G.; MacQuarrie, K.; Aagaard, T. The effect of wind gusts, moisture content and fetch length on sand transport on a beach. *Geomorphology* **2005**, *68*, 115–129. [CrossRef]
28. Bauer, B.O.; Davidson-Arnott, R.G.D. A general framework for modeling sediment supply to coastal dunes including wind angle, beach geometry, and fetch effects. *Geomorphology* **2003**, *49*, 89–108. [CrossRef]
29. De Vries, S.; Arens, S.; De Schipper, M.A.; Ranasinghe, R. Aeolian sediment transport on a beach with a varying sediment supply. *Aeolian Res.* **2014**, *15*, 235–244. [CrossRef]
30. De Vries, S.; Vries, J.V.T.D.; Van Rijn, L.; Arens, S.; Ranasinghe, R. Aeolian sediment transport in supply limited situations. *Aeolian Res.* **2014**, *12*, 75–85. [CrossRef]
31. Gutierrez, B.T.; Plant, N.G.; Thieler, E.R. A Bayesian network to predict coastal vulnerability to sea level rise. *J. Geophys. Res. Space Phys.* **2011**, *116*, 1–15. [CrossRef]
32. Da Silva, G.M.; Hesp, P.A. Coastline orientation, aeolian sediment transport and foredune and dunefield dynamics of Moçambique Beach, Southern Brazil. *Geomorphology* **2010**, *120*, 258–278. [CrossRef]
33. Hesp, P.A.; Smyth, T.A.G. Surfzone-Beach-Dune interactions: Flow and Sediment Transport across the Intertidal Beach and Backshore. *J. Coast. Res.* **2016**, *75*, 8–12. [CrossRef]
34. Jackson, N.L.; Nordstrom, K.F. Effects of Time-dependent Moisture Content of Surface Sediments on Aeolian Transport Rates Across a Beach, Wildwood, New Jersey, USA. *Earth Surf. Process. Landf.* **1997**, *22*, 611–621. [CrossRef]
35. Zhang, W.; Schneider, R.; Kolb, J.; Teichmann, T.; Dudzinska-Nowak, J.; Harff, J.; Hanebuth, T.J. Land–sea interaction and morphogenesis of coastal foredunes—A modeling case study from the southern Baltic Sea coast. *Coast. Eng.* **2015**, *99*, 148–166. [CrossRef]
36. Walker, I.J.; Davidson-Arnott, R.G.D.; Bauer, B.O.; Hesp, P.A.; Delgado-Fernandez, I.; Ollerhead, J.; Smyth, T.A. Scale-dependent perspectives on the geomorphology and evolution of beach-dune systems. *Earth Sci. Rev.* **2017**, *171*, 220–253. [CrossRef]
37. Hoonhout, B.; De Vries, S. A process-based model for aeolian sediment transport and spatiotemporal varying sediment availability. *J. Geophys. Res. Earth Surf.* **2016**, *121*, 1555–1575. [CrossRef]
38. Janssen, T. Aeolian Transport on a Beach: Testing the AeoLIS Aeolian Sediment Transport Model Against the Observed Recovery of Fire Island. Master's Thesis, Delft University of Technology, Delft, The Netherlands, 2016. Available online: https://repository.tudelft.nl/islandora/object/uuid:79d1805e-4a90-404f-bece-c3801eed5d6f?collection=education (accessed on 30 November 2020).
39. Hesp, P. Foredunes and blowouts: Initiation, geomorphology and dynamics. *Geomorphology* **2002**, *48*, 245–268. [CrossRef]
40. Morton, R.A.; Paine, J.G.; Gibeaut, J.C. Stages and durations of post-storm beach recovery, southeastern Texas Coast, USA. *J. Coast. Res.* **1994**, *10*, 884–908.
41. Short, A.D. Three Dimensional Beach-Stage Model. *J. Geol.* **1979**, *87*, 553–571. [CrossRef]
42. Wright, L.; Short, A. Morphodynamic variability of surf zones and beaches: A synthesis. *Mar. Geol.* **1984**, *56*, 93–118. [CrossRef]
43. Short, A.; Hesp, P. Wave, beach and dune interactions in southeastern Australia. *Mar. Geol.* **1982**, *48*, 259–284. [CrossRef]
44. McLean, R.; Shen, J.-S. From Foreshore to Foredune: Foredune Development Over the Last 30 Years at Moruya Beach, New South Wales, Australia. *J. Coast. Res.* **2006**, *221*, 28–36. [CrossRef]
45. Pender, D.; Karunarathna, H. A statistical-process based approach for modelling beach profile variability. *Coast. Eng.* **2013**, *81*, 19–29. [CrossRef]
46. Nordstrom, K.F. Coastal Dunes. In *Coastal Environments and Global Change*; John Wiley and Sons: London, UK, 2015; pp. 178–193.
47. Lalimi, F.Y.; Silvestri, S.; Moore, L.J.; Marani, M. Coupled topographic and vegetation patterns in coastal dunes: Remote sensing observations and ecomorphodynamic implications. *J. Geophys. Res. Biogeosci.* **2017**, *122*, 119–130. [CrossRef]
48. Bitton, M.C.A.; Hesp, P.A. Vegetation dynamics on eroding to accreting beach-foredune systems, Florida panhandle. *Earth Surf. Process. Landf.* **2013**, *38*, 1472–1480. [CrossRef]
49. Durán, O.; Moore, L.J. Vegetation controls on the maximum size of coastal dunes. *Proc. Natl. Acad. Sci. USA* **2013**, *110*, 17217–17222. [CrossRef]

50. Zarnetske, P.L.; Ruggiero, P.; Seabloom, E.W.; Hacker, S.D. Coastal foredune evolution: The relative influence of vegetation and sand supply in the US Pacific Northwest. *J. R. Soc. Interface* **2015**, *12*, 20150017. [CrossRef]
51. Nordstrom, K.F.; Jackson, N.L.; Korotky, K.H.; Puleo, J.A. Aeolian transport rates across raked and unraked beaches on a developed coast. *Earth Surf. Process. Landf.* **2010**, *36*, 779–789. [CrossRef]
52. Zarnetske, P.L.; Hacker, S.D.; Seabloom, E.W.; Ruggiero, P.; Killian, J.R.; Maddux, T.B.; Cox, D. Biophysical feedback mediates effects of invasive grasses on coastal dune shape. *Ecology* **2012**, *93*, 1439–1450. [CrossRef] [PubMed]
53. Van Puijenbroek, M.E.B.; Limpens, J.; De Groot, A.V.; Riksen, M.J.; Gleichman, M.; Slim, P.A.; Van Dobben, H.F.; Berendse, F. Embryo dune development drivers: Beach morphology, growing season precipitation, and storms. *Earth Surf. Process. Landf.* **2017**, *42*, 1733–1744. [CrossRef]
54. Hugenholtz, C.H.; Wolfe, S.A. Biogeomorphic model of dunefield activation and stabilization on the northern Great Plains. *Geomorphology* **2005**, *70*, 53–70. [CrossRef]
55. Lesser, G.; Roelvink, J.; Van Kester, J.; Stelling, G. Development and validation of a three-dimensional morphological model. *Coast. Eng.* **2004**, *51*, 883–915. [CrossRef]
56. Mickey, R.C.; Long, J.W.; Dalyander, P.S.; Jenkins, R.L.; Thompson, D.M.; Passeri, D.L.; Plant, N.G. *Development of a Modeling Framework for Predicting Decadal Barrier Island Evolution*; US Geology Survey: Reston, VA, USA, 2020; p. 46.
57. Mickey, R.C.; Godsey, E.; Dalyander, P.S.; Gonzalez, V.M.; Jenkins, R.L.; Long, J.W.; Thompson, D.M.; Plant, N.G. *Application of Decadal Modeling Approach to Forecast Barrier Island Evolution, Dauphin Island, Alabama*; US Geology Survey: Reston, VA, USA, 2020; p. 45.
58. Passeri, D.L.; Dalyander, P.S.; Long, J.W.; Mickey, R.C.; Jenkins, R.L.; Thompson, D.M.; Plant, N.G.; Godsey, E.S.; Gonzalez, V.M. The Roles of Storminess and Sea Level Rise in Decadal Barrier Island Evolution. *Geophys. Res. Lett.* **2020**, *47*. [CrossRef]
59. Mickey, R.C.; Dalyander, P.S.; McCall, R.; Passeri, D.L. Sensitivity of Storm Response to Antecedent Topography in the XBeach Model. *J. Mar. Sci. Eng.* **2020**, *8*, 829. [CrossRef]
60. Brigham, E.O. *The Fast Fourier Transform and Its Applications*; Prentice Hall: Upper Saddle River, NJ, USA, 1998.
61. Watkins, A.D. A Synthesis of Alabama Beach States and Nourishment Histories. Master's Thesis, University of Alabama, Tuscaloosa, AL, USA, 2011. Available online: http://acumen.lib.ua.edu/u0015/0000001/0000723/u0015_0000001_0000723.pdf (accessed on 30 November 2020).
62. Douglass, S.L. Beach erosion and deposition on Dauphin Island, Alabama, USA. *J. Coast. Res.* **1994**, *10*, 306–328.
63. De Velasco, G.G.; Winant, C.D. Seasonal patterns of wind stress and wind stress curl over the Gulf of Mexico. *J. Geophys. Res.* **1996**, *101*, 18127–18140. [CrossRef]
64. Dzwonkowski, B.; Park, K. Influence of wind stress and discharge on the mean and seasonal currents on the Alabama shelf of the northeastern Gulf of Mexico. *J. Geophys. Res.* **2010**, *115*, 12052. [CrossRef]
65. Enwright, N.M.; Borchert, S.M.; Day, R.H.; Feher, L.C.; Osland, M.J.; Wang, L.; Wang, H. *Barrier Island Habitat Map and Vegetation Survey—Dauphin Island, Alabama, 2015*; US Geological Survey: Reston, VA, USA, 2017; p. 17.
66. Thompson, D.M.; Dalyander, P.S.; Long, J.W.; Plant, N.G. *Correction of Elevation Offsets in Multiple Co-Located Lidar Datasets*; US Geological Survey: Reason, VA, USA, 2017.
67. Morton, R.A. Historical Changes in the Mississippi-Alabama Barrier-Island Chain and the Roles of Extreme Storms, Sea Level, and Human Activities. *J. Coast. Res.* **2008**, *246*, 1587–1600. [CrossRef]
68. Froede, C.R. The Impact that Hurricane Ivan (16 September 2004) Made across Dauphin Island, Alabama. *J. Coast. Res.* **2006**, *223*, 561–573. [CrossRef]
69. Froede, C.R. Changes to Dauphin Island, Alabama, Brought about by Hurricane Katrina (29 August 2005). *J. Coast. Res.* **2008**, *4*, 110–117. [CrossRef]
70. Morton, R.A.; Miller, T.L.; Moore, L.J. *National Assessment of Shoreline Change: Part 1, Historical Shoreline Changes and Associated Coastal Land Loss Along the U.S. Gulf of Mexico*; US Geological Survey: Reston, VA, USA, 2004. Available online: https://pubs.usgs.gov/of/2004/1043/ (accessed on 30 November 2020).
71. Plant, N.G.; Freilich, M.H.; Holman, R.A. Role of morphologic feedback in surf zone sandbar response. *J. Geophys. Res. Space Phys.* **2001**, *106*, 973–989. [CrossRef]
72. Christiansen, M.B.; Davidson-Arnott, R.G.D. Rates of Landward Sand Transport over the Foredune at Skallingen, Denmark and the Role of Dune Ramps. *Geogr. Tidsskr. Dan. J. Geogr.* **2004**, *104*, 31–43. [CrossRef]
73. Houser, C. Synchronization of transport and supply in beach-dune interaction. *Prog. Phys. Geogr. Earth Environ.* **2009**, *33*, 733–746. [CrossRef]

74. Gutierrez, B.T.; Plant, N.G.; Thieler, E.R.; Turecek, A.M. Using a Bayesian network to predict barrier island geomorphologic characteristics. *J. Geophys. Res. Earth Surf.* **2015**, *120*, 2452–2475. [CrossRef]
75. Wolner, C.W.; Moore, L.J.; Young, D.R.; Brantley, S.T.; Bissett, S.N.; McBride, R.A. Ecomorphodynamic feedbacks and barrier island response to disturbance: Insights from the Virginia Barrier Islands, Mid-Atlantic Bight, USA. *Geomorphology* **2013**, *199*, 115–128. [CrossRef]
76. McCall, R.; Vries, J.V.T.D.; Plant, N.; Van Dongeren, A.; Roelvink, J.; Thompson, D.; Reniers, A. Two-dimensional time dependent hurricane overwash and erosion modeling at Santa Rosa Island. *Coast. Eng.* **2010**, *57*, 668–683. [CrossRef]

Journal of
Marine Science and Engineering

MDPI

Article

Investigating Changes in Aeolian Sediment Transport at Coastal Dunes and Sand Trapping Fences: A Field Study on the German Coast

Christiane Eichmanns * and Holger Schüttrumpf

Institute of Hydraulic Engineering and Water Resources Management (IWW), RWTH Aachen University, Mies-van-der-Rohe-Straße 17, 52056 Aachen, Germany; schuettrumpf@iww.rwth-aachen.de
* Correspondence: eichmanns@iww.rwth-aachen.de

Received: 6 November 2020; Accepted: 7 December 2020; Published: 10 December 2020

Abstract: For the restoration and maintenance of beach and dune systems along the coast, knowledge of aeolian sediment transport and its interaction with coastal protection measures is required. As a nature-based solution, sand trapping fences can be an integral part of coastal protection measures initiating foredune development. There are few detailed studies on aeolian sediment transport rates on coastal dunes and sand trapping fences available to date. Thus, in this work, we present the results of field experiments conducted at the beach, coastal dune, and sand trapping fence on the East Frisian island Langeoog. The vertical sediment flux profile was measured by vertical mesh sand traps, and saltiphones measured the instantaneous sediment transport. A meteorological station was set up to obtain wind data. On the beach, dune toe, and dune crest, the stationary wind profile can be described well by the law of the wall. Saturated aeolian sediment transport rates on the beach and dune toe were predicted by widely used empirical models. Between the sand trapping fence, these empirical transport models could not be applied, as no logarithmic wind profile existed. The upwind sediment supply reduced after each brushwood line of the sand trapping fence, thereby, leading to increased deviation from the saturated conditions.

Keywords: aeolian sediment transport; wind flow; field experiments; nature-based solutions; sand trapping fences

1. Introduction

Coastal protection measures on the East Frisian Islands in the German North Sea generally involve coastal dunes, coastal dikes, and associated coastal protection structures [1,2]. The main roles of the coastal dunes on the East Frisian Islands are to provide flood protection for flood-risk areas against storm surges and to secure the existence of these islands [2–5]. These islands form a natural barrier system that reduces the wave energy during storm surges on the mainland coast in Lower Saxony. The islands are, therefore, maintained to reduce flood risks for the hinterland. The maintenance of the coastal dunes at the German North Sea coast generally involves sand nourishment at the coastal dunes, beach, and foreshore [2,6]. Over the past decades, sand nourishments have been an effective method for ensuring coastal protection of the island, to bridge temporary phases of sedimentary deficit caused by sandbar relocations [2,7]. The planting of vegetation and installing sand trapping fences to stabilize the natural dynamics of the dune system represent an additional component of coastal protection on the East Frisian Islands [8].

Nature-based solutions in coastal engineering such, as installing sand trapping fences aim to use natural processes and natural resources to contribute to coastal protection. Nature-based solutions are becoming increasingly popular due to the growing environmental awareness among the population, the increasing economic pressure on coastal areas, and the need for multidisciplinary approaches to meet the challenges of coastal protection. As natural processes can be a part of coastal protection measures, nature-based solutions could be a cost-effective alternative in the implementation of coastal infrastructure [7,9–11].

Sand trapping fences are widely installed along barrier islands to enhance the formation of foredunes after storm surges [12–14]. By installing sand trapping fences on the seaward dune slope close to the dune toe, the onshore wind velocity as well as the wind velocity parallel to the dune are reduced, and sediment can accumulate downwind and between the individual brushwood lines of a sand trapping fence. Using these sand trapping fences systematically, aeolian sedimentation processes can contribute to coastal protection by initiating and facilitating the establishment of the dune toe [15,16].

Both natural as well as anthropogenically modified coastal dunes offer protection against storm surges and sea-level rising. Coastal dunes generally erode due to hydrodynamic forces during storm surges [17,18]. The process of dune erosion and wave–dune interaction is complex in terms of the highly dynamic interaction of soil and water but has been studied extensively in large-scale and small-scale flumes [17,19–21], while the understanding of the preservation of coastal dunes, in general, is still limited [7,22].

Coastal dunes grow predominantly due to aeolian sediment transport processes [23].Therefore, to support the restoration and maintenance of beach and dune systems, accurate knowledge of aeolian sediment transport processes and the effectiveness of coastal protection measures to strengthen coastal dunes is required [3,22,24]. Reliable prediction of sediment transport rates requires an understanding of the fundamental geomorphological processes causing changes in topography, aeolian ripple formation, aeolian ripple evolution, and saltation processes [25].

Generally, a good empirical knowledge of aeolian sediment transport rates in deserts exists [26,27]; however, the prediction of aeolian sediment transport rates in coastal areas is a major challenge due to the variety of influencing factors, such as the vegetation, moisture, wind force, and bed roughness [26,28–30]. Numerous authors have developed prediction models for saturated aeolian sediment transport rates for dry sand under equilibrium conditions in wind tunnel experiments [23,27,31–34].

However, the aeolian sediment transport rates in wind tunnel experiments often differ from the measured sediment transport rates in the field [22,27,35]. The natural systems in coastal areas are very complex due to temporal-spatially changing conditions, which are generally neglected in most transport models, but strongly affect sediment transport rates [7,27,28,36–39]. In addition, to date there are few detailed studies available on aeolian sediment transport rates at different locations on the beach and/or in the coastal dunes [7,22]. In particular, the design of the sand trapping fences and their effect on aeolian sediment transport models have rarely been investigated [12,16,40].

Therefore, the objective of this work was to investigate the changes in aeolian sediment transport at coastal dunes and sand trapping fences. The five research goals were:

(1) Describing the wind profile, shear velocity, and aerodynamic roughness length for the beach, dune toe, and dune crest of the foredunes as well as before and between the sand trapping fence.
(2) Quantifying the vertical sediment flux profile for these measurement locations.
(3) Evaluating the performance of aeolian sediment transport models in comparison to the measured sediment transport rates under consideration of prevailing boundary conditions.
(4) Evaluating the downwind evolution of instantaneous sediment transport at these measurement locations.
(5) Describing the influence of sand trapping fences on aeolian sediment transport.

2. Study Site

In Figure 1 the location of the study site in Germany (a) belonging to the federal state Lower Saxony as part of the East Frisian Islands (b) is shown. The study site in the northern part of the East Frisian island Langeoog (c) is shown in Figure 1d.

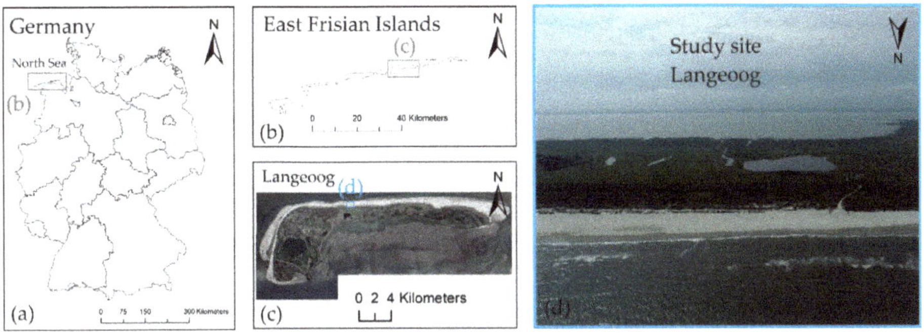

Figure 1. (**a**) Location of the study site in Germany (adapted from [41], with permission from © GeoBasis-DE/ BKG, 2020) belonging to the (**b**) East Frisian Islands, federal state Lower Saxony (adapted from [41], with permission from © GeoBasis-DE/ BKG, 2020) as part of the (**c**) island Langeoog (adapted from [42] with permission from © dl-de/by-2-0, 2020), as well as the location of the (**d**) study site in the Northern part of Langeoog.

The East Frisian Islands are located in the German North Sea and form a barrier island system that developed in the Holocene. Seven permanently inhabited sandy barrier islands stretch over ~90 km and are dissected by six main tidal inlets [2,6]. They are separated from the mainland (distance to mainland ~3–20 km) by a large muddy intertidal area with tidal flats. The Holocene sea level rise, the abundant sediment availability, large-scale embankments, and storm surges have contributed significantly to the development and shape of these barrier islands [43,44]. Up to now, the longshore migration of the intertidal sandbars in the coastal foreland from west to east in irregular periods of decades due to the tides, currents, wave loads, and storm surges caused a significant change in the sediment budget for the East Frisian Islands and resulted in a temporary supply or deficit of sand for these islands [1,2,43,45].

The East Frisian island Langeoog covers an area of ~20 km^2 and has ~20.3 km long coastal dunes stretching from the southwestern side of the island northward toward the east. No grey coastal protection structures are present. From the seaward side to inland, Langeoog's morphology consists of a natural sandy beach followed by foredunes and older dune landscapes with dunes up to 20 m. In the southwest, Langeoog has extensively large areas with salt marshes [2,6,46,47].

Langeoog can be categorized as a mesotidal barrier island. Tides in this area are semi-diurnal with a tidal range of around T_R ~2.7 m at the port entrance [47–49]. At each tidal cycle, the tidal basins with intertidal areas and supratidal salt marshes are filled and subsequently emptied [50]. The main wave direction is from southwest to northwest [51].

The climate is temperate [52] with an annual average temperature of ~11 °C (Norderney, January 2015 to December 2019) and an annual average rainfall of ~754 mm (Norderney, January 2015 to December 2019). The wind direction are mainly from southwest to northwest characterized by wind velocities up to ~30 m/s [53].

Figure 2 shows a close-up aerial photograph of the sand trapping fence. The sand trapping fence was installed from 18 May 2020 to 19 May 2020 by the Lower Saxony Water Management, Coastal Protection and Nature Conservation Agency and was made from locally available brushwood that was put into the sand and arranged in lines parallel and orthogonal to the coastal dune [2,8]. In Figure 2a, an aerial drone photograph and (b) a close-up of the sand trapping fence at the study site of Langeoog are shown.

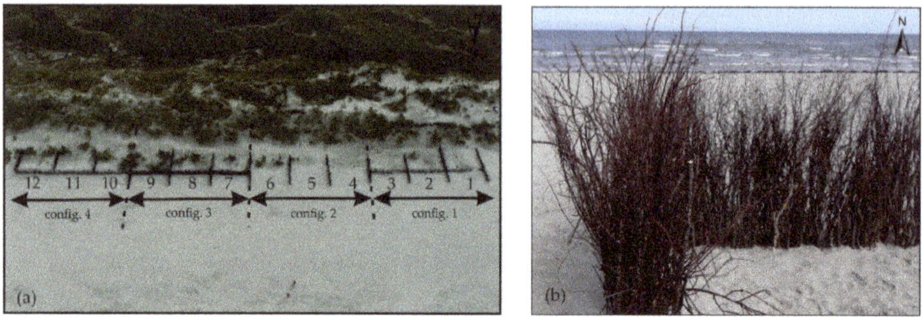

Figure 2. (**a**) Aerial drone photograph of the sand trapping fence with different configurations and (**b**) a close-up of the sand trapping fence consisting of brushwood on the study site of Langeoog (20 May 2020).

When the brushwood bushes were delivered, they were bundled by wires. As far as possible, the wires were removed from the bundles of brushwood. The investigations were conducted in front of field 1, in field 1, and in field 2 (all part of configuration 1). These fields were characterized by three bundles of brushwood per running meter orthogonal and two bundles of brushwood per running meter parallel to the coastal dune [54]. Currently, the design of the protruding brushwood height as well as the porosity of the sand trapping fences as well as their position in relation to the beach profile, and the length of the resulting fields between the sand trapping fences on the East Frisian Islands are based on empirical knowledge.

Figure 3 shows the top view of the study site with all four measurement locations: the beach, the area close to the dune toe, the area on the dune crest of the foredunes, and the sand trapping fence at the dune toe. The seaward side of the foredune is in part protected by a sand trapping fence.

Figure 3. Overview of the different measurement locations: beach, dune toe, dune crest, and sand trapping fence at the study site Langeoog. The aerial photograph was obtained on 19 May 2020.

All four measurement locations differ a lot in their prevailing boundary conditions. A closer visual impression of the locations (a) beach, (b) dune toe, (c) dune crest, and (d) sand trapping fence is given in Figure 4a.

Figure 4. Closer visual impression of the measurement location at the (**a**) beach, (**b**) dune toe, (**c**) dune crest on the foredune, and (**d**) sand trapping fence (position 1) on Langeoog, Germany. Blue arrows indicate the respective wind direction during measurements at each measurement location.

For the measurement location beach, the sand was relatively wet (M = 7.08–10.38 Vol%). A few shell fragments were identified on the beach, see Figure 4a. Close to the dune toe, the sand was relatively dry with a moisture content of maximum M = 1.8 Vol% and without shells or shell fragments, see Figure 4b. Many small sand ripples were recognized moving downwind. On the dune crest, dune grass of varying heights (with a medium height of about 0.5 m) and irregularities in the topography of the dune crest area of about 2 m were present, see Figure 4c. In front of the sand trapping fence, sand was relatively wet with a maximum moisture value of M = 8.73 Vol%. Sand was deposited in the form of a large, rippled sandbar, see Figure 4d, in front of the sand trapping fence. In the sand trapping fence, the moisture content of the sand varied between M = 1.80–8.90 Vol%, and no sand ripples were present.

3. Methodology

3.1. Experimental Setup and Instrumentation

The field experiments began immediately after the completion of the installation of the sand trapping fence and were conducted from 19 May 2020 to 24 May 2020. They were focused on the aeolian transport processes and rates in correlation with the wind regime. The mean sediment transport rates were measured with three modified vertical mesh sand traps constructed according to Sherman et al. (2014) [55], covering a height of z = 0.3 m above the surface, see Figure 5a. The vertical mesh sand traps consisted of six rectangular aluminium tubes arranged one above the other. Each rectangular aluminium tube was b = 0.1 m wide, h = 0.05 m high, and l = 0.25 m long, with edges e = 2 mm thick. To enable sand to be trapped, each tube was covered with a bag made of nylon mono filament with openings of $size$ = 50 µm. A long nylon bag l = 0.75 m was used for the lowest tube, for all other heights the nylon back was l = 0.5 m long. A threaded rod and rubber bands fixed the nylon bags to the frame.

This design of the vertical mesh sand traps was used in field experiments by Campos (2018) [56]. It is a cost effective measurement tool with good efficiency results [55].

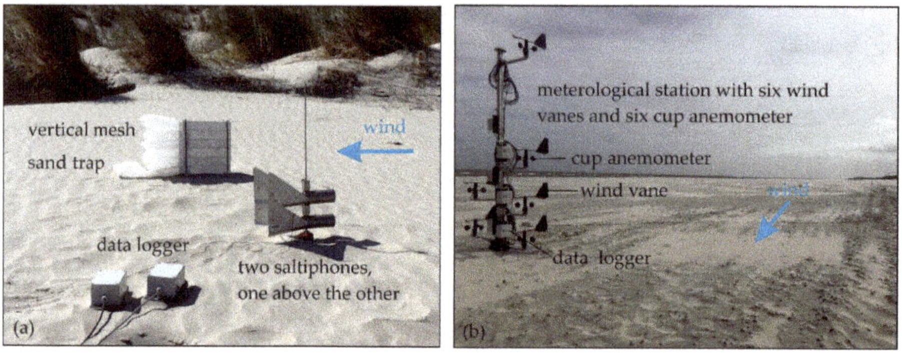

Figure 5. Measuring equipment at the beach of Langeoog, Germany, during the field campaign on 23 May 2020. (**a**) Modified vertical mesh sand trap as well as an array of two saltiphones with respective data loggers, (**b**) meteorological measuring station with six cup anemometers, six wind vanes, and six data loggers. Blue arrows indicate the respective wind direction during measurements.

Six saltiphones from the company Eijkelkamp, see also Figure 5a, were used to measure the instantaneous sediment transport. The saltiphones consist of a microphone, installed in a stainless-steel housing. Each sediment grain hitting the microphone is counted. The maximum cumulative number of impacts per second is generally limited to 1000. The overall measurement accuracy is given with 5% [57]. The data of the saltiphones were recorded by a data logger that was built by the IWW. Two wind vanes on the back of the tube ensured that the microphone was positioned in the wind direction. The saltiphones were calibrated by checking that all six saltiphones gave nearly the same amount of counted grains hitting the microphone for a defined sediment probe. For this purpose, a weighed mass of quartz sand was poured into the housing of each saltiphone at a defined distance. Detailed information about the principle of saltiphones is given by Spaan and van der Abeele (1991) [58] and Goossens et al. (2018) [59].

In this field campaign, a total of three saltiphone stations, each station consisting of two saltiphones, were used. They were placed in such a way that the centre of the microphones was arranged at $z = 0.05$ m and $z = 0.15$ m above the bed surface. Thereby, the instantaneous sediment transport at different altitudes could be estimated. However, in weak wind conditions, it can be possible that the transport layer is so low that most particles pass underneath the lowest sensor [22].

A meteorological station, see also Figure 5b, was set up to obtain the wind velocity and wind direction at the study site.

The sensors OMEGA WMS-23 consisted of a cup anemometer to register the wind velocity and a wind vane to measure the wind direction relative to north. The wind sensors were mounted on a vertical bar at predefined heights of $z = 0.13, 0.17, 0.45, 0.59, 0.91$, and 1.95 m and the wind vanes were at 0.24, 0.28, 0.56, 0.70, 1.02, and 2.06 m, respectively. The wind data were recorded with a measuring rate of $f = 10$ Hz. The measuring accuracy of the wind vane is stated with ± 3° and for the wind anemometer it is the larger value of ± 3% or 0.45 m/s [60].

An unmanned aerial vehicle (UAV, manufacturer DJI Phantom 4 with real time kinematics) was used to obtain ortho-image data, which was referenced by four ground control points. Ortho-images taken by the drone achieved an accuracy of 1 cm + 1 parts-per-million (ppm) of distance between the rover and base horizontally and 1.5 cm + 1 ppm of distance between the rover and base vertically [61]. The exact position of all measuring equipment and reference points were registered additionally using a global navigation satellite system JAVAD GNSS Receiver SigmaD with the same accuracy horizontally and vertically as the drone [62].

A sedimentological analysis according to the specifications in ISO 17892-4 [63] was conducted to determine the grain size distribution at the study site. The spatially variable moisture contents at the vertical mesh sand traps, at the saltiphones, and at the meteorological station were measured by a moisture content sensor Theta-Probe ML3 to quantify the volumetric water content of the sand up to 6 cm beneath its surface [64].

In Figure 6, an aerial photograph of the measurement locations with the instrumental setup is shown. For all measurement locations, except for the dune crest, the same measuring equipment was used. On the dune crest, no sediment transport took place during the conducted experiments; thus, only wind velocity and wind direction were measured at this location. The windblown sand grains were recorded with three saltiphone measuring stations and three vertical mesh sand traps. For the beach and the dune toe, the wind tower was placed 1.0 m windward to the left, and the vertical mesh sand traps were placed 0.5 m windward to the right of the saltiphones.

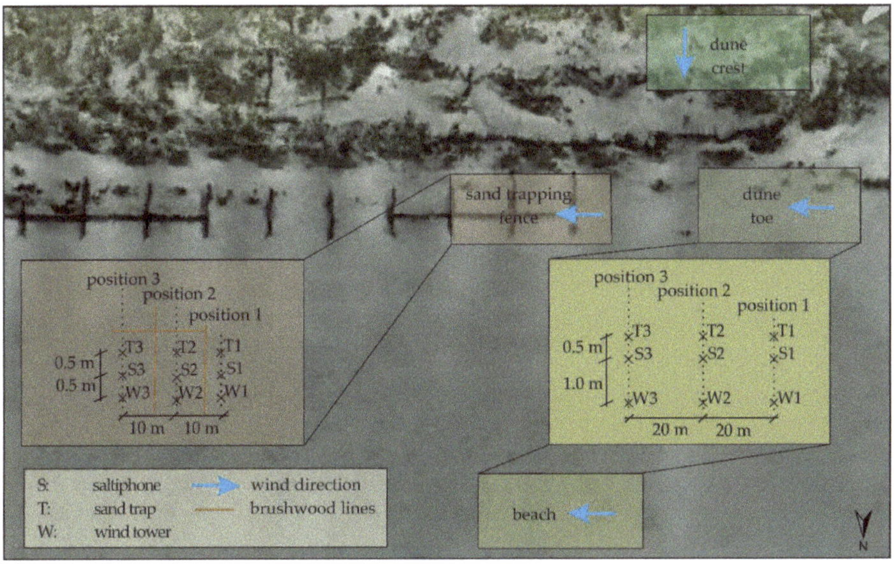

Figure 6. Aerial photograph for the measurement locations: beach, dune toe, dune crest (no sediment transport), and sand trapping fence with the arrangement of the vertical mesh sand traps, the saltiphones, and the wind towers. Blue arrows indicate the respective wind direction during measurements at each measurement location.

The distance between each measuring position was 20 m. The arrangement of the instrumentation for the sand trapping fence differed with regard to Figure 6. Between the brushwood lines of the sand trapping fence the distance between wind tower and saltiphone was reduced to 0.5 m, and the distance between the individual positions was reduced to 10 m. The measuring location sand trapping fence, position 1, was located close to the dune toe in front of the sand trapping fence windward, where the wind was not yet influenced by the brushwood lines. In field 1 (position 2) and field 2 (position 3) the wind regime was influenced by the brushwood lines under the prevailing wind conditions from west during the experiments.

Every 20 min, the wind tower was moved to the next measuring position. Meanwhile, the saltiphones and vertical mesh sand traps remained at their initial location. This procedure was repeated between three and five times starting at the first position again.

3.2. Analysis Method

3.2.1. Wind Velocity, Shear Velocity, and Aerodynamic Roughness Length

Saturated aeolian sediment transport rates can be described by sediment transport models as a function of the shear velocity [7,22,23,25,28,65]. In most models, it is assumed that the sand is dry, the wind velocity and the sediment transport rates are steady and a logarithmic velocity profile over a flat surface is developed [37,55].

In our data set, the wind velocity profile was measured by six cup anemometers at predefined heights. An appropriate averaging time must be selected to gain reliable values for the shear velocity and aerodynamic roughness length. To account for larger boundary layer turbulences at higher altitudes, averaging over longer time scales is required [66]. However, the potential to choose shorter averaging time scales has been recognized when the highly variable aeolian sediment transport is the focus of investigation. As many researches stated [25,65,67], short averaging time scales (~10 s) allow for resolving short-term variations in surface roughness and aeolian sediment transport due to changing wind conditions. For example, Strypsteen et al. (2019) [22] and Ellis et al. [68] (2009) chose time scales for averaging of less than 1 min. In this study, 10 s as the averaging time scale was chosen.

The obtained wind data were filtered. As a criterion for valid wind measurement data, gusts defined by the German Weather Service were excluded. Therefore, the measured wind velocity must not exceed the measured 10-min average wind velocity by at least 5.0 m/s in a time interval greater than 3 s and smaller than 20 s [53]. To obtain stationary wind profiles, the following conditions were defined: The coefficient of variance of each 10-second-interval at each measurement point must not exceed the selected maximum value of $v \leq 0.15$ and the gradient of acceleration of two successive wind profiles over the selected time interval at each measurement point must not be greater than $b \leq 0.5$ m/s^2.

Under neutral atmospheric stability conditions, the wind velocity distribution over the viscous sub boundary layer for aerodynamically rough surfaces can be described [23] with the law of the wall as follows:

$$u_z = \frac{u_*}{\kappa} \cdot \ln \frac{z}{z_0}, \tag{1}$$

where u_z [m/s] is the wind velocity at height z [m], κ [-] is the Kármán constant (here: 0.41), u^* [m/s] is the shear velocity, and z_0 [m] is the aerodynamic roughness length [69,70].

The parameters u^* and z_0 were determined from experimental data through a regression analysis of the measured wind velocities at defined heights over the selected time interval of 10 s. A least-square fit of the wind velocities u [m/s] to the natural logarithm of the wind measurement height ln(z) was applied. The shear velocity u^* [m/s] results from the slope of the regression line of u [m/s] on ln(z) multiplied by the Kármán constant κ [-]. The aerodynamic roughness length is equal the exponential function of the following term: the quotient of the negative y-axis interception and the slope of the regression line u [m/s] on ln(*z*) [25,65,71]. Only wind profiles that had a coefficient of determination greater than $R^2 > 0.90$ were retained for further analysis to ensure only stationary logarithmic wind profiles were analyzed in the following.

With increasing drag and lift force on the particle, there is a critical value at which grain movement is initiated. In nature, this depends on the surface moisture, surface roughness, vegetation, shells, surface slope, and fetch effect [7,35,39,72,73]. The so-called critical shear stress u^{*t} is defined as [18]:

$$u^{*t} = \mathrm{A} \cdot \sqrt{\left(\frac{\rho_s}{\rho_a} - 1\right) \cdot g \cdot d_{50}}, \tag{2}$$

where A [-] is an empirical constant, ρ_a [kg/m^3] is the density of air (here: 1.2 kg/m^3), and ρ_s [kg/m^3] is the density of sand grains (here: 2650 kg/m^3). The gravitational acceleration is given with g [m/s^2] (here: 9.81 m/s^2), and d_{50} [μm] is the mean particle size (here: 210 μm). The empirical constant is assumed as A = 0.11. This value takes into account the effect of cohesion, but not the influence of the protective layer of shells or the increasing cohesion due to higher moisture contents [74–76]. With the given parameters, the critical shear velocity at the study site of Langeoog was calculated as u^{*t} = 0.24 m/s.

3.2.2. Vertical Sediment Flux Profile and Potential Aeolian Sediment Transport Rates

The vertical mesh sand traps, see also Figure 5a, were exposed to sediment flow during a defined exposure time, and, afterward, the collected sand was weighed and divided by the inlet area and measurement duration to calculate the mean sediment flux rate. The vertical distribution of sediment transport can be described by an empirical exponential decay function [22,68,77]:

$$q_z = q_0 \cdot \exp(-\beta \cdot z), \tag{3}$$

where q_z [kg/m²/s] is the sediment transport rate at a predefined height z [m], q_o [kg/m²/s] is the extrapolated saltating sediment mass transport at the surface, and the decay rate β [1/m] as a measure of the vertical concentration gradient. β and q_o can be obtained by a regression analysis to the data of the field experiments [78]. The regression provides generally high correlations of the coefficients, as they are only weakly related to physical aeolian parameters, such as grain size, roughness, and shear velocity. Therefore, a wide range of fitted coefficient values can be found in literature [22,78].

The total mass transport can be calculated by integrating Equation (3) as follows:

$$Q_s = \int_0^\infty q_z \, dz = \frac{q_0}{\beta}, \tag{4}$$

where Qs [kg/m/s] is the total mass transport by saltation. It is assumed that saltating particles start moving immediately above the surface (z = 0 m) [78].

In Table 1, five representative sand transport models related to the shear velocity u^* [m/s], used to calculate the saturated aeolian sand transport rates Qs [kg/m/s] for dry sand, are given. For an overview of all existing models, reference is made to Dong et al. (2003) [79] and Strypsteen (2019) [22]. With the exception of wind velocity, all parameters are considered to be constant over time. C_i [-] are constants derived by wind tunnel and/or field experiments from different authors. D (here: 250 µm) [29] is the reference diameter of the sediment as a standard value for dune sand.

Table 1. Summary of five chosen representative sand transport models for predicting saturated aeolian sand transport rates for dry sand (modified after [22,79]).

Author	Refer-ence	Equation	Description/Assumptions	Equation No.
Bagnold (1937)	[80]	$Q_s = \begin{cases} C_B \cdot \frac{\rho_a}{g} \cdot \sqrt{\frac{d_{50}}{D}} \cdot u_*^3, & u_* > u_{*t} \\ 0, & u_* \leq u_{*t} \end{cases}$	C_B = 2 is chosen for naturally graded sand and the equation is only valid for grain sizes between 0.1–1.0 mm, when the threshold of motion is exceeded.	(5)
Kawamura (1951) andWhite (1979)	[81,82]	$Q_s = C_{KW} \cdot \frac{\rho_a}{g} \cdot (u_* - u_{*t})(u_* + u_{*t})^2$	C_{KW} = 2.78 is chosen for naturally graded sand.	(6)
Bagnold (1954)	[23]	$Q_s = C_B \cdot \frac{\rho_a}{g} \cdot \sqrt{\frac{d_{50}}{D}} \cdot (u_* - u_{*t})^3$	Bagnold corrected his model, introducing the term for threshold shear velocity.	(7)
Hsu (1974)	[83]	$Q_s = C_{Hsu} \cdot \frac{\rho_a}{g} \cdot u_{*t} \cdot \left(u_*^2 - u_{*t}^2\right)$ $C_{Hsu} = \frac{\exp(-0.47 + 4.97\, d_{50}) \cdot 10^{-4}}{10}$	Hsu related the predicted sand transport to the third power of the friction Froude number.	(8)
Kok et al. (2012) and Durán et al. (2011)	[31,84]	$Q_s = C_{DK} \cdot \frac{\rho_a}{g} \cdot u_{*t} \cdot \left(u_*^2 - u_{*t}^2\right)$	C_{DK} = 5 is chosen.	(9)
Van Rijn (2019)	[35]	$Q_s = \begin{cases} C_B \cdot \sqrt{\frac{d_{50}}{D}} \cdot \frac{\rho_a}{g} \cdot \left(u_*^3 - u_{*t}^3\right), & u_* > u_{*t} \\ 0, & u_* \leq u_{*t} \end{cases}$	C_B = 2 is chosen for naturally graded sands.	(10)

The moisture content on a beach is subject to large spatial and temporal variations [35,36,39]. With increasing moisture content, the cohesion between the sand particles rises; therefore, the critical shear velocity is increased. Even low moisture contents (M ~0.5 Vol%) can reduce the transport rates significantly, except when the wind velocity is intense. Saturated moisture content can reduce the instantaneous transport rate by a factor of $m_{moisture}$ = 2 – 10 [30,36,38,76]. Therefore, to predict the sediment transport of wet sand the following equation is used:

$$Q_{s,wet} = \frac{Q_{s,dry}}{m_{moisture}}, \tag{11}$$

where $Q_{s,wet}$ [kg/m/s] is the predicted transport rate for wet sand, and $Q_{s,dry}$ [kg/m/s] is the saturated aeolian sand transport rates for dry sand.

4. Results and Discussion

According to the conducted sieve analyses, the median grain size D_{50} ~218 µm was chosen as an average value for the measurement locations of the beach, dune toe, and sand trapping fence. The sieve analyses results from the measurement location of the dune crest were excluded for determining the median grain size as no sediment transport was measured on the dune crown. For more details regarding grain size distribution, d_{10} and d_{90} are also provided representing the grain sizes for 10% and 90% sieve retainment: D_{10} ~150 µm, D_{90} ~302 µm. The mean size was equal to D_m ~209 µm, and the standard deviation was σ ~212 µm.

During the whole field campaign, the sand at the study site was almost continuously slightly wet with varying moisture contents around an average value of M = 6 Vol% up to a maximum moisture content of M = 14.5 Vol% (average values for all locations). The weather was sunny with a maximum temperature of T = 18 °C. The wind was strong over the selected 10-second-interval with an average wind velocity around u_2 = 16.6 m/s at a height of z = 2.0 m up to stormy conditions with a maximum of u_2 = 24.7 m/s. The wind direction came mainly from the west, which lead to coast parallel sediment transport. Aeolian sand transport occurred in the form of streamers. The average shear velocity was almost always above the threshold of movement of u^{*t} = 0.24 m/s based on Equation (2). However, aeolian sediment transport was not measured throughout the whole field campaign.

4.1. Wind Velocity Profiles, Shear Velocities, and Aerodynamic Roughness Length

In this Section 4.1, the typical results of the measured velocity profiles and corresponding regression lines to determine the shear velocity and the roughness length for the four different measurement locations, the beach, dune toe, dune crest, and between the sand trapping fences at the study site, are presented. In Figure 7, the wind profiles to determine the shear velocity and roughness length parameters for the locations (a) beach and (b) dune toe are shown. The blue dots show the measured wind velocity at a predefined elevation, the arrows show the wind direction—in this example, the wind comes from the west, and the red line shows the regression line.

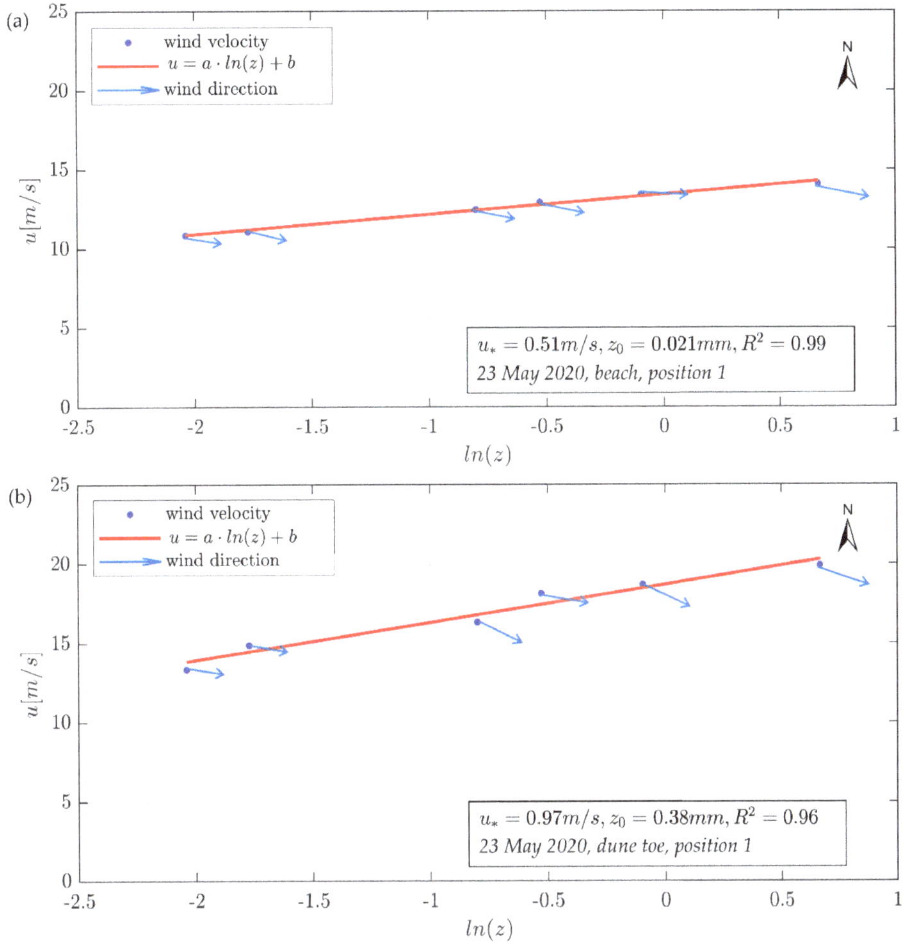

Figure 7. Representative example of a velocity profile for the locations (**a**) beach and (**b**) dune toe on the study site of Langeoog. The blue dots show the measured wind velocity, the blue vectors show the wind direction, and the red line indicates the regression line, see Equation (1).

The regression for all evaluated wind data fits the location beach and dune toe well with an overall coefficient of determination of R^2 = 0.95 (u^* = 0.93 m/s) and R^2 = 0.88 (u^* = 0.43 m/s). The lower coefficient of determination of the regression at the dune toe can be explained by the larger unevenness in the topography compared to the beach. Thereby, it is proven for the locations of the dune toe and beach that the wind profile can be displayed appropriately by the law of the wall, see Equation (1).

In Figure 8a, the wind profile for the location dune crest and, in Figure 8b, the wind profile for the location sand trapping fence in field 1 (position 2) are shown. Regression analyses for position 1 and position 3 were evaluated, but are not shown in the following Figure 8.

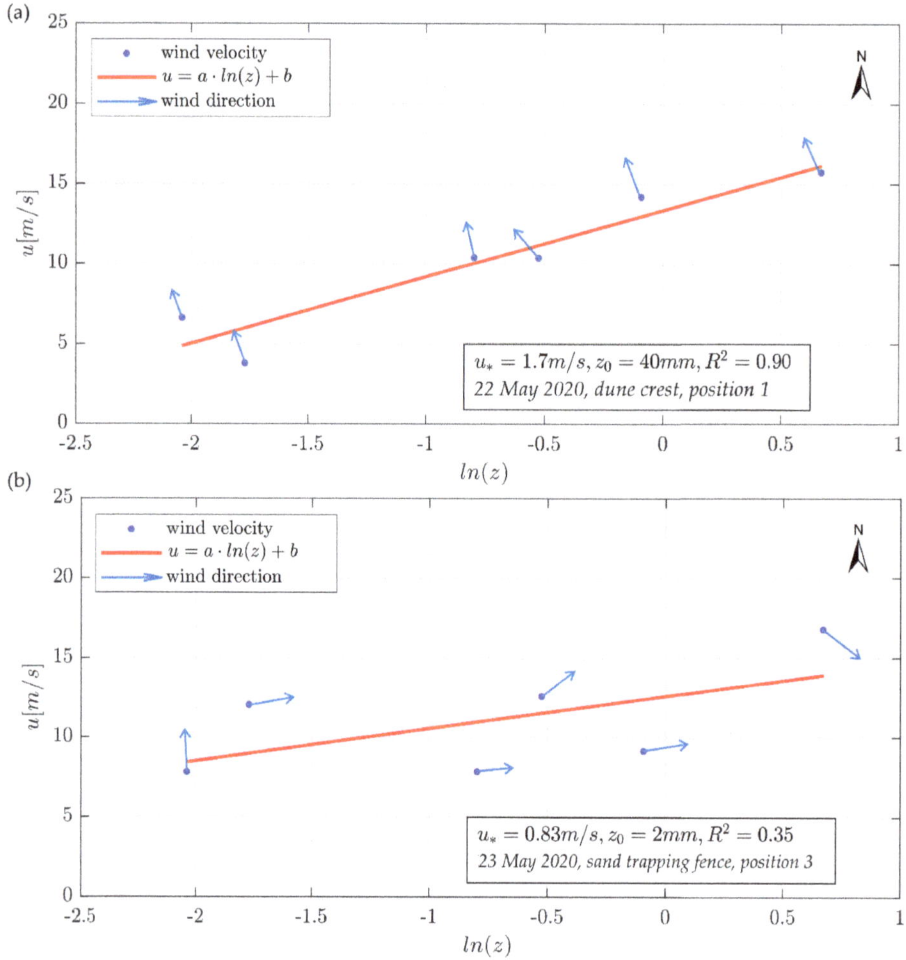

Figure 8. Representative example of a velocity profile to determine the shear velocity and roughness length for the locations (**a**) dune crest and (**b**) sand trapping fence (position 2) on the study site of Langeoog. The blue dots show the measured wind velocity, the blue vectors show the wind direction, and the red line indicates the regression line, see Equation (1).

The regression analysis of all wind data fit the measuring points for the location dune crest on the foredunes with an overall determination coefficient of R^2 = 0.82 (u^* = 0.65 m/s). Due to the strong wind, the dune grass bent on the coastal dune to varying degrees depending on the wind velocity, resulting in different roughness lengths. Even with the same vegetation cover and height, the wind flow characteristics can be different [85]. In front of the sand trapping fence at position 1, see also Figure 6, the overall coefficient of determination was R^2 = 0.89 (u^* = 0.92 m/s). The measuring location of the sand trapping fence, position 1, was located close to the dune toe in front of the sand trapping fence windward, so that the wind was not yet influenced by the sand trapping fence.

Between the sand trapping fence at location 2 and location 3, the overall coefficient of determination was significantly lower with R^2 = 0.34 and R^2 = 0.38, respectively. On the basis of the curve fitting, between the sand trapping fence, no stationary logarithmic velocity profiles were present. In between the sand trapping fence, the wind directions and the wind velocities varied considerably, see also the wind vectors in Figure 8b. We assumed that turbulent fluctuations took place [16,86], and therefore the effects of short-time variations in the surface roughness associated with large eddies cannot be displayed appropriately by a one-dimensional wind profile according to Equation (1).

In Figure 9, the relationship between the calculated shear velocity u^* [m/s] and the aerodynamic roughness length z_0 [mm] at the different measurement locations are shown. The black line indicates the regression for the beach, dune toe, and sand trapping fence (position 1) with the following equation:

$$z_0 = a \cdot u_*^b. \quad (12)$$

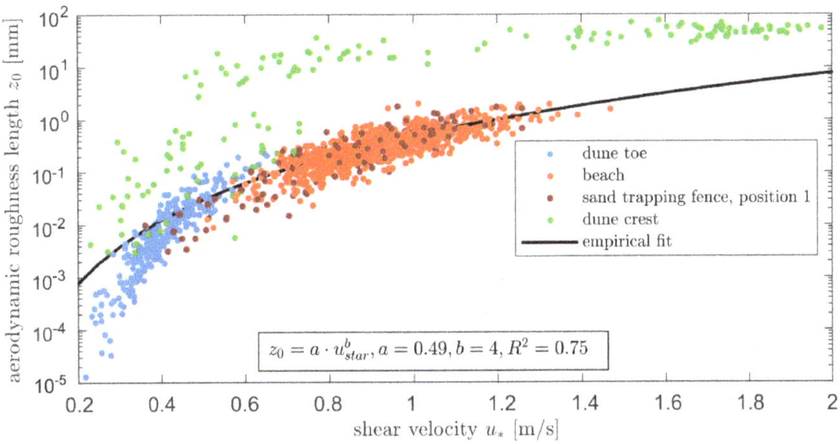

Figure 9. Regression line for the calculated shear velocity and aerodynamic roughness length for the measurement locations of the dune toe, beach, and sand trapping fence (position 1) and calculated aerodynamic roughness length for the location dune crest (not part of the regression).

The regression parameters are defined as $a = 0.49$ and $b = 4$ with a determination coefficient of $R^2 = 0.75$. The dune crest data is not implemented in the regression line.

The values for the aerodynamic roughness length close to the dune toe lie between $z_0 = 2.6 \cdot 10^{-6}$–0.23 mm, on the beach $z_0 = 0.009$–2.03 mm, and in front of the sand trapping fence between $z_0 = 0.003$–1.79 mm. The results of the dune crest cannot be compared directly to the results of the dune toe, beach, and sand trapping fence (position 1) due to the presence of vegetation and higher bed roughness on the dune crest (see also Section 2). At the dune crest, the values were significantly higher (~100 times) and were subject to more scatter with aerodynamic roughness lengths of around $z_0 = 0.003$–90.96 mm. However, they follow the same trend as the measurement locations of the dune toe, beach, and sand trapping fence (position 1). The values are in a similar range as other field experiments for estimating the aerodynamic roughness length over a coastal dune field [85].

The aerodynamic roughness length increased with the increasing shear velocity. The aerodynamic roughness length increased as a power law with the increasing shear velocity [22,25].

4.2. Vertical Sediment Flux Profile

For the measurement locations of the beach, dune toe, and sand trapping fence (position 1 and position 2), the vertical mesh sand traps over $z = 0.3$ m above the surface were analyzed in the following. In Figure 10, a representative example of the vertical distribution of sediment transport fluxes derived from the vertical mesh sand traps is shown for the measurement location of the beach. The trapped sediment over the height is described using an exponential decay curve as described in Equation (3).

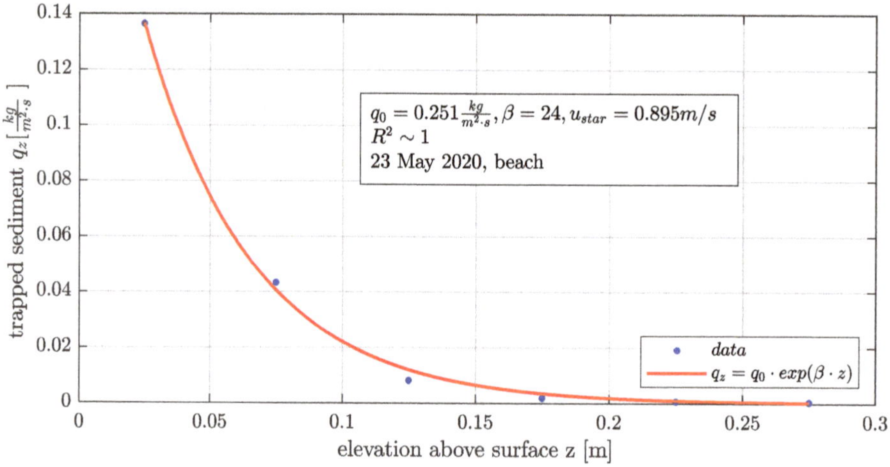

Figure 10. Representative example of the vertical distribution of sediment transport flux on the beach derived from the vertical mesh sand trap for the measurement location of the beach and the best fit exponential decay function.

The sand was caught predominantly in heights of less than $z \leq 0.05$ m. When the wind was really strong with u^*~1 m/s, sand was caught in heights over $z > 0.15$ m. This is in accordance with the literature reporting that most of the sediment transport in coastal areas occurs in a thin layer $z < 0.05$ m above the surface [22].

In Tables 2 and 3, the results for the trapped sediment from the vertical mesh sand traps with the total mass flux Q_s [kg/m/h] over the measuring duration t [min], with the decay rate $ß$ [1/m] and the extrapolated saltating sediment mass transport at the surface q_0 [kg/m^2/h], and the correlation coefficients R^2 [-] of the regression analysis are shown. The corresponding shear velocity u^* [m/s], the moisture content M [Vol%], and the aerodynamic roughness length z_0 [mm] from other analyses are also shown. The transport fluxes varied between $Q_s = 0.65 - 57.03$ kg/m/h with shear velocities up to a maximum of $u^* = 1.01$ m/s. The measuring duration varied between t = 19–105 min.

Table 2. Results of the total mass flux over the measuring duration, decay rate, extrapolated saltating sediment mass transport at the surface, correlation coefficients of the regression analysis of the vertical wind profile (rounded values), shear velocity, moisture content, and aerodynamic roughness length.

Location	No.	Q_s [kg/m/h]	t [min]	$ß$ [1/m]	q_0 [kg/m^2/h]	R^2 [-]	u^* [m/s]	M [Vol%]	z_0 [mm]
Beach	1	27.83	49	35	966.70	1.00	1.05	8.67	0.64
	2	50.72	53	17	839.68	0.97	0.99	8.57	0.52
	3	47.62	83	13	610.52	0.88	0.96	10.38	0.43
	4	37.29	46	24	904.85	1.00	0.89	7.47	0.33
	5	27.38	43	39	1075.91	1.00	0.88	8.54	0.39
	6	57.03	58	29	1676.80	1.00	0.97	8.28	0.47
	7	43.44	27	39	1696.28	1.00	0.90	7.08	0.34
	8	22.77	19	46	1048.21	1.00	0.85	7.82	0.28
Dune toe	9	2.70	105	48	122.33	0.91	0.42	1.10	0.03
	10	1.34	105	63	83.69	0.88	0.42	1.53	0.03
	11	0.65	105	24	13.76	0.91	0.42	1.80	0.03

Table 3. Results of the total mass flux over the measuring duration, decay rate, extrapolated saltating sediment mass transport at the surface, correlation coefficients of the regression analysis of the vertical wind profile (rounded values), shear velocity, moisture content, and aerodynamic roughness length.

Location	No.	Qs [kg/m/h]	t [min]	$ß$ [1/m]	q_0 [kg/m²/h]	R^2 [-]	u^* [m/s]	M [Vol%]	z_0 [mm]
Dune crest	n/a, no sediment transport took place								
Sand trapping fence 1	12	11.92	69	39	462.06	1.00	0.90	8.73	0.36
	13	6.42	65	42	268.87	1.00	0.62	8.67	0.14
	14	39.62	20	36	1416.88	1.00	0.97	5.67	0.32
	15	6.47	80	37	199.05	1.00	0.97 *	5.67	0.32 *
	16	5.24	80	34	157.17	1.00	1.01	5.22	0.78
	17	21.74	80	31	662.45	1.00	1.01 *	3.43	0.78 *
Sand trapping fence 2	18	0.89	60	33	25.77	0.93	**	10.30	**
	19	2.45	60	36	79.53	1.00	**	4.77	**
	20	4.80	60	60	314.97	1.00	1.01 **	3.65	0.78 **
Sand trapping fence 3	too little sediment transport to conduct regression analysis, no vertical wind profile according to the law of the wall present								

* no wind data available, u^* and z_0 assumed from data of the closest time interval; ** no wind data available between the sand trapping fence.

The regression analysis of the transport fluxes showed correlation coefficients around $R^2 = 0.88$–1.00. The regression parameters varied with $ß = 29$–63/m and $q_0 = 25.77$–1696.28 kg/m²/h.

The decay rate generally increases for a constant saltating sediment mass transport at the surface with higher wind velocity or less moisture content, indicating that more sand is transported at higher elevations above the surface [56]. Considerable variation was present when comparing the decay rate and saltating sediment mass transport at the surface for different measurement locations, where the prevailing boundary conditions, like the moisture content and shear velocities, varied. Close to the dune toe, the wind regime was of medium intensity, and the sand was relatively dry, resulting in lower saltation heights, compared to the beach and sand trapping fence, when it was relatively wet; however, high wind velocities occurred. The presence of slightly wet sand increased the threshold of motion for sand, resulting in a lower total sediment flux.

Aeolian sediment transport is highly variable in both space and time. During one event, the trapped sand in two vertical mesh sand traps standing next to each other can vary a lot. This happens, if wind forces and sand transport occur in the form of streamers resulting in spatial-temporally changing threshold shear velocity [22,26,35,78].

At the measurement location of the sand trapping fence, position 3, there was too little sediment caught in the vertical mesh sand trap over the defined time interval. Thus, the vertical sediment flux profile cannot be described with the approach according to Equation (3).

The literature gives a wide range of fitting coefficients and different methods for trapping sand above the surface [22,68,78,79]. The studies at the Belgian coast at Mariakerke and Koksijde are described in the following to compare the results of this work with other authors. Strypsteen (2019) [22] measured the trapped sand with Modified Wilson And Cool Samplers (MWAC) on a dry beach, where the bottles were placed at the heights of $z = 6.5$, 13.5, 21, and 28.5 cm above the surface. The results were analyzed with the potential decay function as an average value out of 12 measurement locations on the beach. With an average shear velocity between $u^* = 0.29 - 0.54$ m/s and transport fluxes between $Q_s = 7.2 - 121.8$ kg/m/h, the decay rate varied between $ß = 14.06 - 28.57$/m. The measuring time varied between minutes up to days depending on the intensity of the sediment transport [87].

Campos (2018) [56] measured the sediment transport fluxes with vertical mesh sand traps on the Belgian coast at Koksijde. With an average shear velocity between $u^* = 0.40$–0.53 m/s and transport fluxes between $Q_s = 38.4$–88 kg/m/h, the decay rate varied between $ß = 18.13$–23.72/m. In the study, each run was performed nine times with measurement durations between t = 16 min up to t = 38 min. Little to no moisture was present [56].

These two selected field campaigns measured significantly higher total mass flux but lower shear velocities and decay rates. This could be explained by the presence of slightly wet sand on the study site Langeoog, increasing the threshold of motion for the sand, resulting in a lower total sediment flux (see also Section 2) for higher shear velocities. Further reasons for these deviations may be that Campos (2018) [56] used shorter time intervals for trapping the sand in the vertical mesh traps. The trap efficiency of the measurement tool MWAC, stated as the ratio of total particle mass passing through the air per defined area to the actual value, varies between 42%–120% [59,88,89]. For the vertical mesh sand traps, there was good efficiency reported; however, the efficiency decreased for longer time periods [55,56]. The MWAC sand trap began measuring at $z \geq 6.5$ cm above the surface, whereas the vertical sand trap began measuring right above the sand surface.

4.3. Observed Aeolian Sediment Transport Rates at the Measurement Location of the Beach

In Figure 11, the theoretical comparison between the field data of Langeoog beach, dune toe, sand trapping fence (position 1), and five chosen transport models derived by Bagnold (1937) [80], Kawamura (1951) [82], Bagnold (1954) [23], Kok et al. (2012) [31], and van Rijn (2019) [35] (see also Table 1) are shown. The Bagnold (1954) [23] model results in the smallest transport rates compared to the other prediction models, especially for smaller shear velocities. The orange dots show the results of the location beach, the blue dots of the location dune toe, and the brown dots of the location sand trapping fence (location 1).

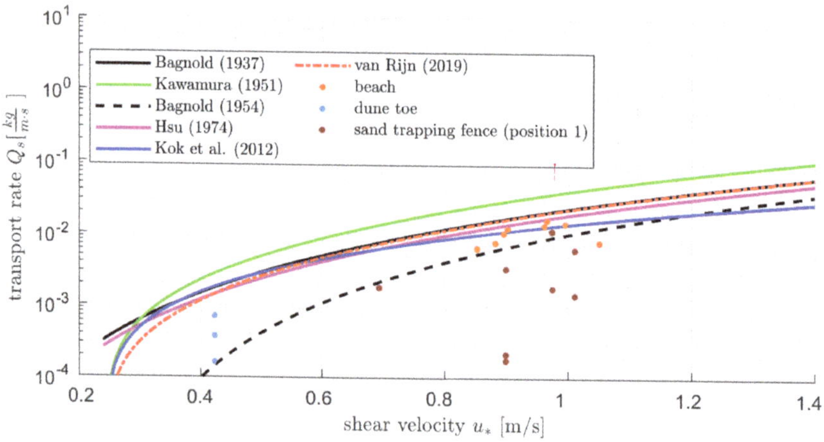

Figure 11. Comparison between the measured sediment transport rates on Langeoog and the predicted sediment transport rates for five different transport models.

The measured data from Langeoog beach can be described well with the predicted transport models if the moisture content as a reduction factor is considered. For the location dune toe and sand trapping fences (position 1) the results scattered more but still showed good results. We could not clearly evaluate which equation fit the results best as only high wind velocities with shear velocities around $u^* = 0.70 - 1.05$ m/s were recorded, except for $u^* = 0.42$ m/s, where the results scattered a lot. For further analysis, smaller shear velocities are required to indicate the best fit regression line.

An application of the transport models from Table 1 is only useful for the entrance boundary condition (upwind the sand trapping fence, position 1). Between the sand trapping fences, the transport models cannot be used as the sediment supply from above was too small and a logarithmic wind profile no longer existed (see Section 4.1).

4.4. Downwind Evolution of Saltiphone Data for the Measurement Location Beach

In Figure 12, an example of the wind and saltiphone data on the beach on 23 May 2020 for the time interval 12:35 pm to 13:10 pm is shown. This is a representative example out of four repeating measurements on the beach. The calculated shear velocities from the wind profiles and the mean wind direction are shown as an average value over 10 s. The critical shear velocity is also given, see Figure 12a. Additionally, the 95% quantiles of the saltation impacts over 10-s time intervals are shown. In Figure 12a, the black line represents the calculated shear velocity, the dotted red line indicates the critical shear velocity (u^{*t} = 0.24 m/s, see Section 3.2.1), whereas the light orange line shows the mean wind direction. Figure 12b shows the saltiphone data for position 1 and (c) for position 2, with (d) for position 3 in the downwind direction. The blue bars represent the counted sand grains of the lower saltiphones, and the orange bars are for the upper saltiphones. In terms of the wind conditions, the experiments at the beach were intense with an average wind velocity of u_2 = 19.2 m/s (u^* = 0.88 m/s), resulting in saltation impacts up to s = 750 impacts/10 s. The wind came from the west (wind direction dr = 274°) with slight wind velocity fluctuations.

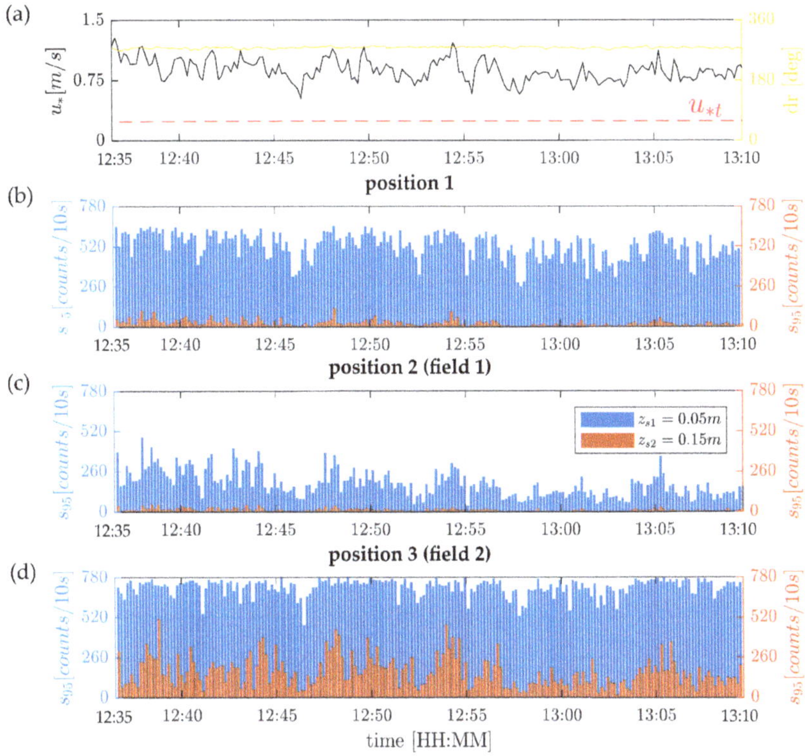

Figure 12. Time series of the (**a**) shear velocity, critical shear velocity, mean wind direction, and (**b–d**) 95% quantiles of saltation impacts over 10-s time intervals for the measurement location of the beach conducted on 23 May 2020.

The shear velocity was always above the critical shear velocity during the whole experiment, showing continuous sediment transport for all three positions. In general, as the number of counted sediment grains on the lower saltiphone increased, so did the number of counted sediment grains on the upper saltiphone. This suggests that the height of the saltating layer increased. At position 2 (20 m downwind), lower impacts were measured compared to position 1, followed by an increase of the saltation impacts at position 3 (40 m downwind). The smaller number of impacts at the second location can likely be attributed to a higher moisture content or more shells, protecting the beach against the erosion of sand grains, resulting in higher critical shear velocities [35,90,91].

We assumed that, between location 1 and location 2, there was a sediment sink due to the high moisture contents, whereas, between position 2 and position 3, there was a sediment source with lower moisture contents. During the whole day, the moisture content was relatively high with values around M = 3.3–14.5 Vol%, showing almost saturated values. However, the moisture content was very variable in time and space. The higher moisture content resulted in higher critical shear velocities compared to the dune toe, where the sand was relatively dry.

4.5. Downwind Evolution of Saltiphone Data for the Measurement Location of the Dune Toe

In Figure 13, a representative example of the wind and saltiphone data close to the dune toe on 19 May 2020 for the time interval 12:15 pm to 13:05 pm is shown. This is an example out of three repeated measurements of the dune toe.

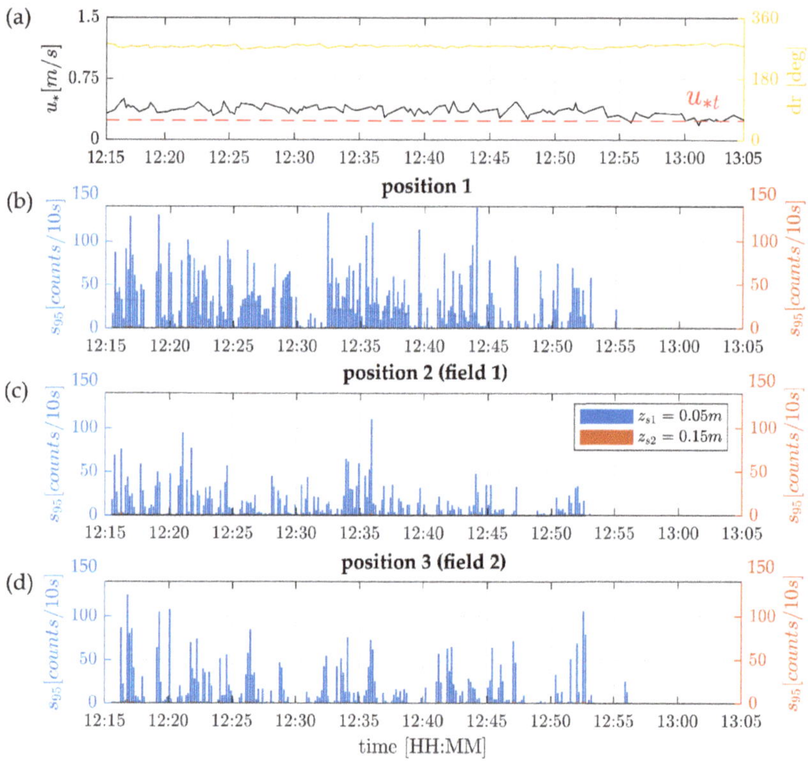

Figure 13. Time series of the (**a**) shear velocity, critical shear velocity, mean wind direction, and (**b–d**) 95% quantiles of saltation impacts over 10-s time intervals for the measurement location of the dune toe conducted on 19 May 2020.

The average wind velocity was u_2 = 11.6 m/s, and the average shear velocity was u^* = 0.36 m/s. The wind was relatively steady coming from west (dr = 277°), parallel to the dune. During the time interval shown in Figure 13, the shear velocity was mostly slightly above the critical shear velocity, whereby, around 12:56 pm and 13:02 pm, the critical shear velocity was undercut, and no sediment transport was further measured. The results, therefore, indicate that the calculated critical shear velocity was a good approximation of the threshold for the initiation of sediment transport.

The lower saltiphones recorded more impacts than the upper saltiphones. As only sporadic saltation impacts were measured by the upper saltiphones, they are hard to distinguish in Figure 13; see, for example, Figure 13b around 12:17 pm. The number of impacts measured by the upper saltiphones increased due to a wind peak. This shows that most of the sediment transport occurred below $z \leq 0.15$ m. This also agrees with the results of the vertical mesh sand traps close to the dune toe, where at heights of $z \geq 0.15$ m above the surface little sediment were caught (see also Section 4.2).

Time intervals with increased saltation impacts corresponded temporally for all measuring positions downwind. Most grain impacts were recorded at position 1, followed by position 2 (20 m downwind) and position 3 (40 m) approximately in the same order of magnitude. The peaks in the wind velocity correspond with the peaks in the counted particles by the saltiphones.

Little spatial variations in sediment transport gradients are found downwind and reflect the spatially varying availability of sand related to the surface roughness and moisture. The moisture content of the sand close to the dune was relatively low with average values around M ~1.1–1.8 Vol% and, therefore, played a minor role for supply limiting factors. However, due to the medium intense wind velocities, the bed ripples moved downwind, causing a high spatial variability of the aerodynamic roughness length (see also Figure 4). Generally, these ripples occur with wind velocities of $u = 5 - 10$ m/s, and, with higher wind velocities, these ripples are gradually smoothed out [67].

4.6. Downwind Evolution of Saltiphone Data Influenced by Sand Trapping Fences

In Figure 14, a representative example of the time series on 24 May 2020 from 09:45 am to 10:08 am for the measurement location sand trapping fence is shown. This is a representative example out of thirteen repeating measurements. Position 1 is in front of the sand trapping fence and is treated as a boundary condition for incoming sediment that accumulates at the brushwood lines of the sand trapping fence. Position 2 (field 1) and position 3 (field 2) are in between the sand trapping fences at 10 m and 20 m, respectively, downwind (see Figure 6). All three locations belong to the sand trapping configuration 1, see Figure 2a. The average wind velocity at 2 m of height was $u_2 = 20.8$ m/s with a few fluctuations. The shear velocity was $u^* = 0.94$ m/s and always above the critical shear velocity. The wind direction was relatively steady coming from the west ($dr = 262°$), which implies parallel wind to the dune.

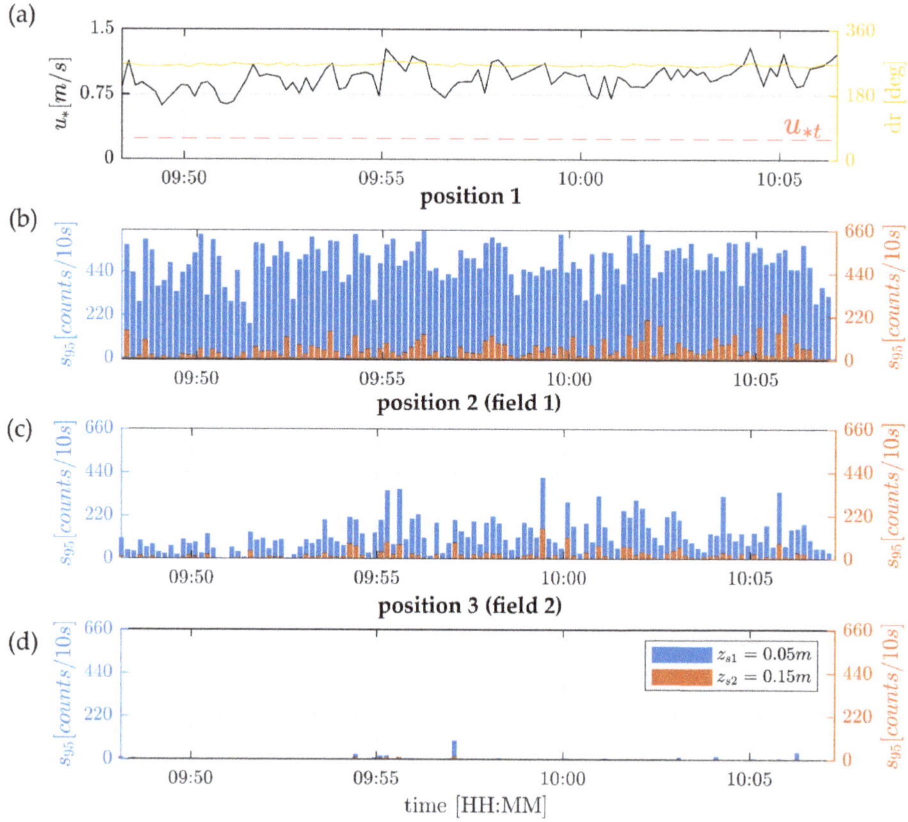

Figure 14. Time series of the (**a**) shear velocity, critical shear velocity, mean wind direction, and (**b–d**) 95% quantiles of saltation impacts over 10-s time intervals for the measurement location of the sand trapping fence conducted on 24 May 2020.

The number of counted particles decreased in the downwind direction behind the first line of brushwood of the sand trapping fence. In the first field of the sand trapping fence, less sediment was transported than in front of the sand trapping fence. In the second field, even less sediment was transported. With the measured wind conditions and the given time interval, the sand transport between the sand trapping fence could only be measured up to the second sand trap field. In field 2, the counted saltiphones were almost not measurable or zero.

In Figure 15, 95% quantiles of the counted sand grains s_{95} [counts/10 s] at position i relative to the boundary condition $s_{95,position\ 1}$ [counts/10 s] in percentage over the distance downwind y [m] through the sand trapping fence for different shear velocities are shown. The 95%-quantiles of the saltation impacts were averaged over a 60-min time interval with approximately the same shear velocity. Position 1 (upwind of the sand trapping fence) was taken as the boundary condition with an input of 100% saltation impacts. In field 1 and field 2, the saltiphone data are shown relative to the boundary condition (position 1). To illustrate the decrease in the saltation impacts before and between the sand trapping fence in a simplified way, the saltation impacts are shown constant over the distance downwind y_1 = 0–10 m, y_2 = 10–20 m, and y_3 =20–30 m—although in reality, this is not the case.

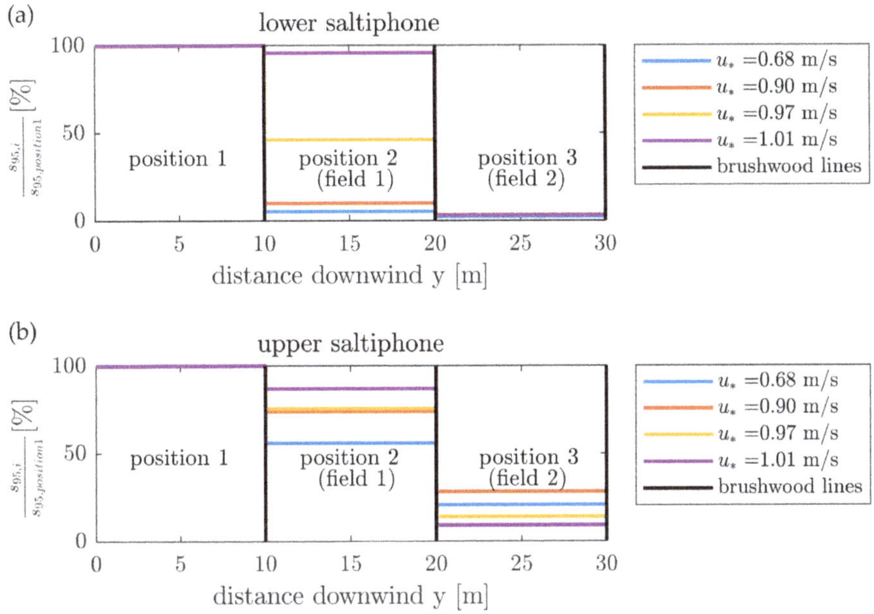

Figure 15. Comparison of the downwind evolution of the saltiphone data before and behind the sand trapping fence for different shear velocities for (**a**) the lower saltiphone and (**b**) the upper saltiphone.

For all shear velocities between $u^* = 0.68 - 1.01$ m/s, the counted saltation impacts were significantly reduced between the sand trapping fences, indicating the deposition of sand in the distance downwind. In field 1 (position 2), the sediment transport for the lower saltiphone was reduced between 5–95% compared to position 1, and, in field 2 (position 3), the sediment transport was reduced between 96–98% compared to position 1. For the upper saltiphone, the sediment transport in field 1 (position 2) was reduced between 13–44% compared to position 1, and, in field 2 (position 3), the sediment transport was reduced between 72–91% compared to position 1. At high shear velocities of up to $u^* = 1.01$ m/s, the sand transport in field 1 was only slightly reduced for both saltiphone heights. For the upper saltiphone, the saltation impact was less reduced compared to the lower saltiphone. Due to the installation of the bundled brushwood, the sand trapping fence was relatively dense directly above the bed surface. This indicates that more sediment grains hit the brushwood and were deposited close to the bed surface.

The investigation at the sand trapping fence were conducted directly after the installation of the sand trapping fence, and, therefore, the fields caught relatively little sand. Sediment was caught predominantly by the first two brushwood lines, where sediment transport could also be measured. This implies that the first two brushwood lines of this type of sand trapping fence configuration (see also Figure 2), with the westerly longshore wind and the actual amount of trapped sand in between the sand trapping fence, were the most effective brushwood lines to trap sand. However, on Langeoog, many of the elderly installed sand trapping fences are already completely filled with sand, showing that not only the first two brushwood lines exposed to longshore wind but also the brushwood lines that are not in an exposed wind position trap sand effectively. This can be explained by the fact that strong winds from the west over north to south can also transport sediment to the sand trapping fence, where it can accumulate at the brushwood lines orthogonal and parallel to the coastal dune. In addition, with a larger amount of trapped sediment between the individual brushwood lines and longshore wind, other fields (not directly exposed to the longshore wind) can be filled with sediment as well.

5. Conclusions

This work presents the results of field experiments conducted on the coast of the East Frisian island Langeoog, analyzing wind and sediment transport data at the beach, coastal dune, and sand trapping fence. The following conclusions can be drawn from the results presented in this work:

(1) The logarithmic wind profiles for the beach, the area close to the dune toe, the dune crest of the foredunes, and upwind of the sand trapping fence (position 1) can be well described by the law of the wall. In between the sand trapping fence, no stationary logarithmic velocity profile was present. Here, the wind directions and the wind velocities varied considerably over the distance from the bottom for any given point in time. We assumed that turbulent fluctuations took place, and therefore the effects of short-time variations in surface roughness associated with large eddies cannot be displayed appropriately by a one-dimensional wind profile. The values for the aerodynamic roughness length close to the dune toe, on the beach, and in front of the sand trapping fence can be described as a function of the shear velocity. At the dune crest, where vegetation was present, the values were ~100 times higher. With increasing aerodynamic roughness length, the shear velocity also increased.

(2) The vertical sediment flux profile of aeolian sand transport for the measurement locations of the beach, dune toe, and sand trapping fence (position 1 and 2) can be described well with an empirical exponential decay function. The regression parameters exponential decay rate and the extrapolated saltation flux at the surface were subject to considerable variation for different measurement locations. Close to the dune toe, the wind regime was of medium intensity, and the sand was relatively dry. On the beach and between the sand trapping fence, high wind velocities occurred, and the sand was relatively wet, resulting in higher critical shear velocities and lower total sediment fluxes.

(3) The sediment caught in the vertical mesh sand traps divided through the inlet area and the measurement duration could well describe the sediment transport rates on the coastal area. The results are described well by the transport models derived by Bagnold (1937) [80], Kawamura (1951) [82], Bagnold (1954) [23], Kok et al. (2012) [31], and van Rijn (2019) [35], considering the moisture content as a further reduction factor. Between the sand trapping fences, these transport models cannot be applied as no logarithmic wind profile existed and the upwind sediment supply was reduced after each brushwood line, thereby, leading to increased deviation from the saturated conditions.

(4) The saltiphones are a good measurement tool to determine the moment sediment transport is initiated and, thus, for determining the value of the critical shear velocity. Small spatial variations in the sediment transport gradients were found downwind close to the dune toe. They reflected the spatially varying availability of erodible sand, which is related to the surface roughness and moisture. However, due to the medium intense wind velocities, the bed ripples were moving downwind, causing a high spatial variability of aerodynamic roughness length. At the measurement location of the beach (position 2, 20 m downwind), a smaller number of impacts were measured compared to position 1, followed by an increase of the saltation impacts at position 3 (40 m downwind). We assumed that, between location 1 and location 2, a sediment sink was present due to the high moisture contents, whereas, between position 2 and 3, we assumed a sediment source was present due to the lower moisture contents. During the whole field campaign, the moisture content was high resulting in higher critical shear velocities.

(5) Sediment transport was significantly reduced by the sand trapping fence. The sand transport was highest upwind of the sand trapping fence. Downwind, in field 1 and field 2, the sediment transport for the lower and upper saltiphones was reduced. In field 2, the sediment transport was almost not measurable or zero.

This study provided the first insights into aeolian sediment transport at different locations on the coastal dune and the influence of sand trapping fences on aeolian sediment transport. We strongly recommend to extend the knowledge of the influences of sand trapping fences on aeolian sediment transport in standardized wind tunnel experiments to gain more quantitative data. To evaluate the influence of sand trapping fences on dune development for longer timescales, aerial drone investigations of the coastal dune should be conducted on the study site Langeoog. For this purpose, an analysis of a series of digital elevation models at the coastal dune and at the sand trapping fence will be performed to investigate the dune volume changes around the fence depending on each sand trapping configuration. The aim of the future investigation is to evaluate the efficiency of each sand trapping configuration under different prevailing wind conditions.

Author Contributions: Conceptualization, C.E. and H.S.; methodology, C.E.; formal analysis, C.E.; investigation: C.E., data processing, C.E., resources, H.S.; original draft preparation, C.E.; review and editing, H.S.; visualization, C.E.; supervision, H.S.; project administration, H.S.; and funding acquisition, C.E., H.S., and others. All authors have read and agreed to the published version of the manuscript.

Funding: This research was funded by the German Federal Ministry of Education and Research (BMBF) within the project ProDune (grant number 03KIS125) that was initiated in the framework of the German Coastal Engineering Research Council (KFKI).

Acknowledgments: The authors thank the project partner Lower Saxony Water Management, Coastal Protection and Nature Conservation Agency for their help in attaining permission to conduct the field experiments and for sharing their expertise in the project support group. The authors thank the administration of the National Park Wadden Sea of Lower Saxony for the permission to conduct the field experiments on the East Frisian Islands. We, furthermore, thank the technical staff and student assistants of the IWW for their help in conducting the field experiments.

Conflicts of Interest: The authors declare no conflict of interest. The funders had no role in the design of the study; in the collection, analyses, or interpretation of data; in the writing of the manuscript, or in the decision to publish the result.

References

1. Peters, K.; Pohl, M. Kuratorium für Forschung im Küsteningenieurwesen (German Coastal Engineering Research Council). Die Küste (The Coast), Issue 88, EAK 2002, *Empfehlungen für die Ausführung von Küstenschutzwerken durch den Ausschuss für Küstenschutzwerke der Deutschen Gesellschaft für Geotechnik e.V. und der Hafenbautechnischen Gesellschaft e.V. (Recommendations for Coastal Protection Structures. Working group 'Coastal Protection Structures' as a Joint Commitee of the German Geotechnical Society and the German Port Technology Association)*; Westholsteinische Verlagsanstalt Boyens & Co.: Heide, Germany, 2020; ISBN 978-3-939239-54-0.
2. Niedersächsischer Landesbetrieb für Wasserwirtschaft, Küsten- und Naturschutz (Lower Saxony Water Management, Coastal Protection and Nature Conservation Agency). *Generalplan Küstenschutz Niedersachsen: Ostfriesische Inseln, Küstenschutz Band 2*; Niedersächsischer Landesbetrieb für Wasserwirtschaft, Küsten- und Naturschutz (Lower Saxony Water Management, Coastal Protection and Nature Conservation Agency): Lower Saxony, Germany, 2010.
3. Short, A.D.; Hesp, P.A. Wave, Beach and Dune Interactions in Southeastern Australia. *Mar. Geol.* **1982**, *48*, 259–284. [CrossRef]
4. Nordstrom, K.F.; Jackson, N.L.; Korotky, K.H.; Puleo, J.A. Aeolian transport rates across raked and unraked beaches on a developed coast. *Earth Surf. Process. Landforms* **2011**, *36*, 779–789. [CrossRef]
5. Ranwell, D.; Boar, R. *Coast Dune Management Guide*; Institute of terrestrial Ecology: Wallingford, UK, 1986.
6. Hillmann, S.; Blum., H.; Thorenz, F. *National Analysis—Germany Lower Saxony*; Niedersächsischer Landesbetrieb für Wasserwirtschaft, Küsten- und Naturschutz (Lower Saxony Water Management, Coastal Protection and Nature Conservation Agency): Lower Saxony, Germany, 2019.
7. De Vries, S. Physics of Blown Sand and Coastal Dunes. Ph.D. Thesis, Delft University of Technology, Delft, The Netherlands, 2013.
8. Pye, K.; Saye, S.; Blott, S. *Sand Dune Processes and Management for Flood and Coastal Defence: Part 4: Techniques for Sand Dune Management*; R&D Technical Report FD1392/TR 2007; Environment Agency: Bristol, UK, 2007.
9. Scheres, B.; Schüttrumpf, H. Nature-Based Solutions in Coastal Research—A New Challenge for Coastal Engineers? In *APAC 2019*; Viet, N.T., Xiping, D., Tung, T.T., Eds.; Springer Singapore: Singapore, 2020; pp. 1383–1389. ISBN 978-981-15-0290-3.

10. Sutton-Grier, A.; Gittman, R.; Arkema, K.; Bennett, R.; Benoit, J.; Blitch, S.; Burks-Copes, K.; Colden, A.; Dausman, A.; DeAngelis, B.; et al. Investing in Natural and Nature-Based Infrastructure: Building Better Along Our Coasts. *Sustainability* **2018**, *10*, 523. [CrossRef]
11. Morris, R.L.; Konlechner, T.M.; Ghisalberti, M.; Swearer, S.E. From grey to green: Efficacy of eco-engineering solutions for nature-based coastal defence. *Glob. Chang. Biol.* **2018**, *24*, 1827–1842. [CrossRef]
12. Ruz, M.-H.; Anthony, E.J. Sand trapping by brushwood fences on a beach-foredune contact: The primacy of the local sediment budget. *Zeitschrift für Geomorphologie Suppl. Issues* **2008**, *52*, 179–194. [CrossRef]
13. Miller, D.; Thetford, M.; Yager, L. Evaluation of Sand Fence and Vegetation for Dune Building Following Overwash by Hurricane Opalon Santa Rosa Island, Florida. *J. Coast. Res.* **2001**, *17*, 936–948, West Palm Beach (Florida).
14. Mendelssohn, I.A.; Hester, M.W.; Monteferrante, F.J.; Talbot, F. Experimental Dune Building and Vegetative Stabilization in a Sand-Deficient Barrier Island Setting on the Louisiana Coast, USA. *J. Coast. Res.* **1991**, *7*, 137–149, Fort Lauderdale (Florida).
15. Grafals-Soto, R.; Nordstrom, K. Sand fences in the coastal zone: Intended and unintended effects. *Environ. Manag.* **2009**, *44*, 420–429. [CrossRef]
16. Li, B.; Sherman, D.J. Aerodynamics and morphodynamics of sand fences: A review. *Aeolian Res.* **2015**, *17*, 33–48. [CrossRef]
17. van Thiel de Vries, J.S.M. *Dune Erosion during Storm Surges*; IOS Press: Amsterdam, The Netherlands, 2009; ISBN 1607500418.
18. Hesp, P. Foredunes and blowouts: Initiation, Geomorphology and Dynamics. *Geomorphology* **2002**, *48*, 245–268. [CrossRef]
19. van Rijn, L.C. Coastal erosion and control. *Ocean Coast. Manag.* **2011**, *54*, 867–887. [CrossRef]
20. Vellinga, P. Beach and Dune Erosion during Storm Surges. Ph.D. Thesis, Delft University of Technology, Deltares, Delft, The Netherlands, 1986. Publication 372, Delft Hydraulics.
21. D'Alessandro, F.; Tomasicchio, G.R. Wave–dune interaction and beach resilience in large-scale physical model tests. *Coast. Eng.* **2016**, *116*, 15–25. [CrossRef]
22. Strypsteen, G. Monitoring and Modelling Aeolian Sand Transport at the Belgian Coast. Ph.D. Thesis, KU Leuven, Faculty of Engineering Technology, Leuven, Belgium, 2019.
23. Bagnold, R.A. *The Physics of Blown Sand and Desert Dunes*, 2nd ed.; Methuen: London, UK, 1954.
24. Arens, S.M. Rates of aeolian transport on a beach in a temperate humid climate. *Geomorphology* **1996**, *17*, 3–18. [CrossRef]
25. Field, J.P.; Pelletier, J.D. Controls on the aerodynamic roughness length and the grain-size dependence of aeolian sediment transport. *Earth Surf. Process. Landforms* **2018**, *43*, 2616–2626. [CrossRef]
26. Baas, A.C.W.; Sherman, D.J. Spatiotemporal Variability of Aeolian Sand Transport in a Coastal Dune Environment. *J. Coast. Res.* **2006**, *225*, 1198–1205. [CrossRef]
27. Sherman, D.J.; Li, B.; Ellis, J.T.; Farrell, E.J.; Maia, L.P.; Granja, H. Recalibrating aeolian sand transport models. *Earth Surf. Process. Landforms* **2013**, *38*, 169–178. [CrossRef]
28. Nickling, W.G.; Davidson-Arnott, R.G.D. Aeolian Sediment Transport on beaches and Coastal Sand dunes. In Proceedings of the Symposium on Coastal Sand Dunes 1990, Guelph, ON, Canada, 12–14 September 1990; pp. 1–35.
29. Nickling, W.G.; McKenna Neumann, C. Aeolian Sediment Transport. In *Geomorphology of Desert Environments*; University of Guelph: Guelph, ON, Canada, 2009.
30. Davidson-Arnott, R.; Ollerhead, J.; Hesp, P.A.; Walker, I.J. Spatial and Temporal Variability in Intensity of Aeolian Transport on a Beach and Foredune 2003. Available online: https://www.semanticscholar.org/paper/SPATIAL-AND-TEMPORAL-VARIABILITY-IN-INTENSITY-OF-ON-Davidson-Arnott-Ollerhead/c72d9f66c802cb33dbbbb64fb7b8056aac107696 (accessed on 25 November 2020).
31. Kok, J.F.; Parteli, E.J.R.; Michaels, T.I.; Karam, D.B. The physics of wind-blown sand and dust. *Rep. Prog. Phys.* **2012**, *75*, 106901. [CrossRef] [PubMed]
32. Bagnold, R.A. The movement of desert sand. *Geogr. J.* **1935**, *85*, 342–365. [CrossRef]
33. Zingg, A.W. Wind tunnel studies of the movement of sedimentary material: Soil Conservation Service, Manhattan, Kansas. In Proceedings of the 5th Hydraulic Conference, Iowa City, IA, USA, 17 August 1953.
34. Owen, P.R. Saltation of uniform grains in air. *J. Fluid Mech.* **1964**, *20*, 225–242. [CrossRef]

35. van Rijn, L.C. Aeolian Transport Over a Flat Sediment Surface. Aeolian Transp. 2019. Available online: https://www.leovanrijn-sediment.com/papers/Aeoliansandtransport2018.pdf (accessed on 25 November 2020).
36. Bauer, B.O.; Davidson-Arnott, R.G.D.; Hesp, P.A.; Namikas, S.L.; Ollerhead, J.; Walker, I.J. Aeolian sediment transport on a beach: Surface moisture, wind fetch, and mean transport. *Geomorphology* **2009**, *105*, 106–116. [CrossRef]
37. Bauer, B.O.; Davidson-Arnott, R.G.D. A general framework for modeling sediment supply to coastal dunes including wind angle, beach geometry, and fetch effects. *Geomorphology* **2003**, *49*, 89–108. [CrossRef]
38. Davidson-Arnott, R.G.D. Instantaneous and Mean Aeolian Sediment Transport Rate on Beaches. *J. Coast. Res.* **2009**, *56*, 297–301.
39. Davidson-Arnott, R.G.D.; Yang, Y.; Ollerhead, J.; Hesp, P.A.; Walker, I.J. The effects of surface moisture on aeolian sediment transport threshold and mass flux on a beach. *Earth Surf. Process. Landforms* **2008**, *33*, 55–74. [CrossRef]
40. Ning, Q.; Li, B.; Ellis, J.T. Fence height control on sand trapping. *Aeolian Res.* **2020**, *46*, 100617. [CrossRef]
41. Bundesamt für Kartographie und Geodäsie (Federal Agency for Cartography and Geodesy). Geodaten der deutschen Landesvermessung. Bundesamt für Kartographie und Geodäsie (Federal Agency for Cartography and Geodesy), Leipzig, Germany. © GeoBasis-DE/ BKG (2020) Referring to the Licence Text. Available online: http://sg.geodatenzentrum.de/web_public/nutzungsbedingungen.pdf (accessed on 25 November 2020).
42. Landesamt für Geoinformation und Landesvermessung Niedersachsen. Auszug aus den Geodaten des Landesamtes für Geoinformation und Landesvermessung Niedersachsen. Landesamt für Geoinformation und Landesvermessung Niedersachsen, Germany. © dl-de/by-2-0 (2020) Referring to the Licence Text. Available online: www.govdata.de/dl-de/by-2-0 (accessed on 25 November 2020).
43. de Groot, A.V.; Oost, A.P.; Veeneklaas, R.M.; Lammerts, E.J.; van Duin, W.E.; van Wesenbeeck, B.K. Tales of island tails: Biogeomorphic development and management of barrier islands. *J. Coast. Conserv.* **2017**, *21*, 409–419. [CrossRef]
44. Reise, K. Coast of change: Habitat loss and transformations in the Wadden Sea. *Helgol. Mar. Res.* **2005**, *59*, 9–21. [CrossRef]
45. Oost, A.P.; Hoekstra, P.; Wiersma, A.; Flemming, B.; Lammerts, E.J.; Pejrup, M.; Hofstede, J.; van der Valk, B.; Kiden, P.; Bartholdy, J.; et al. Barrier island management: Lessons from the past and directions for the future. *Ocean Coast. Manag.* **2012**, *68*, 18–38. [CrossRef]
46. Hacker, E.; Johannsen, R. *Ingenieurbiologie, 1. Aufl*; UTB: Stuttgart, Germany, 2011; ISBN 9783838533322.
47. Thorenz, F. Coastal Flood Defence and Coastal Protection along the North Sea Coast of Niedersachsen; Die Küste (The Coast) No. 74. 2008. Available online: https://hdl.handle.net/20.500.11970/101601 (accessed on 25 November 2020).
48. Bundesamt für Seeschifffahrt und Hydrographie. *Gezeitenkalender 2020. Hoch- und Niedrigwasserzeiten für die Deutsche Bucht und deren Flussgebiete*; Bundesamt für Seeschifffahrt und Hydrographie No. 2117: Hamburg/Rostock, Germany, 2019; ISBN 978-3-86987-924-6.
49. Hayes, M.O. *Barrier Island Morphology as a Function of Tidal and Wave Regime: Barrier Islands from the Gulf of St. Lawrence to the Gulf of Mexico*; Academic Press: New York, NY, USA, 1979; pp. 1–27.
50. Niemeyer, H.D. Long Term Morphodynamical Development of the East Frisian Island and Coast. In Proceedings of the 24th International Conference on Coastal Engineering, Kobe, Japan, 23–28 October 1994.
51. Hagen, R.; Freund, J.; Plüß, A.; Ihde, R.; Validierungsdokument EasyGSH-DB Nordseemodell. Teil: UnTRIM2—SediMorph—UnK. Bundesanstalt für Wasserbau, Karlsruhe, Germany; 2019. Available online: https://doi.org/10.18451/k2_easygsh_1 (accessed on 25 November 2020).
52. Kottek, M.; Grieser, J.; Beck, C.; Rudolf, B.; Rubel, F. World Map of the Köppen-Geiger climate classification updated. *METZ* **2006**, *15*, 259–263. [CrossRef]
53. Deutscher Wetterdienst. Climata Data Center. Wind, Perception and Temperature. 2020. Available online: Ftp://opendata.dwd.de/climate_environment/CDC/ (accessed on 25 November 2020).
54. Niedersächsischer Landesbetrieb für Wasserwirtschaft, Küsten- und Naturschutz (Lower Saxony Water Management, Coastal Protection and Nature Conservation Agency). *Documentation of the Configuration of the Installed Sand Trapping Fence on Langeoog*; Niedersächsischer Landesbetrieb für Wasserwirtschaft, Küsten- und Naturschutz (Lower Saxony Water Management, Coastal Protection and Nature Conservation Agency): Lower Saxony, Germany, 2020.

55. Sherman, D.J.; Swann, C.; Barron, J.D. A high-efficiency, low-cost aeolian sand trap. *Aeolian Res.* **2014**, *13*, 31–34. [CrossRef]
56. Campos, L.A.D. Quantification Methods for Aeolian Sand Transport on Beaches. Ph.D. Thesis, University of Twente, Enschede, The Netherlands, 2018.
57. Eijkelkamp Soil & Water. Manual, Saltiphone. 2019. Available online: https://en.eijkelkamp.com/products/sensors-monitoring_uk/saltiphone.html (accessed on 25 November 2020).
58. Spaan, W.P.; van den Abeele, G.D. Wind borne particle measurements with acoustic sensors. *Soil Technol.* **1991**, *4*, 51–63. [CrossRef]
59. Goossens, D.; Nolet, C.; Etyemezian, V.; Duarte-Campos, L.; Bakker, G.; Riksen, M. Field testing, comparison, and discussion of five aeolian sand transport measuring devices operating on different measuring principles. *Aeolian Res.* **2018**, *32*, 1–13. [CrossRef]
60. Omega Engineering. Wind Monitor Stations: Product Specifications. Available online: https://in.omega.com/pptst/WMS-20.html (accessed on 25 November 2020).
61. DJI Phantom 4 RTK. Product Specifications. Available online: https://www.dji.com/de/phantom-4-rtk/info (accessed on 25 November 2020).
62. Javad. *Data Sheet GNSS Receiver Sigma-3*. 2019. Available online: http://download.javad.com/manuals/hardware/Sigma_Operators_Manual.pdf (accessed on 25 November 2020).
63. ISO. *ISO 17892-4:2016 Geotechnical Investigation and Testing—Laboratory Testing of Soil—Part 4: Determination of Particle Size Distribution*; German Version EN ISO 17892-4:2016; Beuth Verlag GmbH: Berlin, Germany, 2016.
64. DeltaT Devices. ML3 Soil Moisture Sensor. Product Specifications. 2020. Available online: https://www.ugt-online.de/ (accessed on 25 November 2020).
65. Namikas, S.L.; Bauer, B.O.; Sherman, D.J. Influence of averaging interval on shear velocity estimates for aeolian transport modeling. *Geomorphology* **2003**, *53*, 235–246. [CrossRef]
66. Vickers, D.; Mahrt, L. The Cospectral Gap and Turbulent Flux Calculations. *J. Atmos. Oceanic Technol.* **2003**, *20*, 660–672. [CrossRef]
67. Belly, P.Y. *Sand Movement by Wind*; Technichal Memorandum No. 1; US Army Corps of Engineers, Coastal Engineering Research Center: Washington, DC, USA, 1964.
68. Ellis, J.T.; Li, B.; Farrell, E.J.; Sherman, D.J. Protocols for characterizing aeolian mass-flux profiles. *Aeolian Res.* **2009**, *1*, 19–26. [CrossRef]
69. Nikuradse, J. *Strömungsgesetze in Rauhen Rohren [Flow Laws in Rough Pipes]*; NACA: Washington, DC, USA, 1950.
70. Vithana. The Effect of Stone Protrusion on the Incipient Motion of Rock Armour under the Action of Regular Waves. Ph.D. Thesis, University College London, London, UK, 2013.
71. Bergeron, N.E.; Abrahams, A.D. Estimating shear velocity and roughness length from velocity profiles. *Water Resour. Res.* **1992**, *28*, 2155–2158. [CrossRef]
72. Delgado-Fernandez, I. A review of the application of the fetch effect to modelling sand supply to coastal foredunes. *Aeolian Res.* **2010**, *2*, 61–70. [CrossRef]
73. Hoonhout, B.; de Vries, S. Field measurements on spatial variations in aeolian sediment availability at the Sand Motor mega nourishment. *Aeolian Res.* **2017**, *24*, 93–104. [CrossRef]
74. Shao, Y.; Lu, H. A simple expression for wind erosion threshold friction velocity. *J. Geophys. Res. Space Phys.* **2000**, *105*, 22437–22443. [CrossRef]
75. Han, Q.; Qu, J.; Liao, K.; Zhu, S.; Zhang, K.; Zu, R.; Niu, Q. A wind tunnel study of aeolian sand transport on a wetted sand surface using sands from tropical humid coastal southern China. *Environ. Earth Sci.* **2011**, *64*, 1375–1385. [CrossRef]
76. van Rijn, L.C.; Strypsteen, G. A fully predictive model for aeolian sand transport. *Coast. Eng.* **2020**, *156*, 103600. [CrossRef]
77. Poortinga, A.; Keijsers, J.G.S.; Maroulis, J.; Visser, S.M. Measurement uncertainties in quantifying aeolian mass flux: Evidence from wind tunnel and field site data. *PeerJ* **2014**, *2*, e454. [CrossRef] [PubMed]
78. Bauer, B.O.; Davidson-Arnott, R.G.D. Aeolian particle flux profiles and transport unsteadiness. *J. Geophys. Res. Earth Surf.* **2014**, *119*, 1542–1563. [CrossRef]
79. Dong, Z. Height profile of particle concentration in an aeolian saltating cloud: A wind tunnel investigation by PIV MSD. *Geophys. Res. Lett.* **2003**, *30*, 820. [CrossRef]
80. Bagnold, R.A. The size-grading of sand by wind. *Proc. R. Soc. Lond. Ser. A Math. Phys. Sci.* **1937**, *163*, 250–264.

81. White, B.R. Soil transport by wind on Mars. *J. Geophys. Res. Space Phys.* **2013**, 4643–4651. [CrossRef]
82. Kawamura, R. Study on sand movement by wind. *Rep. Inst. Sci. Technol.* **1951**, *5*, 95–112.
83. Hsu, S.A. Coastal Studies Institute; Louisiana State University Baton Rouge; Louisiana 70803. Computing Eolian Sand Transport from Routine Weather Data. In Proceedings of the 14th International Conference on Coastal Engineering, Copenhagen, Denmark, 24–28 June 1974.
84. Durán, O.; Claudin, P.; Andreotti, B. On aeolian transport: Grain-scale interactions, dynamical mechanisms and scaling laws. *Aeolian Res.* **2011**, *3*, 243–270. [CrossRef]
85. Levin, N.; Ben-Dor, E.; Kidron, G.J.; Yaakov, Y. Estimation of surface roughness (z_0) over a stabilizing coastal dune field based on vegetation and topography. Earth Surf. Process. *Landforms* **2008**, *33*, 1520–1541. [CrossRef]
86. Gillies, J.A.; Etyemezian, V.; Nikolich, G.; Glick, R.; Rowland, P.; Pesce, T.; Skinner, M. Effectiveness of an array of porous fences to reduce sand flux: Oceano Dunes, Oceano CA. *J. Wind Eng. Ind. Aerodyn.* **2017**, *168*, 247–259. [CrossRef]
87. Strypsteen, G.; Montreuil, A.-L.; Rauwoens, P. Aeolian Sand Transport at the Belgian Coast: Field Campaigns and first Results. Coastal Dynamics. 2017. Available online: http://coastaldynamics2017.dk/onewebmedia/117_strypsteen.pdf (accessed on 25 November 2020).
88. Goossens, D.; Offer, Z.; London, G. Wind tunnel and field calibration of five aeolian sand traps. *Geomorphology* **2000**, *35*, 233–252. [CrossRef]
89. Poortinga, A.; van Minnen, J.; Keijsers, J.; Riksen, M.; Goossens, D.; Seeger, M. Measuring fast-temporal sediment fluxes with an analogue acoustic sensor: A wind tunnel study. *PLoS ONE* **2013**, *8*, e74007. [CrossRef] [PubMed]
90. van der Wal, D. Effects of fetch and surface texture on aeolian sand transport on two nourished beaches. *J. Arid. Environ.* **1998**, *39*, 533–547. [CrossRef]
91. van der Wal, D. The Impact of the Grain-Size Distribution of Nourishment Sand on Aeolian Sand Transport. *J. Coast. Res.* **1998**, *14*, 620–631.

Publisher's Note: MDPI stays neutral with regard to jurisdictional claims in published maps and institutional affiliations.

Journal of
Marine Science and Engineering

MDPI

Article

Use of Nanosilica for Increasing Dune Erosion Resistance during a Sea Storm

Elisa Leone [1,*], Nobuhisa Kobayashi [2], Antonio Francone [3], Samuele De Bartolo [1], Davide Strafella [1], Felice D'Alessandro [4] and Giuseppe Roberto Tomasicchio [1]

1 Department of Engineering for Innovation, University of Salento, EUMER Campus Ecotekne, Via Monteroni, 73100 Lecce, Italy; samuele.debartolo@unisalento.it (S.D.B.); davide.strafella@unisalento.it (D.S.); roberto.tomasicchio@unisalento.it (G.R.T.)
2 Center for Applied Coastal Research, University of Delaware, Newark, DE 19716, USA; nk@udel.edu
3 Department of Civil, Environmental, Land, Construction and Chemistry (DICATECh), Polytechnic of Bari, Via Orabona 4, 70125 Bari, Italy; antonio.francone@poliba.it
4 Department of Environmental Science and Policy, University of Milan, Via Celoria 2, 20133 Milano, Italy; felice.dalessandro@unimi.it
* Correspondence: elisa.leone@unisalento.it

Abstract: Dune recovery interventions that integrate natural, sustainable, and soft solutions have become increasingly popular in coastal communities. In the present study, the reliability of an innovative non-toxic colloidal silica-based solution for coastal sand dunes has been verified for the first time by means of laboratory experiments. An extensive experimental campaign aimed at studying the effectiveness of the use of nanosilica has been conducted in the 2D wave flume of the EUMER laboratory at the University of Salento (Italy). The study was first based on a horizontal seabed and then a cross-shore beach-dune profile was drawn similar to those generally observed in nature. Detailed measurements of wave characteristics and observed bed and cross-shore beach-dune profiles were analyzed for a wide range of wave conditions. In both cases, two sets of experiments were carried out. After the first set of experiments performed resembling the native conditions of the models composed with natural sand, the effects of the injection of the mineral colloidal silica-based grout were investigated. The observations show that mineral colloidal silica increases the mechanical strength of non-cohesive sediments reducing the volume of dune erosion, thus improving the resistance and longevity of the beach-dune system.

Keywords: erosion; vulnerability; eco-recovery; colloidal nanosilica; sediment transport; laboratory experiments; longevity; resistance

Citation: Leone, E.; Kobayashi, N.; Francone, A.; De Bartolo, S.; Strafella, D.; D'Alessandro, F.; Tomasicchio, G.R. Use of Nanosilica for Increasing Dune Erosion Resistance during a Sea Storm. *J. Mar. Sci. Eng.* **2021**, *9*, 620. https://doi.org/10.3390/jmse9060620

Academic Editor: Patrick Hesp

Received: 7 May 2021
Accepted: 31 May 2021
Published: 3 June 2021

1. Introduction

The continuously growing human population along the coasts of the world will aggravate the impact of human activities on the coastal environments [1]. Therefore, defensive and restoration activities of coastal areas will become increasingly important [2]. Sandy dunes constitute the natural barrier protecting coastal environments against flooding due to storm surge and wave impacts. The process of defense and restoration consists in preventing degradation and assisting the recovery of an ecosystem that has been damaged or destroyed with respect to its health, integrity, and sustainability [3–5]. In the last decades, several soft and hard engineering solutions dealing with coastal dune erosion prevention have been proposed [6,7]. Despite the results in terms of erosion reduction obtained by the installation of engineering structures, a significant alteration of geomorphological processes is rather unavoidable. To improve traditional construction techniques that used rock, concrete, and steel, new alternative techniques are proposed, using geosynthetic materials for revetments, scour protection, and reclamation [8]. In this regard, planting vegetation and/or constructing plant fences along the backshore was found to be a more natural solution. Installation of planted vegetation limits the mobility of the sand caused by aeolian forces or wave action [9,10]. Nevertheless, a recurrent issue that is addressed in restoration using plants is counteracting the negative impact of depressed biodiversity and even loss of endemic species [11,12]. Moreover, great care must be taken when dealing with changes in geomorphological processes. The collective movement of solid particles, known as sediment transport, is a complex problem related to flow conditions, the nature of the sediment, and the cohesive forces between particles. Historical overviews of the influence of wave conditions, sediment size, sediment gradation, etc., have been extensively studied, for example, by Jonsson [13], Madsen and Grant [14], Sleath [15], Nielsen [16], Fredsøe and Deigaard [17], Van Rijn [18], Soulsby [19], and cross-shore wave mechanics and sediment transport on beaches has been investigated by Nairn and Southgate [20], Kobayashi and Johnson [21], Dean and Dalrymple [22], Tomasicchio et al. [23,24], D'Alessandro and Tomasicchio [25], D'Alessandro et al. [26], Sancho et al. [27], and Brunone and Tomasicchio [28]. These works show the wave-induced bottom shear stress as an essential parameter for calculation of beach erosion and accretion rate and wave height attenuation.

In the present study, the use of nanosilica for coastal sand dune strengthening has been evaluated in the 2D wave flume of EUMER at Unisalento, Lecce (www.eumer.eu, accessed on 10 September 2019). The method concerns the injection of a colloidal silica-based solution into the sand in order to obtain a consolidated layer and reduce the erosion process due to wave action. Testing the performance of the use of nanosilica included two stages. In the first stage, wave-driven sand transport was measured on a mobile sand bed for a wide range of flow conditions. The bed with natural well-sorted fine sand was compared with the bed with the silica-based consolidating material, which changed the cohesive properties of the sand. Observing the behavior of both the natural and consolidated sand has clarified the relationship between the sand properties and hydrodynamic loadings (e.g., bed shear stress). Thereafter, physical model tests on the morphological development of the beach-dune system under irregular waves were conducted with the aim of observing and comparing the behavior of natural and consolidated sand. The conducted experiments, including setup, data analysis, and comparison of the data are presented in the following sections.

2. Laboratory Facility and Nanosilica Grout

2.1. Experimental Facility

In the framework of the INNODUNE project, physical model tests were conducted in a 2D wave flume at the EUMER lab, Campus Ecotekne, Lecce (Figure 1). The flume is 45 m long, 1.5 m wide, and 2 m deep. The wave generation is performed by a single paddle equipped with an active wave absorption system allowing tests to be run without the effect of spurious model-induced reflections. Froude's law was selected as the proper criterion to down-scale the hydrodynamic parameters; however, on fine-grained sand beaches, the geometric scaling leading to a clay-size sediment were avoided to prevent the inception of cohesive effects [29]. Therefore, the grain size of the sediment in the flume is not scaled by the geometric scale, resulting in the same sand in the model [30]. The sediment is well-sorted fine sand with a median diameter d_{50} of 0.245 mm. The measured density of sediment grains, ρ_s, and porosity, ε, of the sand are 2.6 g/cm^3 and 0.4, respectively. The experimental program involved measurements of time-varying water level and bed profile. A wave gauge system from HR Wallingford acquired the water level with an accuracy of the order of 0.1 mm [31], enabling both statistical and spectral analysis of the waves through the HR DAQ data acquisition and analysis software program.

The cross-shore profile is observed with the laser profiler described by HR Wallingford. The system consists of a vertical position controlled device driven by a small electric motor with a mechanical transmission actuating a rod. At the lower end of the rod, a low-powered laser distance sensor mounted inside a waterproof housing is provided. The position of the probe is measured by encoders on the moto shafts, leading to the observation of the bed height, even rather inclined and irregular profile, with an accuracy of the order of

0.1 mm [32]. The bed profile observation is performed by the horizontal motion of the vertical servo controlled device placed in a carriage.

Figure 1. Top view (**left**) and side view (**right**) of 2D wave flume at EUMER lab, Unisalento.

2.2. Nanosilica-Based Grout

The additive to consolidate the sand consists of a non-toxic aqueous suspension with dispersed nanometric particles of colloidal silica (Figure 2). These particles are made of quartz (SiO_2) with a tetrahedral distribution of the atoms which makes the mixture completely mineral, highly reactive, and more stable than other quartz-based crystalline compounds. Before injecting nanosilica into the sand, an accelerator, i.e., a solution of sodium chloride (10% NaCl in water), is added to the suspension.

Figure 2. Aqueous suspension of colloidal nanosilica.

The solution acts as an activator, triggering a gelification process that induces the suspension to become a gel. Depending on the dosage, the accelerator adjusts the product workability time [33]. Nanosilica, before gelation, has a very low viscosity (very close to water: about 10 mPa·s compared to 1 mPa·s), which means a very high permeation and penetration capacity.

After mixing the components, the dispersion of the nanosilica-based grout was carried out by a pressure pulverizer at 20 °C, penetrating into the surface layer of the sand. After spraying the grout and completion of the gelling time, a homogeneous consolidation of the sand was observed. The result consisted of a pore-filling process and an increase in cohesion of the sand; nevertheless, the nanosilica-based material did not interfere with germination or seedling growth because permeability was ensured. On a control volume, different amounts of nanosilica were tested, defined by a percentage of the grout relative to the voids in the sand. In particular, tested amounts were 5%, 2%, and 1% which resulted in a consolidated sand layer of different depth, h_c, and porosity ε (Table 1). In Figure 3, samples of consolidated layers are shown.

Table 1. Characteristics of consolidated layers of each percentage of nanosilica.

Percentage of Nanosilica	h_c [cm]	ε [-]
5%	3.0	0.20
2%	1.5	0.24
1%	0.8	0.25

Figure 3. Samples of consolidated layers for different percentages of nanosilica: 5%, 2%, and 1%.

3. Horizontal Bed Experiments

The first part of the experimental campaign was carried out on a horizontal seabed. The experiment was conducted because the wave boundary layer on a horizontal bed is well understood. The bottom shear stress on natural and consolidated sands can be estimated using existing wave boundary layer models. In the wave flume, two concrete blocks with a triangular transversal geometry and 1:3 slope were positioned at a distance of 2 m from each other. The blocks were 0.3 m high, 0.9 m long, and as wide as the width of the flume. The 2-m gap between the blocks was filled with sand. A picture of the first physical model is shown in Figure 4.

Figure 4. Horizontal sand bed layout.

One wave gauge (WG1) was located as close as possible to the wave generator and three wave gauges (WG2–WG4) were used to evaluate the wave conditions over the horizontal seabed and to separate incident and reflected waves in the wave flume. The bed profile was observed with the laser profiler for an extension of 1.5 m on the seabed to avoid the edge effects due to the slopes. A side view of the full experimental setup is shown in Figure 5.

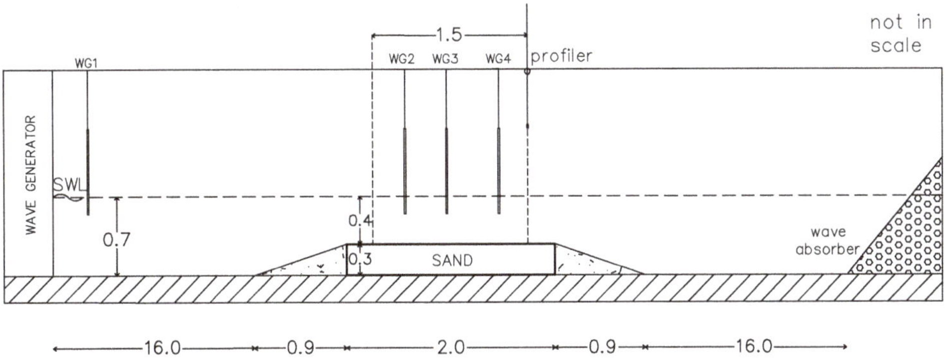

Figure 5. Side view of experimental setup of the horizontal seabed model.

Non-breaking wave conditions were adopted. Regular and irregular wave tests were conducted on the horizontal seabed. The incident and reflected wave parameters were obtained using the theory of Mansard and Funke [34]. The reflection coefficient, k_r, defined as the ratio between the reflected and incident wave height was less than 0.25. Table 2 lists the incident wave characteristics in sequence. In particular, tests marked with HN (Horizontal Natural) refer to a movable bottom of natural sand without adding colloidal nanosilica. Tests marked with HC (Horizontal Consolidated) refer to a condition of sand with a percentage of nanosilica calculated so as to achieve a consolidated sand layer of 5 cm. The regular wave conditions are defined by wave height, H, and wave period, T. The irregular wave conditions are based on a Jonswap spectrum with peak enhancement factor, γ, equal to 3.3, and defined by energy-based significant wave height, H_{m0}, and peak period, T_p. Three water depths above the bed, h, were chosen, starting from 0.4 m increased by 0.1 m up to 0.6 m. The water depth in front of the wave generator was (h + 0.3) m. Two tests of regular waves and two tests of irregular waves were carried out for each h. The duration of wave attacks was defined considering the time required for bed form to develop evenly along the seabed [35]. In particular, for regular waves, a number of waves were specified to be generated; under irregular waves, the bed forms may grow slowly, especially under low velocities that are not much above threshold. Some exploratory experiments were conducted to determine the necessary durations. In both cases, the total duration also includes the time the wave takes to reach the model and return to the wave generator which, for convenience, has been associated with the duration of 30 waves. This calculated time was excluded during wave data analysis [35].

Wave parameters are subject to changes due to wave transformation and wave height may decrease, due to wave energy loss. The intensity of wave energy dissipation depends on the incident wave height, water depth, and bottom characteristics. Using the wave conditions acquired in HC and HN tests, the amplitude of the wave orbital velocity, U_w, and one half of the horizontal orbital excursion, A_w, just above the bottom boundary layer for regular waves were calculated as [19]:

$$U_w = \frac{\pi H}{T \sinh(kh)}$$
$$A_w = \frac{U_w T}{2\pi} \quad (1)$$

where *sinh* is the hyperbolic sine, $k = \frac{2\pi}{L}$ is the wave number, L is the wavelength, and h the water depth over the seabed. For irregular waves, H and T are replaced by H_{m0} and T_p.

Table 2. Incident of regular and irregular wave characteristics on the horizontal seabed.

Test Group	Test Name	Wave Type	h [m]	H/H_{m0} [m]	T/T_p [s]	Duration [s]
HN	HN1	Regular	0.40	0.116	1.00	180
	HN2		0.40	0.138	1.20	216
	HN3		0.50	0.153	1.79	324
	HN4		0.50	0.219	1.99	360
	HN5		0.60	0.195	1.79	324
	HN6		0.60	0.220	2.19	396
	HN7	Irregular	0.40	0.095	1.48	8145
	HN8		0.40	0.118	1.79	2574
	HN9		0.50	0.146	2.01	2460
	HN10		0.50	0.183	2.42	1752
	HN11		0.60	0.182	2.15	1606
	HN12		0.60	0.206	2.41	1752
HC	HC1	Regular	0.40	0.118	1.00	180
	HC2		0.40	0.141	1.20	216
	HC3		0.50	0.160	1.80	324
	HC4		0.50	0.227	2.00	360
	HC5		0.60	0.209	1.80	324
	HC6		0.60	0.231	2.19	396
	HC7	Irregular	0.40	0.097	1.48	8145
	HC8		0.40	0.120	1.80	2574
	HC9		0.50	0.153	2.02	2460
	HC10		0.50	0.189	2.46	1752
	HC11		0.60	0.188	2.15	1606
	HC12		0.60	0.211	2.42	1752

Bed form conditions responded to increasing wave energy for HN2–12. The measured ripple height and wavelength were estimated about 1 cm and 10 cm, respectively. In the remaining tests (HN1 and HC1–12) the seabed configuration did not change in response to the wave attack, remaining flat.

4. Movement of Natural and Consolidated Sands

An estimate of the threshold of motion can be given by the Shields parameter, θ_0, in terms of the ratio of the force exerted by the bed shear stress acting on a grain on the bed, to the submerged weight of the grain counteracting this [19]. The bed shear stress, τ_w, represents the most important hydrodynamic property of waves for sediment transport purposes. It is usually obtained from U_w and the bottom friction factor, f_w. Acting on the bed, τ_w is made up of contributions due to the skin friction, τ_{ws}, produced by (and acting on) the sediment grains and the form drag, τ_{wf}, produced by the pressure field associated with the flow over ripples and/or larger features on the bed. If the bed is flat and sediment transport is not intense, the distinction between total and skin friction quantities is not necessary. It is important to be aware that only the skin friction contribution acts directly on the sediment grains, and the skin friction is normally used to calculate the threshold of motion, bedload transport, and reference concentration or pick-up rate for grains in suspension. On the other hand, it is the total bed shear stress that corresponds to the overall resistance of the flow and determines the turbulence intensities which influence the diffusion of suspended sediment higher in the water column. In literature, several expressions of f_w were found with a great variability in terms of Reynolds number, $Re = \frac{A_w U_w}{\nu}$, where ν is the kinematic viscosity, and the relative roughness, $\frac{A_w}{k_N}$, where k_N is the Nikuradse equivalent bed roughness. k_N can be written as sum of grain-related

k_{Ns} and form drag k_{Nf} components. Among the available formulas, the formulation by Soulsby [19] was adopted to calculate f_w. This is a choice taken by many authors [36–38]. The formulation by Soulsby does not depend on the wave Reynolds number, and a rough turbulent bed is assumed. Therefore, θ_0, τ_w, and f_w are defined by:

$$\begin{gathered}\theta_0 = \frac{\tau_w}{g\ (\rho_s - \rho) d_{50}} \\ \tau_w = \frac{1}{2}\rho f_w {U_w}^2 \\ f_w \ depends \ on \ \frac{A_w}{K_N}\end{gathered} \tag{2}$$

where g is the acceleration due to gravity, ρ is the density of the water, and the density and the median diameter of the sand are ρ_s =2.6 g/cm^3 and d_{50} =0.245 mm in this study.

In the present work, K_N is evaluated in the form $K_N = 2.5\, d_{50}$ proposed by Nielsen (1992) [16].

According to Shields, the sediment movement may be observed when a critical value of bed shear stress is reached with a typical value of θ_0 is of about 0.05 [39]. Figure 6 shows the calculated θ_0 for each test considering only the skin friction shear stress contribution which acts directly on the sediment grains.

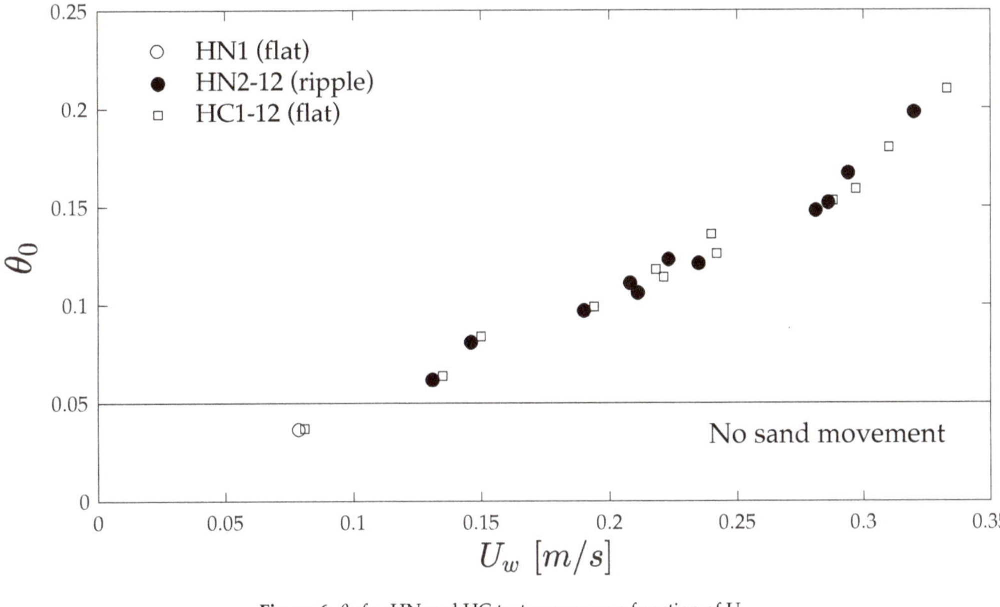

Figure 6. θ_0 for HN and HC test groups as a function of U_w.

In HN tests, the results were consistent with Shield's theory. In tests with θ_0 greater than 0.05, mobilization at the bottom was observed. Only in the case of HN1, the critical value not reached. On the other hand, in HC tests, although the threshold θ_0 is reached, mobility at the bottom was not observed. In particular, the largest value of θ_0 was observed in the HC4 regular wave test with a duration of 360 s corresponding to a consolidated sand resistance force per unit area of 0.81 N/m^2. For irregular waves, θ_0 became as large as 0.16 in the HC10 with a duration of 1752 s, reaching a consolidated sand resistance force per unit area of 0.61 N/m^2.

This experiment indicated the practical utility of nanosilica grout but, since its performance against breaking waves is unknown, new experiments using breaking waves were carried out.

5. Dune Erosion Experiments

A second part of the experimental campaign consisted of a model beach-dune system located at 23.3 m from the wave generator. The model in the flume had a length of 6.9 m. The beach had a slope of 1/10 by an extension of 5.2 m and a slope of 1/50 by an extension of 0.2 m. The dune length was 1.5 m with a seaward slope of 1/1.5. The dune crest length was 0.5 m, and the dune height above the toe was 0.35 m. These slope gradients are typically observed along the Italian coastline. The still water level (SWL) was chosen in such a way that the part of the beach with a slope of 1/10 was mostly submerged and the part with a slope of 1/50 was fully emerged. Waves were measured using wave gauges WG1-WG7. A side view of the beach-dune system as built with the instrumentation is shown in Figure 7. The origin along the onshore coordinate is taken at WG2. WG1 was located as near as possible to the wave generator to check the accuracy of the wave generation and wave absorption capability of the wave maker. The remaining gauges measured the free surface elevations from the offshore and wave shoaling zones and for reflection analysis following the theory of Mansard and Funke [34]. Given the bathymetry, WG2-WG3-WG4 were placed before the beach and WG5-WG6-WG7 before the dune. In Table 3, the onshore coordinate X (m) and water depth below SWL at each of the seven WGs are listed.

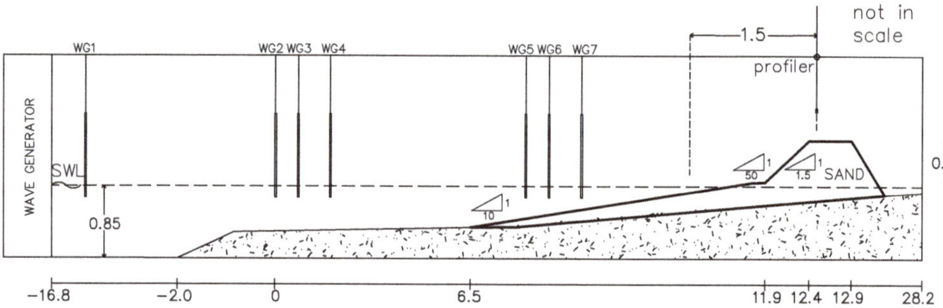

Figure 7. Side view of experimental setup of beach-dune system model.

Table 3. Onshore coordinate X and water depth below SWL at WG1-WG7.

Wave Gauge (WG)	WG1	WG2	WG3	WG4	WG5	WG6	WG7
Onshore X [m]	−15.8	0	0.4	0.9	7.0	7.3	7.7
Depth [m]	0.85	0.53	0.53	0.52	0.44	0.41	0.38

The cross-shore profile of the beach-dune system is measured with the laser profiler in the zone of noticeable profile changes x = 11.0 m–12.5 m (1.5 m onshore distance). This choice stems from the fact that the nanosilica-based grout was sprayed from the toe up to the crest of the dune to evaluate the increase of the erosion resistance exclusively on the dune face.

Two sets of tests were carried out allowing the observation of the dune erosion process and comparison with the behavior of the system with the nanosilica grout. After the first set of tests, performed resembling the native conditions of the beach-dune system made of natural well-sorted fine sand, the effects of spraying the colloidal silica-based grout over the face of the dune were investigated. Seven dune tests (D) using natural (N) and consolidated (C) sands were carried out. The wave conditions are defined by H_{m0} and T_p at WG 2–4 outside the surf zone and the wave steepness $s = \frac{2\pi}{g}\frac{H_{m0}}{T_p^2}$ is between 0.01

and 0.07. Starting from time t = 0, the initial profile was exposed to four runs composed of 230, 230, 330, and 330 waves, respectively. The beach profile was measured after each run [35]. Again, 30 waves were excluded during wave data analysis for each run. Table 4 summarizes the incident wave characteristics in the dune erosion experiments.

Reflection coefficients appear relatively large because of the steep dune slope, typically observed along the Italian coastlines and exposed to waves that are not too large. Table 4 shows similar wave characteristics for the corresponding test, DNj and DCj, with j = 1–7. The differences are less than 10% for H_{m0}, 15% for T_p, and 20% for k_r.

Table 4. Incident irregular wave H_{m0}, T_p, $s = \frac{H_{m0}}{(gT_p^2/2\pi)}$, and reflection coefficient k_r at WG 2–4.

Test	Run	H_{m0} [m]	T_p [s]	s [-]	k_r [-]	Test	Run	H_{m0} [m]	T_p [s]	s [-]	k_r [-]
DN1	1	0.101	1.17	0.047	0.203	DC1	1	0.104	1.17	0.048	0.189
	2	0.104	1.14	0.051	0.199		2	0.104	1.20	0.047	0.207
	3	0.098	1.19	0.044	0.198		3	0.100	1.19	0.045	0.217
	4	0.097	1.19	0.044	0.188		4	0.100	1.19	0.045	0.227
DN2	1	0.116	1.56	0.030	0.257	DC2	1	0.116	1.56	0.031	0.253
	2	0.115	1.56	0.030	0.236		2	0.116	1.56	0.031	0.260
	3	0.114	1.57	0.030	0.235		3	0.114	1.57	0.030	0.272
	4	0.114	1.54	0.031	0.235		4	0.114	1.53	0.031	0.280
DN3	1	0.118	1.99	0.019	0.299	DC3	1	0.121	1.99	0.020	0.337
	2	0.118	1.99	0.019	0.279		2	0.121	2.02	0.019	0.344
	3	0.117	2.16	0.016	0.258		3	0.120	2.21	0.016	0.321
	4	0.117	2.27	0.015	0.238		4	0.120	2.21	0.016	0.325
DN4	1	0.123	2.49	0.013	0.402	DC4	1	0.124	2.42	0.014	0.388
	2	0.123	2.45	0.013	0.401		2	0.123	2.42	0.013	0.397
	3	0.122	2.55	0.012	0.379		3	0.122	2.47	0.013	0.386
	4	0.122	2.53	0.012	0.363		4	0.122	2.47	0.013	0.386
DN5	1	0.155	1.23	0.065	0.256	DC5	1	0.152	1.23	0.064	0.259
	2	0.154	1.23	0.065	0.269		2	0.152	1.23	0.064	0.265
	3	0.150	1.19	0.068	0.247		3	0.149	1.19	0.068	0.267
	4	0.151	1.19	0.068	0.254		4	0.146	1.19	0.066	0.251
DN6	1	0.188	1.52	0.052	0.299	DC6	1	0.202	1.53	0.055	0.283
	2	0.188	1.52	0.052	0.294		2	0.202	1.53	0.055	0.295
	3	0.184	1.52	0.051	0.289		3	0.196	1.54	0.053	0.305
	4	0.183	1.54	0.049	0.284		4	0.195	1.56	0.051	0.312
DN7	1	0.188	1.99	0.030	0.259	DC7	1	0.208	1.99	0.034	0.278
	2	0.188	1.99	0.030	0.267		2	0.208	1.99	0.034	0.294
	3	0.185	2.27	0.023	0.235		3	0.208	1.99	0.034	0.297
	4	0.185	2.27	0.023	0.231		4	0.199	1.99	0.032	0.244

6. Erosion of Natural and Consolidated Dunes

In the beach-dune model, incident wave characteristics observed between the offshore and shoaling zones were analyzed to examine the cross-shore wave transformation. In the DN and DC tests, H_{m0}, T_p, and k_r at WG 2–4 and WG 5–7 were compared and shown in Figure 8 for all the tests with the four runs.

In most of the tests, the wave showed no significant change from WG 2–4 to WG 5–7, except for the DN6-DC6 and DN7-DC7 in which H_{m0} decreased due to wave breaking. Comparing the results between the DN and DC tests, k_r increased along the flume in the cases of consolidated sand. The above tendencies are associated with the application of nanosilica to the dune, which limits the erosion of the steep dune face. This behavior is most evident in test DC7, where k_r increased from about 0.28 at WG 2–4 to about 0.36 at WG 5–7 because of the decrease of H_{m0} due to wave breaking.

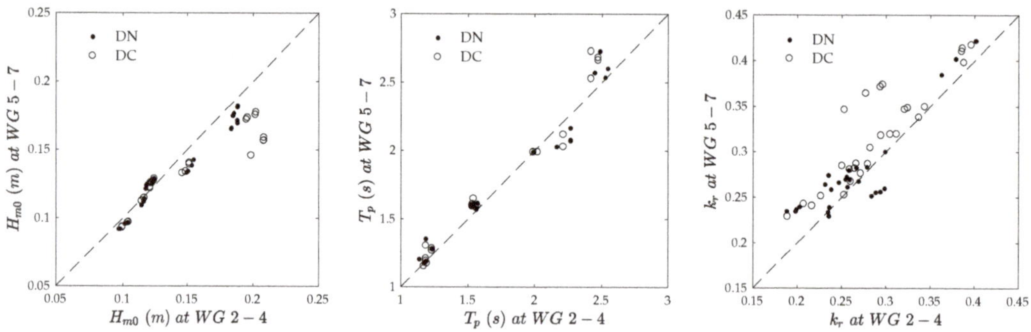

Figure 8. Comparison of wave characteristics of WG 2–4 and W G5–7 for DN and DC tests.

The erosion process of the beach-dune system was observed. Before each new test, the initial conditions of the beach-dune system were restored. It is important to note that it is difficult to exactly reconstruct the same model geometry at the start of each test. Although some geometric parameters were unidentical for all tests (e.g., dune height), the deviation is small enough that they can be considered constant. The process of dune face erosion in the DN tests changed as the test continued. Initially, the slope of the dune face was 1/1.5 (vertical/horizontal). Many waves reach and erode the dune face. The scarped dune front becomes steeper until it is nearly vertical. Then, after a first series of waves within the first run, a different erosion mechanism starts. The dune face only retreats under waves impacting it. Episodically, the dune face collapses and big lumps of sediment fall or slide down the dune face on the beach in front of the dune. This sand is picked up by swash and inner surf zone processes that move the sediment further seaward extending the new foreshore. At this stage of the erosion process, the number of waves impacting the dune is certainly lower, consequently the erosion rate is reduced. In the DC tests, the injection of the nanosilica protected the surface layer of the dune exposed to run up, so the erosion process did not take place in the first series of waves, as in the DN tests. The dune face remained unchanged for a higher number of waves. A front and side view of the cross-shore profile after four runs of DN4 and DC4 is shown in Figure 9 in order to appreciate the effect of nanosilica on the dune face.

The cross-shore profiles of the beach-dune system were evaluated by the laser profiler attached to a mobile platform with x-axis defined positively onshore from x = 11 m up to x = 12.5 m (see Figure 7). Before the start of each test as well as after every wave run, profiles were recorded along the centerline. The profiler velocity was set in order to have a horizontal resolution of 0.2 mm. Figure 10 presents the comparison of the observed cross-shore profiles of DN4 and DC4.

For the DN4 test, a progressive erosion of the dune face from the first run up to the end of the test was observed. An offshore transport was noticeable from the crest, leading to sand being deposited on the toe of the dune and onto the underlying beach. This is compounded by onshore sand transport increasing sand accumulation at the toe area. Focusing on the DC4 test, it is evident that the cross-shore profile of the dune face did not changed as a result of the injection of the nano-silica based grout. Only the onshore transport was observed, leading to accumulation (of a lower magnitude than DN4) at the toe of the dune and erosion in the inner surf zone.

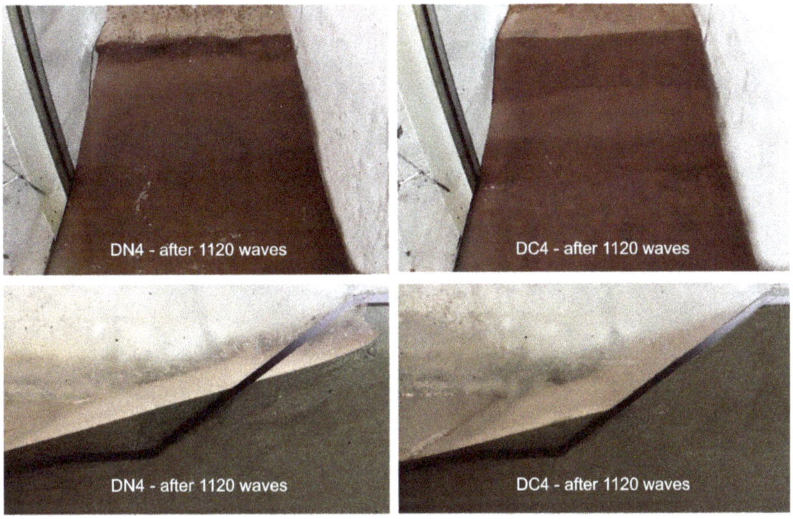

Figure 9. Front view (**up**) and side view (**down**) of cross-shore profile after four runs of DN4 and DC4.

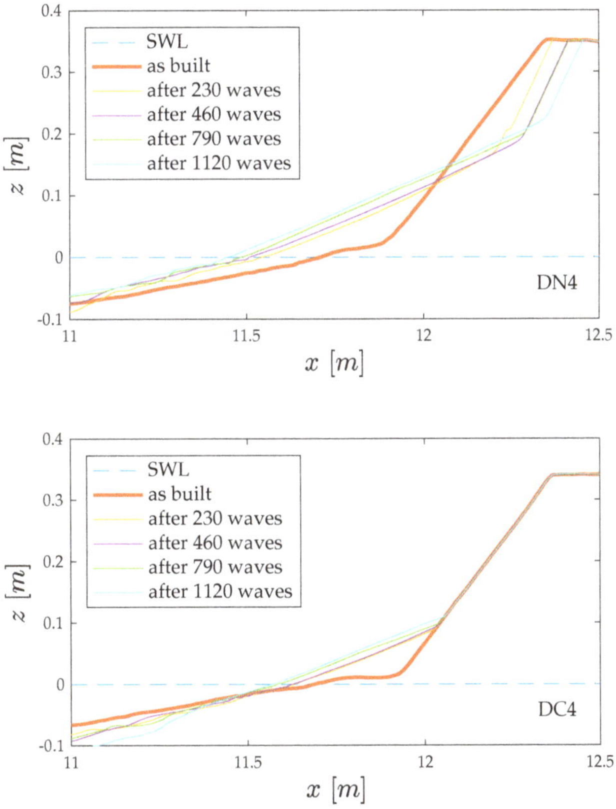

Figure 10. Measured cross-shore profiles after 230, 460, 790, and 1120 waves for tests DN4 (**up**) and DC4 (**down**).

Erosion progression of the dune was analyzed in order to quantify the effect of the nanosilica-based grout on the dune surface. The eroded area, A_e, after each run of the DN and DC tests was normalized by the square of H_{m0} observed at WG 2–4. Table 5 lists the sequence of the normalized area $A_* = \frac{A_e}{H_{m0}^2}$ and the wave steepness s based on H_{m0} and T_p at WG 2–4 for the DN and DC tests.

Figure 11 shows A_* as a function of s for tests DN1-DN7 for increasing number of waves. Empirical formula for dune erosion can be found for $s = 0.012 - 0.068$ by fitting an exponential function of the observed data for each number of waves as:

$$A_{*,n} = a(N)\, e^{-b(N)*s} \quad (3)$$

where the subscript n indicates that the formula is valid for tests with natural sand, N is the number of waves from 230 to 1120, and a and b are the exponential function coefficients listed in Table 6.

Table 5. Wave steepness $s = \frac{H_{m0}}{(gT_p^2/2\pi)}$ at WG 2–4 and measured A_e and $A_* = \frac{A_e}{H_{m0}^2}$ for DN and DC tests.

Test	Run	s [-]	A_e [m^2]	A_* [-]	test	Run	s [-]	A_e [m^2]	A_* [-]
DN1	1	0.047	0.0018	0.175	DC1	1	0.048	0.0001	0.009
	2	0.051	0.0019	0.177		2	0.047	0.0001	0.009
	3	0.044	0.0020	0.209		3	0.045	0.0001	0.010
	4	0.044	0.0022	0.235		4	0.045	0.0002	0.020
DN2	1	0.030	0.0045	0.336	DC2	1	0.031	0.0001	0.007
	2	0.030	0.0059	0.443		2	0.031	0.0001	0.007
	3	0.030	0.0081	0.625		3	0.030	0.0002	0.015
	4	0.031	0.0087	0.671		4	0.031	0.0002	0.015
DN3	1	0.019	0.0032	0.229	DC3	1	0.020	0.0001	0.007
	2	0.019	0.0049	0.355		2	0.019	0.0002	0.014
	3	0.016	0.0050	0.365		3	0.016	0.0002	0.014
	4	0.015	0.0068	0.497		4	0.016	0.0004	0.028
DN4	1	0.013	0.0150	0.985	DC4	1	0.014	0.0001	0.007
	2	0.013	0.0208	1.370		2	0.013	0.0001	0.007
	3	0.012	0.0227	1.525		3	0.013	0.0001	0.007
	4	0.012	0.0247	1.673		4	0.013	0.0002	0.013
DN5	1	0.065	0.0026	0.108	DC5	1	0.064	0.0001	0.003
	2	0.065	0.0035	0.149		2	0.064	0.0001	0.004
	3	0.068	0.0037	0.166		3	0.068	0.0002	0.009
	4	0.068	0.0047	0.207		4	0.066	0.0002	0.009
DN6	1	0.052	0.0049	0.139	DC6	1	0.055	0.0001	0.002
	2	0.052	0.0069	0.195		2	0.055	0.0002	0.005
	3	0.051	0.0101	0.300		3	0.053	0.0002	0.005
	4	0.049	0.0152	0.452		4	0.051	0.0101	0.265
DN7	1	0.030	0.0116	0.327	DC7	1	0.034	0.0001	0.002
	2	0.030	0.0121	0.342		2	0.034	0.0001	0.002
	3	0.023	0.0186	0.541		3	0.034	0.0001	0.002
	4	0.023	0.0227	0.667		4	0.032	0.0123	0.312

Based on the analysis of the test results, the effect of T_p on dune erosion in terms of A_* was evident. In particular, A_* was larger for lower values of s, confirming its dependency on T_p in accordance with van Gent et al. [40].

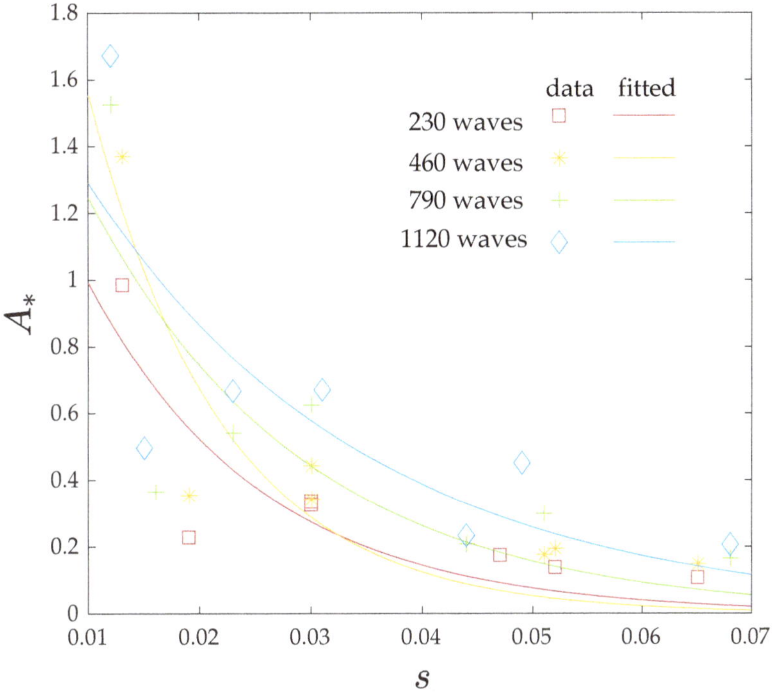

Figure 11. Normalized eroded area A_* after 230, 460, 790, and 1120 waves as a function of wave steepness s for tests DN1-DN7.

Table 6. a and b as a function of N for tests DN1-DN7.

N	a	b
230	1.88	64.0
460	3.60	84.1
790	2.09	51.8
1120	1.93	40.1

For tests DC1-DC7, a significant reduction in A_* was observed. The erosion threshold, in terms of A_*, is 0.05 except for DC6 and DC7, which exhibited a higher value for N = 1120. Figure 12 shows A_* as a function of s in the DN and DC tests for increasing N. The difference in terms of A_* between the DN and DC tests is significant because the erosion in the DC tests was significantly reduced; nevertheless, erosion was observed at DC6 and DC7, between waves 790 and 1120. The present experiments suggest the limitations of the nanosilica grout in terms of erosion resistance and durability because of the fairly sudden failure of the nanosilica grout.

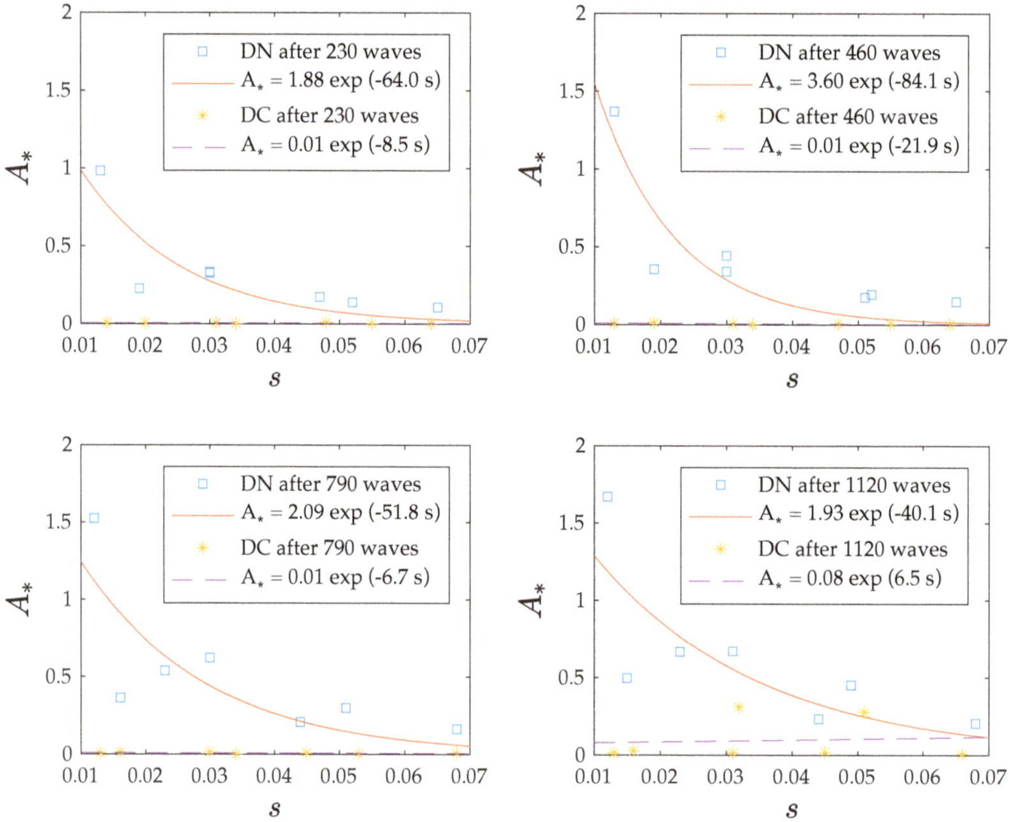

Figure 12. A_* as a function of s in the DN and DC tests for N = 230 (**upper left**), 460 (**upper right**), 790 (**lower left**), and 1120 (**lower right**).

The general empirical formula, for both DN and DC tests, is expressed as:

$$A_* = R\,a(N)\,e^{-b(N)*s} \tag{4}$$

where R is defined as the reduction factor for the consolidated sand:

$$\begin{aligned} R &= 1 && \text{For natural sand} \\ R &= \frac{A_*}{a\,e^{(-b\,s)}} && \text{For consolidated sand} \end{aligned} \tag{5}$$

Equation (5) implies that $R < 1$ in the DC tests accounts for the effect of increased erosion resistance. The estimated $a\,e^{(-b\,s)}$ using Table 6 and R using A_* in DC tests are listed in Table 7. A significant reduction in erosion was confirmed, with a threshold in terms of R, of 0.28 except for DC6 and DC7 for the number of waves equal to 1120. The reduction factor of R will need to be predicted for actual field applications of the nanosilica grout.

Table 7. Wave characteristics for DC tests, estimated $a\,e^{(-b\,s)}$ for DN tests and R for DC tests.

Test	H_{m0} [m]	T_p [s]	s [-]	N	ae^{-bs}	R
DC1	0.104	1.17	0.048	230	0.09	0.10
	0.104	1.20	0.047	460	0.05	0.19
	0.100	1.19	0.045	790	0.21	0.05
	0.100	1.19	0.045	1120	0.33	0.06
DC2	0.116	1.56	0.031	230	0.27	0.03
	0.116	1.56	0.031	460	0.28	0.03
	0.114	1.57	0.030	790	0.45	0.03
	0.114	1.53	0.031	1120	0.56	0.03
DC3	0.121	1.99	0.020	230	0.55	0.01
	0.121	2.02	0.019	460	0.73	0.02
	0.120	2.21	0.016	790	0.91	0.02
	0.120	2.21	0.016	1120	1.08	0.03
DC4	0.124	2.42	0.014	230	0.83	0.01
	0.123	2.42	0.013	460	1.19	0.01
	0.122	2.47	0.013	790	1.12	0.01
	0.122	2.47	0.013	1120	1.18	0.01
DC5	0.152	1.23	0.064	230	0.03	0.11
	0.152	1.23	0.064	460	0.02	0.28
	0.149	1.19	0.068	790	0.06	0.14
	0.146	1.19	0.066	1120	0.12	0.08
DC6	0.202	1.53	0.055	230	0.07	0.02
	0.202	1.53	0.055	460	0.04	0.11
	0.196	1.54	0.053	790	0.15	0.03
	0.195	1.56	0.051	1120	0.27	0.99
DC7	0.208	1.99	0.034	230	0.27	0.01
	0.208	1.99	0.034	460	0.28	0.01
	0.208	1.99	0.034	790	0.63	0.01
	0.199	1.99	0.032	1120	0.77	0.40

7. Conclusions

Dune protection interventions vary greatly in scale and complexity depending primarily on the extent of the existing dune system and on the degree of degradation that has affected the dune system. The effectiveness of the use of nanosilica was investigated with two physical models in the 2D wave flume of the EUMER laboratory at the University of Salento (Italy). The first part of the experimental campaign was carried out on a horizontal seabed to allow us to gain knowledge of the nanosilica properties related to hydrodynamic loadings and effective testing procedures. Compared with a natural loose sand bottom, the injection of colloidal nanosilica consolidated a surface layer of the seabed. The most interesting aspect concerning the first part of the experimental campaign is that the sediment mobility threshold of consolidated sand increased considerably. In the second part of the experimental study, the behavior of colloidal silica-based grout was investigated on a beach-dune system under irregular wave tests. Cross-shore profiles of a beach-dune system observed for seven different wave conditions in the case of natural sand and consolidated sand were compared. It was observed that, with the injection of the consolidating nanosilica-based grout, the cross-shore dune profile under five wave attacks remained practically unchanged; under the other two wave attacks, a delay in the erosion process was observed. The results demonstrate significant advantages in the use of nanosilica to improve erosion resistance of the dune.

From laboratory evidence, the use of nanosilica is found to be promising, although it should include a carefully planned procedure to achieve and maintain strong awareness and support within the broader community, as with any dune rehabilitation and management strategy. In this regard, research in several aspects can be further deepened, such as the influence on aeolian transport, vegetation dynamics related to the deposition and erosion of the sediments at the dunes, and multi-scale consequences on biological components and the overall sediment balance in the coastal region. As already mentioned, several methods can be used to prevent dune erosion depending on the context and scale of degradation; they include the use of sand moving equipment or sand trapping techniques such as dune-forming fences, the spreading of brush matting, and revegetation. Comparing with more traditional solutions, the proposed intervention does not increase the dune size footprint, avoiding further obstacles. Moreover, it also has an economic implication since, paralleled with the cost of the material, which can vary depending on the context of the dune environment to be treated, there is a considerable saving on the costs of implementation of the intervention in terms of human resources and equipment to be used.

It is worth noting that several potential applications could be considered not only in the field of dune protection but also concerning phenomena in which hydrodynamic quantities such as turbulent and mean velocities, bed shear stress, and turbulent stresses undoubtedly govern the processes of entrainment, transport, and deposition [41,42].

A numerical model will need to be developed for practical applications of nanosilica grout and to investigate the behavior of consolidated sand on the beach-dune system across different time and space scales. The numerical model may be calibrated using the small-scale experiments in this study.

Author Contributions: Conceptualization and supervision of the investigation, N.K. and G.R.T.; laboratory investigation, E.L. and D.S.; data analysis and curation, writing—original draft preparation, E.L.; review and editing, A.F., S.D.B., and F.D. All authors have read and agreed to the published version of the manuscript.

Funding: This research was funded by the Regione Puglia through the grant project titled "Sperimentazione di tecnologie innovative per il consolidamento di dune costiere (INNO-DUNECOST)", POR Puglia FESR FSE 2014-2020-Sub-Azione 1.4.B, Contract n. RM5UKM2.

Institutional Review Board Statement: Not applicable.

Informed Consent Statement: Not applicable.

Data Availability Statement: Not applicable.

Acknowledgments: Authors thank the four anonymous reviewers for their fruitful comments.

Conflicts of Interest: The authors declare no conflict of interest.

References

1. Taveira-Pinto, F.; Rosa-Santos, P.; Fazeres-Ferradosa, T. Anthropogenic influences on Integrated Coastal Zone Management. *Rev. Gestão Costeira Integr.* **2020**, *20*, 215–217. [CrossRef]
2. Hamza, W.; Tomasicchio, G.R.; Ligorio, F.; Lusito, L.; Francone, A. A Nourishment Performance Index for Beach Erosion/Accretion at Saadiyat Island in Abu Dhabi. *J. Mar. Sci. Eng.* **2019**, *7*, 173. [CrossRef]
3. Gann, G.D.; McDonald, T.; Walder, B.; Aronson, J.; Nelson, C.R.; Jonson, J.; Hallett, J.G.; Eisenberg, C.; Guariguata, M.R.; Liu, J.; et al. International principles and standards for the practice of ecological restoration. Second edition. *Restor. Ecol.* **2019**, *27*, S1–S46. [CrossRef]
4. Gencarelli, R.; Johnson, B.; Kobayashi, N.; Tomasicchio, G.R. On dune erosion and breaching. In Proceedings of the 5th International Conference of Coastal Structures, Venice, Italy, 2–4 June 2007.
5. Gencarelli, R.; Tomasicchio, G.R.; D'Alessandro, F.; Frega, F. Dune erosion prediction during storm surges. In Proceedings of the International Conference on the Application of Physical Modelling to Port and Coastal Protection—Coastlab 08, Bari, Italy, 2–5 July 2008; Nuova Editoriale BIOS: Cosenza, Italy; pp. 113–116, ISBN 978-88-6093-046-0.
6. Martínez, M.L.; Hesp, P.A.; Gallego-Fernandez, J.B. *Coastal Dune Restoration: Trends and Perspectives*; Springer: Berlin/Heidelberg, Germany, 2013; pp. 323–339.
7. Gencarelli, R.; Tomasicchio, G.R.; Kobayashi, N.; Johnson, B. Effects of Hurricane Isabel along the North Carolina Coastline: Beach Profile Evolution and Dune Erosion. In Proceedings of the 3rd SCACR—International Short Conference on Applied Coastal Research, Lecce, Italy, 2–4 June 2008.
8. Fazeres-Ferradosa, T.; Welzel, M.; Schendel, A.; Baelus, L.; Santos, P.R.; Pinto, F.T. Extended characterization of damage in rubble mound scour protections. *Coast. Eng.* **2020**, *158*, 103671. [CrossRef]

9. Maza, M.; Lara, J.L.; Losada, I.J. Experimental analysis of wave attenuation and drag forces in a realistic fringe Rhizophora mangrove forest. *Adv. Water Resour.* **2019**, *131*, 103376. [CrossRef]
10. Kobayashi, N.; Gralher, C.; Do, K. Effects of Woody Plants on Dune Erosion and Overwash. *J. Waterw. Port Coastal Ocean Eng.* **2013**, *139*, 466–472. [CrossRef]
11. Kutiel, P.; Kutiel, H.; Lavee, H. Vegetation response to possible scenarios of rainfall variations along a Mediterranean–extreme arid climatic transect. *J. Arid. Environ.* **2000**, *44*, 277–290. [CrossRef]
12. Mason, T.; French, K.; Russell, K. Moderate impacts of plant invasion and management regimes in coastal hind dune seed banks. *Biol. Conserv.* **2007**, *134*, 428–439. [CrossRef]
13. Jonsson, I.G. Wave Boundary Layers and Friction Factors. *Coast. Eng.* **1967**, 127–148. [CrossRef]
14. Madsen, O.S.; Grant, W.D. Quantitative Description of Sediment Transport by Waves. *Coast. Eng.* **1977**, 1092–1112. [CrossRef]
15. Sleath, J.F. *Sea Bed Mechanics*; Ocean Engineering; Wiley: New York, NY, USA, 1984.
16. Nielsen, P. *Coastal Bottom Boundary Layers and Sediment Transport*; Tuh Tuck Link; World Scientific: Singapore, 1992; Volume 4.
17. Fredsøe, J.; Deigaard, R. *Mechanics of Coastal Sediment Transport*; World Scientific: Singapore, 1992; Volume 3, p. 369.
18. Van Rijn, L.C. *Principles of Sediment Transport in Rivers, Estuaries and Coastal Seas*; Aqua Publications: Amsterdam, The Netherlands, 1993; Volume 1006, pp. 11–13.
19. Soulsby, R.L. Dynamics of marine sands: A manual for practical applications. *Oceanogr. Lit. Rev.* **1997**, *9*, 947.
20. Nairn, R.B.; Southgate, H.N. Deterministic profile modelling of nearshore processes. Part 2. Sediment transport and beach profile development. *Coast. Eng.* **1993**, *19*, 57–96. [CrossRef]
21. Kobayashi, N.; Johnson, B.D.; Karjadi, E.A. Cross-Shore Sand Transport on Beaches. *Coast. Eng.* **2001**, 3165–3178. [CrossRef]
22. Dean, R.G.; Dalrymple, R.A. *Coastal Processes with Engineering Applications*; Cambridge University Press: Cambridge, UK, 2002.
23. Tomasicchio, G.R.; D'Alessandro, F.; Barbaro, G. Composite modelling for large-scale experiments on wave–dune interaction. *J. Hydraul. Res.* **2011**, *49*, 15–19. [CrossRef]
24. Tomasicchio, G.R.; Sánchez-Arcilla, A.; D'Alessandro, F.; Ilic, S.; James, M.R.; Sancho, F.; Fortes, C.J.; Schüttrumpf, H. Large-scale experiments on dune erosion processes. *J. Hydraul. Res.* **2011**, *49*, 20–30. [CrossRef]
25. D'Alessandro, F.; Tomasicchio, G.R. Wave–dune interaction and beach resilience in large-scale physical model tests. *Coast. Eng.* **2016**, *116*, 15–25. [CrossRef]
26. D'Alessandro, F.; Tomasicchio, G.R.; Musci, F.; Ricca, A. Dune erosion physical, analytical and numerical modelling. In Proceedings of the Coastal Engineering, Santander, Spain, 14 December 2012.
27. Sancho, F.; Abreu, T.; D'Alessandro, F.; Tomasicchio, G.R.; Silva, P.A. Surf hydrodynamics under collapsing coastal dunes. *J. Coast. Res.* **2011**, *64*, 144–148.
28. Brunone, B.; Tomasicchio, G. Wave Kinematics at Steep Slopes: Second-Order Model. *J. Waterw. Port Coast. Ocean Eng.* **1997**, *123*, 223–232. [CrossRef]
29. Paul, M.; Kamphuis, J.; Brebner, A. Similarity of Equilibrium Beach Profiles. *Coast. Eng.* **1972**, 1217–1236. [CrossRef]
30. Henry, P.Y.; Aberle, J. *Hydralab+ Deliverable D8.3 Protocols for Scaling Morphodynamics in Time*; Zendo: Genève, Switzerland, 2018.
31. Wallingford, H.R. Wave Gauge System: User Manual. Available online: http://equipit.hrwallingford.com/products/wave-gauges/wave-gauge-systems-8-channels (accessed on 10 September 2019).
32. Wallingford, H.R. 2D Bed Profiling System for Physical Models. Available online: https://www.hrwallingford.com/expertise/equipment-and-technology/instrumentation-and-software-physical-models (accessed on 10 September 2019).
33. Chieregato, A.; Salazar, C.G.O.; Todaro, C.; Martinelli, D.; Peila, D. *Test di Laboratorio di Iniezione per L'impermeabilizzazione e Consolidamento di Terreni Granulari per Mezzo di Materiali Innovativi*; GEAM (GEoingegneria Ambientale e Mineraria): Milan, Italy, 2014.
34. Mansard, E.; Funke, E. The Measurement of Incident and Reflected Spectra Using a Least Squares Method. *Coast. Eng.* **1980**, 154–172. [CrossRef]
35. Frostick, L.E.; McLelland, S.J.; Mercer, T.G. *Users Guide to Physical Modelling and Experimentation: Experience of the HY-DRALAB Network*; CRC Press: Boca Raton, FL, USA, 2011.
36. Mariotti, G.; Fagherazzi, S. Wind waves on a mudflat: The influence of fetch and depth on bed shear stresses. *Cont. Shelf Res.* **2013**, *60*, S99–S110. [CrossRef]
37. Zhu, Q.; van Prooijen, B.; Wang, Z.; Yang, S. Bed-level changes on intertidal wetland in response to waves and tides: A case study from the Yangtze River Delta. *Mar. Geol.* **2017**, *385*, 160–172. [CrossRef]
38. Shi, B.; Cooper, J.R.; Pratolongo, P.D.; Gao, S.; Bouma, T.J.; Li, G.; Li, C.; Yang, S.; Wang, Y.P. Erosion and Accretion on a Mudflat: The Importance of Very Shallow-Water Effects. *J. Geophys. Res. Oceans* **2017**, *122*, 9476–9499. [CrossRef]
39. Shields, A. Application of Similarity Principles and Turbulence Research to Bed-Load Movement. 1936. Available online: https://resolver.caltech.edu/CaltechKHR:HydroLabpub167 (accessed on 25 November 2019).
40. Van Gent, M.; Steetzel, H.; Boers, M.; Smith, J.M.; Coeveld, E.; Walstra, D.; Van De Graaff, J. Dune erosion tests to study the influence of wave periods. In Proceedings of the 30th ICCE, San Diego, CA, USA, 3–8 September 2006.
41. Zordan, J.; Juez, C.; Schleiss, A.J.; Franca, M.J. Entrainment, transport and deposition of sediment by saline gravity currents. *Adv. Water Resour.* **2018**, *115*, 17–32. [CrossRef]
42. Vahdati, V.J.; Yaghoubi, S.; Torabipour, A.; Correia, J.A.; Fazeres-Ferradosa, T.; Taveira-Pinto, F. Combined solutions to reduce scour around complex foundations: An experimental study. *Mar. Syst. Ocean Technol.* **2020**, *15*, 81–93. [CrossRef]

Journal of
Marine Science and Engineering

MDPI

Article

Role of Storm Erosion Potential and Beach Morphology in Controlling Dune Erosion

Laura Lemke * and Jon K. Miller

Department of Civil, Environmental and Ocean Engineering, Stevens Institute of Technology, Hoboken, NJ 07030, USA; jmiller@stevens.edu
* Correspondence: llemke@stevens.edu

Abstract: Coastal erosion is controlled by two sets of factors, one related to storm intensity and the other related to a location's vulnerability. This study investigated the role of each set in controlling dune erosion based on data compiled for eighteen historical events in New Jersey. Here, storm intensity was characterized by the Storm Erosion Index (*SEI*) and Peak Erosion Intensity (PEI), factors used to describe a storm's cumulative erosion potential and maximum erosive power, respectively. In this study, a direct relationship between these parameters, beach morphology characteristics, and expected dune response was established through a classification tree ensemble. Of the seven input parameters, PEI was the most important, indicating that peak storm conditions with time scales on the order of hours were the most critical in predicting dune impacts. Results suggested that PEI, alone, was successful in distinguishing between storms most likely to result in no impacts (PEI < 69) and those likely to result in some (PEI > 102), regardless of beach condition. For intensities in between, where no consistent behavior was observed, beach conditions must be considered. Because of the propensity for beach conditions to change over short spatial scales, it is important to predict impacts on a local scale. This study established a model with the computational effectiveness to provide such predictions.

Keywords: erosion potential; dune erosion; CART analysis; New Jersey

Citation: Lemke, L.; Miller, J.K. Role of Storm Erosion Potential and Beach Morphology in Controlling Dune Erosion. *J. Mar. Sci. Eng.* **2021**, *9*, 1428. https://doi.org/10.3390/jmse9121428

Academic Editors: Giuseppe Roberto Tomasicchio, Ferdinando Frega and Felice D'Alessandro

Received: 17 November 2021
Accepted: 9 December 2021
Published: 14 December 2021

1. Introduction

1.1. Background

In the absence of human-made features, dunes act as a community's final barrier to direct impact by hydrodynamic forces during coastal events. Spatial variation of surge-induced structural damages during major storms has been related to dune size [1,2]. Because post-storm dune recovery occurs over time scales of years to decades [3–5], communities are left even more vulnerable to impact during subsequent events. Consequently, there is considerable interest in understanding the conditions that lead to the erosion of these protective features.

In general, erosion is controlled by two sets of factors, one related to storm intensity and the other related to a location's vulnerability. Traditionally, storm intensity was characterized by a single parameter such as wind speed [6,7], wave height, or storm surge. In recent decades, however, shortcomings of these methods have been brought to light particularly with regard to characterizing a storm's erosion potential. For this reason, additional parameters were developed that combined multiple factors (i.e., wave conditions; water level; storm duration) known to be drivers of coastal erosion [8–14]. These parameters served as proxies for storm impacts, with qualitative relationships established between erosion potential and observed impacts. Direct quantitative relationships were more challenging to establish because of: (1) difficulties in obtaining spatially varying storm information and quantified impacts; and (2) variation introduced by other site-specific controls of dune erosion.

Site-specific controls describe how vulnerable a particular location is and influence how much of a storm's erosion potential is realized. These controls are often why neighboring coastal communities exhibit different responses to similar storm conditions. As an example, during Hurricane Sandy (New Jersey, United States, 2012), Ortley Beach and Seaside Park experienced very different impacts to their dunes and landward structures despite being separated by just 4 km. In Ortley Beach, the dune was entirely removed (Figure 1), leaving landward structures susceptible to major damage by hydrodynamic forces. However, the dune at Seaside Park remained intact with reduced structural impacts [15]. Because of the locations' proximity, it was assumed that each experienced similar storm conditions (i.e., total water level; wave height). Therefore, the differences between the impacts of the two locations were likely the result of the reduced protective capacity of the beach and dune system at Ortley Beach.

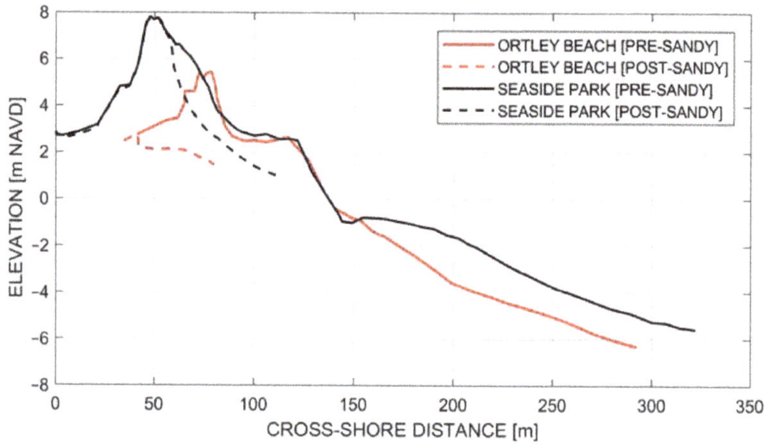

Figure 1. Pre- and post-Sandy beach profiles at Seaside Park (black) and Ortley Beach (red). Data reproduced from The Stockton University Coastal Research Center [16].

The above outcome is similar to that observed during other historical storms such as Hurricane Ivan [17] and those in Australia in 2011 [18] and 2016 [19], where local morphology was found to explain the spatial variation in dune erosion. Pre-storm dune toe and crest elevations, dune width, dune volume, and berm width have all been identified as important controls [17–23]. For a particular combination of wave and water level conditions, these morphological parameters control how much of the wave energy reaches and erodes the dune.

The examples above demonstrate the advantage in estimating beach profile responses to a storm on a local scale (i.e., town by town), if not even smaller (i.e., lot by lot). To provide predictions at such a scale requires computational efficient methods. Traditionally, either 1D or 2D process-based models have been utilized to predict differences in morphological responses to storms. However, as the required number of simulations increases these methods may become inefficient [24]. Computational efficiency is particularly important in the application of forecasting where wave and water level forecasts change on the order of hours and in the application of regional vulnerability assessments, where a large number of storm parameters are tested. For this reason, we turn to data-driven models, techniques shown to be effective in predicting expected outcomes after being trained on a sufficiently large data set.

The present study proposes the development of a data-driven model capable of providing local predictions of storm-induced dune impacts. The model combines a descriptor of storm erosion potential (described in the following section), with site-specific parameters, describing vulnerability, to directly predict local impacts. This approach is viewed as advantageous as it reduces the number of predictor variables by combining storm-related drivers into a known proxy of coastal erosion. Additionally, it separates the role of storm parameters, which are readily available through regional hindcasts and forecasts, from the local morphology, which may be highly variable over small scales and may be less readily available.

1.2. The Storm Erosion Index

Of the storm intensity measures noted in the above section, this study focuses on the Storm Erosion Index (*SEI*), which was introduced by Miller and Livermont [14] and describes a storm's erosion potential by the sum of an instantaneous erosion intensity (*IEI*) over the duration of the storm (t_d):

$$SEI = \sum_{t_d} IEI(t_i) = \sum_{t_d} W_*(t_i) \left[\frac{0.068 H_b(t_i) + S(t_i)}{B + 1.28 H_b(t_i)} \right] \quad (1)$$

where H_b is the depth-limited breaking wave height (H_b = 0.8 h_b), W^* is the width of the active surf zone (approximated as the distance to the breakpoint), B is the berm height, S is the water level height above the mean water line, and t_i is a time index.

Advantages of *SEI* in evaluating erosion potential include its consideration of the three primary storm-related drivers of coastal erosion; mathematical foundation in the physical storm response of an equilibrium beach profile (Equation (1)); applicability to both tropical and extratropical storms; computational efficiency; and ability to serve as a proxy of storm-induced impacts. *SEI* has been shown capable of identifying and classifying major storms [25,26] and of identifying spatial variation in storm intensity [26–28].

Although *SEI* is not a direct predictor of storm impacts (because it does not include information on beach state), Miller and Livermont [14] found that storms with higher *SEI* values typically caused greater erosion. They found that the time-lag between storm occurrence and typically seasonal beach profiles and shoreline measurements made direct comparisons with *SEI* difficult due to the propensity for the beach to recover between surveys. To supplement the comparisons between *SEI* and the measured shoreline changes at Wildwood, New Jersey, comparisons between qualitative post-storm erosion reports collected by the New Jersey Department of Environmental Protection (NJDEP) and *SEI* were examined. The results confirmed that erosional events classified as moderate to severe typically had higher *SEI* values than those classified as slight or minor.

While several studies [14,25,28] focused on the link between *SEI* and erosion, Janssen et al. [27] highlighted the importance of a related parameter, the Peak Erosion Intensity (PEI) as well. PEI, defined by Miller and Livermont [14] as a measure of a storm's maximum instantaneous erosion intensity (*IEI*), considers a storm's wave height, total water level, and the timing of the two maxima. For Hurricane Michael, PEI was shown to capture the surge-induced structural damages which occurred when water levels were high and the affected areas were inundated [27].

The present study serves as an extension of the aforementioned works by establishing a direct relationship between erosion potential and morphological impacts through the development of a data-driven model. Specifically, the objectives of this study are to:

(a) Investigate the role of erosion potential, described by *SEI* and PEI, and beach morphology in controlling observed storm-induced dune erosion, and
(b) Develop a tool which predicts dune erosion in a computationally efficient manner, enabling future forecasts and vulnerability assessments to be performed on a local scale.

2. Study Area and Storm Climate

This study focuses on New Jersey's Atlantic coastline, which has some of the most populated and developed barrier islands in the country [29]. The barrier islands, in addition to the coastal bluffs of central Monmouth County, and the barrier-spit system of Sandy

Hook are fronted by sandy beaches (d_{50} = 0.2 to 0.6 mm). The largest and most persistent dunes tend to be located in the central part of the state in Ocean County (Figure 2).

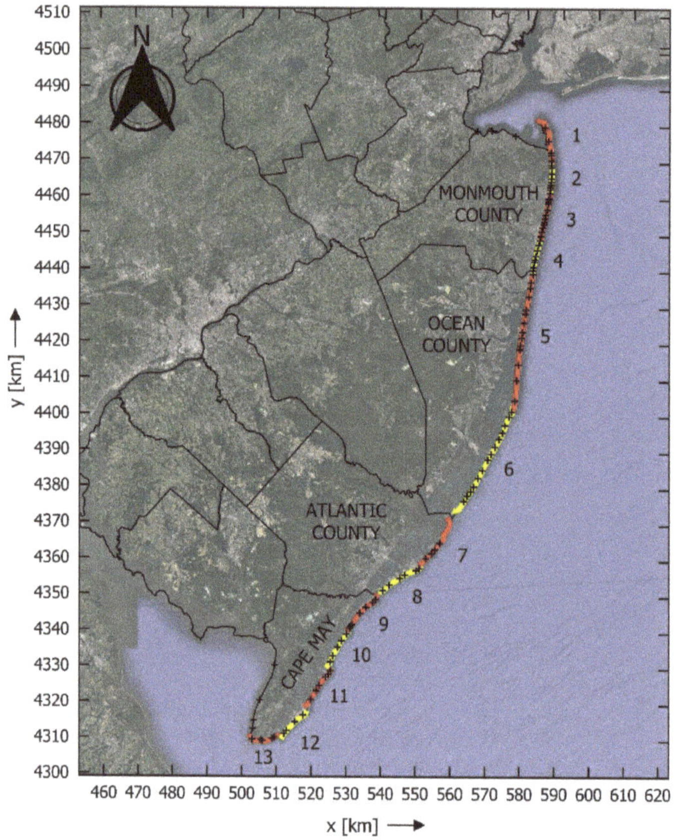

Figure 2. Study area location. Alternating red and yellow lines depict NJ shoreline segments utilized by Lemke and Miller [26] to produce the storm erosion potential climatology. Segments are numbered from north to south, with #1 at Sandy Hook and #13 at Cape May. Black lines depict NJ County borders with the four Atlantic coastal counties labeled accordingly. Black markers depict NJBPN survey locations. Map coordinate system is NAD83/UTM Zone 18N in kilometers.

The region typically experiences three to four storms per year, comprised of both tropical and extratropical events [26]. Events such as Hurricane Sandy (2012), the Veteran's Day Storm (November 2009), and the December 1992 nor'easter represent some of the most extreme events in recent history. Observed impacts during these storms were widespread, with the degree varying by town, arguably based on both spatially varying erosion potential and vulnerability.

The erosion potential of these events and others was captured by Lemke and Miller [26] using the Storm Erosion Index. Specifically, they characterized the intensity of approximately 130 storms occurring between 1980 and 2013 at thirteen shoreline segments within the state (Figure 2) using information extracted from available wave and water level hindcasts. Storm intensities were reported in terms of *SEI*, representative of the cumulative erosion potential of the storm, and PEI, representative of the maximum erosive power of the storm.

Many of the 130 events identified and classified by Lemke and Miller [26] could be described as minor, with *SEI* values an order of magnitude lower than the statewide maximum. In between those minor events, and the upper three events listed above, were those of more moderate and perhaps more typical erosion potential. In contrast to events such as Hurricane Sandy, which has had countless publications dedicated to documenting its impacts, e.g., [1,22,30], those of moderate intensity tend to be less well documented. However, these events have also caused dune erosion, particularly in areas where dunes were more vulnerable. This study aims to capture these events as well, building a model capable of predicting dune erosion due to storms of varying intensities (i.e., moderate and high erosion potentials).

From the climatology, eighteen events were identified as those most likely to have produced dune erosion based on storm intensity alone. These events were ranked in the top twenty storms by either *SEI* or PEI. Table 1 provides a list of these storms as well as their statewide average *SEI* and PEI values. In general, these storms have return periods of greater than one year. The storms capture a range of storm types including high-intensity, short-duration events such as Hurricane Irene (August 2011) and lower-intensity, long-duration events such as the 12 October 2005 storm.

Table 1. Average *SEI* and PEI values, and associated return periods (t_r), for eighteen historical storms in New Jersey. Return periods are based on frequency of occurrence curves created by Janssen [31].

Event	*SEI*	*SEI* t_r [yr]	PEI	PEI t_r [yr]
October 1991	1534	5.2	67.3	5.2
January 1992	1017	2.2	64.4	4.1
December 1992	3326	24	90.8	20
December 1994	869	1.7	57.5	2.3
January 1996	719	1.3	51.1	1.4
February 1998	1434	4.5	67.0	5.0
September 2003	1040	2.3	53.8	1.8
October 12 2005	1779	6.8	54.3	1.8
October 25 2005	866	1.7	60.0	1.4
September 2006	964	2.0	50.3	1.4
November 2007	670	1.2	60.4	2.9
May 2008	998	2.1	53.4	1.7
September 2008	1714	6.4	54.3	1.8
November 2009 (Vets)	2986	18	74.8	9.1
March 2010	1220	3.2	58.7	2.5
September 2010	581	1.0	62.9	3.6
August 2011 (Irene)	788	1.4	73.3	8.2
October 2012 (Sandy)	3056	19	119	49

3. Methodology

3.1. Data Compilation and Preliminary Analysis

To achieve the study objectives listed above, a robust data set of storm characteristics, pre-storm morphological conditions, and observed dune erosion was built leveraging existing long-term records for the region. Spatially varying storm characteristics were extracted for the storm subset (i.e., those listed in Table 1) from the climatology produced by Lemke and Miller [26]. The parameters included *SEI*, PEI, maximum water level (MaxWL), maximum breaking wave height (MaxH_b), storm duration (t_d), and cumulative wave energy (cumE).

Pre-storm morphological data and observed dune erosion were obtained for each storm from the New Jersey Beach Profile Network (NJBPN). The NJBPN, maintained by The Stockton University Coastal Research Center, included profile data available at 110 locations throughout the state (Figure 2). This information was collected annually from 1986 to 1994 and then biannually (spring/fall) thereafter using a total station [32] with an accuracy on the order of centimeters [33]. Unless a storm occurred, dune morphology in these locations was unlikely to change significantly by natural processes in the time between these surveys. Therefore, these data were determined to be sufficient for identifying storm-induced dune erosion.

At each NJBPN location, pre- and post-storm profiles were identified for each storm. Only profiles which included a dune in the pre-storm profile were utilized in the analysis. On average, pre-storm surveys occurred 3 months before storm occurrence, while post-storm surveys occurred 2 months after storm occurrence. It was assumed that observed dune impacts were the result of the identified storm and not the result of multiple smaller storms. Any profiles for which a set of pre- and post-storm surveys bound more than a single significant storm event were not considered in the analysis. Any profiles where a coastal protection project (i.e., beach nourishment) occurred between pre- and post-storm surveys were also excluded.

From each of the pre-storm profiles, a series of morphological parameters characterizing the vulnerability of the profile to storm-induced changes were extracted. A schematic of these parameters is presented in Figure 3 while definitions are provided in Table 2. Automated methods for extracting the dune crest and toe positions were developed following that of Brodie and Spore [34]. Specifically, the dune crest was identified as the maximum seaward observed elevation within the primary dune and dune toe was identified by the seaward maximum in curvature. The dune heel was defined as the landward most position where the slope was greater than 1:20. When possible, a comparison between these extracted features and those identified from aerial imagery taken around the same time was performed as a validation of the extraction methods.

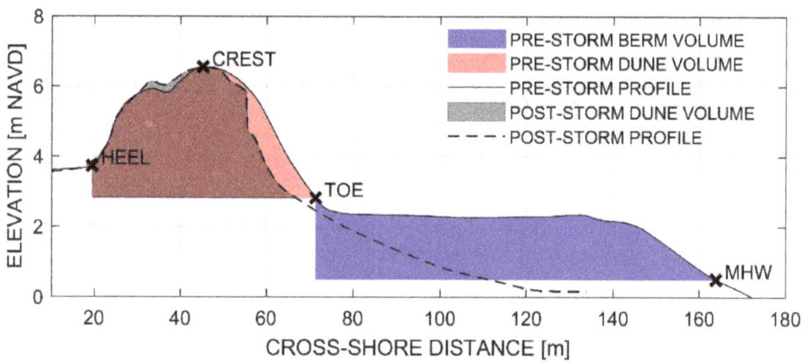

Figure 3. Example of extracted morphologic parameters for surveyed pre- and post-storm profiles.

Table 2. List of extracted morphologic parameters.

Parameter	Definition
Berm width (bwidth)	Horizontal cross-shore distance from dune toe to shoreline (MHW)
Berm volume (mhwvol)	Volume of material seaward of dune toe and above MHW per alongshore unit
Foredune width (fwidth)	Horizontal cross-shore distance from dune crest to dune toe
Foredune volume (fvol)	Volume of material seaward of dune crest and above dune toe per alongshore unit
Dune volume (dvol)	Volume of material seaward of dune heel and above dune toe per alongshore unit
Dune crest elevation (crestz)	Primary dune peak elevation
Dune toe elevation (toez)	Primary dune toe elevation
Dune crest "freeboard" (crestfb)	Height of crest relative to storm maximum water level
Dune toe "freeboard" (toefb)	Height of toe relative to storm maximum water level
Intertidal slope (islope)	End point slope between MHW and MLW
Beach slope (bslope)	End point slope between dune toe and MHW
Foredune slope (fslope)	End point slope between dune crest and dune toe

Storm-induced dune impacts were quantified by comparing the pre- and post-storm profiles. While other impacts including the change in position (i.e., recession and/or reduction) of the dune crest and toe were initially considered, this study focused on volume loss within the pre-storm dune footprint (Figure 3). Volume losses were reported as a percentage of the pre-storm dune volume:

$$dloss(\%) = \frac{A_{pre} - A_{post}}{A_{pre}} \times 100 \tag{2}$$

This parameter was a reliable indicator of overall dune performance. In comparison, relying on dune toe recession or crest reduction alone led to misinterpretations regarding the severity of dune impacts. For example, in several cases, material removed from the foredune was deposited at the dune toe and the toe appeared to move seaward despite net losses in volume and crest height. Additionally, crest heights appeared to remain constant in wide flat dunes despite extensive losses to the foredune.

Dune impacts were classified into one of three classes based on percent loss (Table 3). Damages were classified as Major when volume losses were greater than 40%, Moderate when volume losses were between 5% and 40%, and Minor when volume losses were less than 5%. During Hurricane Sandy, dunes with greater than 40% loss typically exhibited signs of overtopping (i.e., loss of crest elevation; sand deposited landward of the dune). In situations where the dune is overtopped, landward structures may be subject to damage.

Table 3. Classification of storm-induced dune impacts based on quantitative changes.

Damage Class	Definition
Major	Dune volume loss > 40%
Moderate	Dune volume loss 5–40%
Minor	Dune volume loss < 5%

Preliminary analyses were performed on the compiled storm and morphological data prior to use in model development. The range of data was evaluated to confirm that the data set provided sufficient coverage to train and test a data-driven model. Correlations between all parameters, including those describing impacts to the dune, were evaluated through the calculation of a Spearman's rank correlation coefficient. This analysis was performed to highlight relationships between the variables as well as to reduce the number variables used as input to the model. The results and implications of this analysis are discussed in a later section.

3.2. Classification and Regression Tree (CART) Analysis

This study used classification trees to investigate the relationship between the predictor variables (storm and beach characteristics) and the output variable (dune damage class). In general, classification trees, like other data-driven models, rely on identifying patterns within a training data set and using those patterns to predict future outputs. Specifically, classification trees employ a series of binary splits based on the predictor variables to sort data into homogeneous groups based on the output class. Although considered relatively simple, they can reveal underlying interactions between data and identify the most important predictor variables from a large list of those considered. Other advantages to classifications trees include their insensitivity to the underlying distribution of the data; ability to handle missing data; use of numerical and categorical predictor data; and insensitivity to outliers.

For the present study, the compiled data set was split into two subsets, one used to train the model and the other used to independently test the model. The samples for each subset were randomly chosen without replacement with 70% belonging to the training set and 30% belonging to the testing set. A classification tree was generated using the training data to predict the expected dune damage class based on a reduced set of the predictor variables. The classification tree was grown to its full depth and subsequently pruned following the procedures outlined in Breiman et al. [35] and summarized in De'ath and Fabricius [36] and Olden et al. [37]. Pruning was performed to improve the model's performance in predicting the outputs of new samples by generalizing trends and reducing overfitting.

Because of the model's tendency to produce false negatives (i.e., dune damage predicted as class lower than observed), a custom misclassification cost matrix was introduced to penalize the errors associated with false negatives as higher than those associated with false positives. False negatives were considered riskier as they could, in a forecasting application, lead to an underestimation of expected impacts and underprepared communities. Misclassification of an observation into one class lower (i.e., observed Major damage misclassified as Moderate or observed Moderate damage misclassified as Minor) was "c" times worse than a false positive. Misclassification of an observation into two classes lower (i.e., observed Major damage misclassified as Minor) was c^2 worse than a false positive. A sensitivity test was performed to determine the optimal cost coefficient (c).

This first model was found to be unstable, meaning that small changes in the training sample led to differently grown trees and ultimately performance. Instability is recognized as a common drawback to tree models which manifests from a classification tree's inherent tendency to select splits which optimize the model for a particular training sample, even when pruned [38,39]. For this reason, ensembles are often utilized with the aim of reducing the final prediction's variance by averaging the result over many trees. While several methods are available (e.g., boosting; random forests), the present study found that bootstrap aggregation, or "bagging", resulted in the optimal ensemble. This procedure involved taking a specified number of bootstrap samples of size "N" from the training set of size "N" and growing a tree from each sample [40,41]. Performance of the model was evaluated using the resubstitution error of the training set and error of the testing set. Performance was sensitive to the number of trees included in the ensemble and the depth of those trees. Similar to the single classification tree, a custom misclassification cost matrix was introduced to weight the error associated with particular observations as higher than others. Sensitivity tests were performed to determine the optimal configuration (i.e., misclassification cost coefficient (c); number of trees; depth of trees).

4. Results

4.1. Preliminary Analysis

The final data set had 865 sets of observations compiled across eighteen storms between 1991 and 2012. On average, there were 48 sets of observations associated with each storm. Storms with less than this (as few as 20) tended to occur before 2000, while those after 2000 tended to have more (up to 70 profiles each). This was a result of: (1) reduced frequency of surveys prior to 1994; (2) reduced number of profiles surveyed in the state prior to 2000; and (3) the reduced number of profiles with dunes prior to 2000.

The eighteen storms captured a full range of impacts, with losses extending from 0% to 100%. The most impactful storm (Hurricane Sandy) resulted in losses of greater than 40% at nearly half of the profiles (Figure 4). There was a clear skew in the data with 85% of the observed dune impacts being classified as Minor damages, 10% as Moderate, and 5% as Major. Typically, only the most extreme storms resulted in dune damage and even then, only at a portion of the profiles with the clear outlier being Hurricane Sandy.

It should be noted that Figure 4 only reflects profiles where the impacts could be attributed to a single storm. Profiles where there were impacts but it could not be reasonably ascertained which storm caused the erosion based on the pre- and post-storm survey dates were not included. Therefore, it should not be assumed that storms such as the October 1991 or the September 2008 storms resulted in no impacts to the dune in the state. Their impacts simply could not be discerned from those caused by similarly intense storms (e.g., January 1992; May 2008) occurring between pre- and post-storm surveys.

The correlation analysis between all parameters was utilized to identify predictor variables which were highly correlated with one another as well as to identify those correlated with the response variables. This analysis was adapted from that presented by

Beuzen et al. [19]. Figure 5 shows the Spearman rank correlations between all parameters. Important correlations, namely those which had correlations (ρ) greater than 0.4 and were statistically significant to the 95% level ($p < 0.05$), are highlighted by black "X" markers.

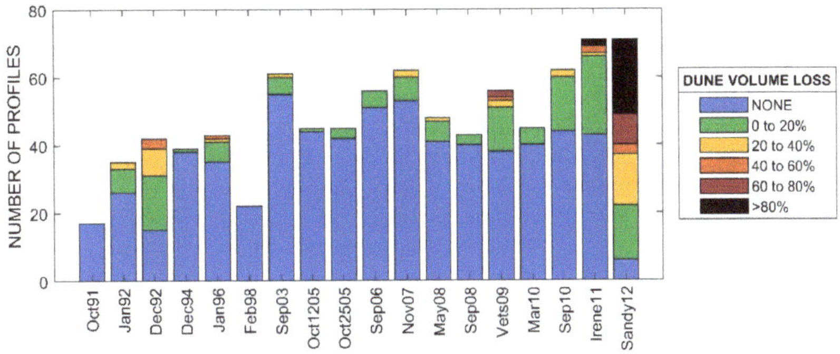

Figure 4. Instances of dune volume loss, as a percentage of pre-storm dune volume, for eighteen storms in New Jersey.

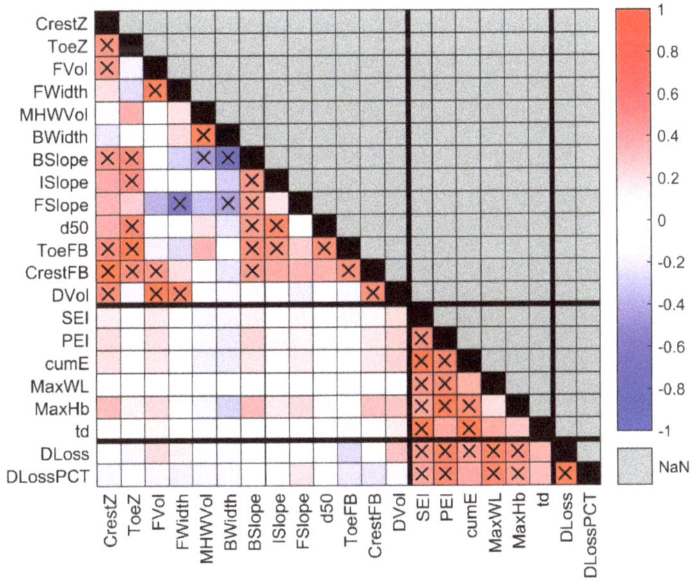

Figure 5. Spearman rank correlations between all predictor and impact variables. Black "X" markers indicate correlations of $|\rho| > 0.4$ and $p < 0.05$. Figure adapted from Beuzen et al. [19].

The storm intensity parameters were found to have strong and statistically significant correlations with the response variables. The strongest correlations were that of *SEI*, PEI, MaxWL, and MaxH$_b$ with values between 0.5 and 0.7. Generally, as storm intensity increased, the observed dune percent loss also increased. However, for all the comparisons there was spread in the data, indicating that a combination of parameters likely controlled dune erosion. This interaction was investigated using the classification trees discussed in the following section.

The correlation results were also utilized to reduce the number of variables used as input and limit redundancy in the model. Many of the predictor variables showed strong and significant correlations with one another. As an example, the volume of sand contained within the berm (MHWVol) was, not surprisingly, highly correlated with both berm width (bwidth) (ρ = 0.83) and berm slope (bslope) (ρ = −0.82). Therefore, only berm volume was considered as input to the classification tree and berm width and slope were initially excluded. Similarly, maximum water level, maximum breaking wave height, and storm duration were not included as they were highly correlated with *SEI* and PEI. These strong and significant correlations were expected as *SEI* and PEI combine wave and water level information in a physically meaningful way to describe storm intensity (see Equation (1)). Individual tests were performed to examine whether including variables initially excluded improved model performance. Results did not show substantial improvement for the classification tree models. The final list of parameters utilized in as input in the model were:

- Crest elevation (CrestZ)
- Toe elevation (ToeZ)
- Berm volume (MHWVol)
- Dune volume (DVol)
- Sediment grain size (d_{50})
- The Storm Erosion Index (*SEI*)
- Peak Erosion Intensity (PEI)

4.2. Prediction of Damage Classes

This section describes two model configurations that were developed to predict the expected damage classes using the predictor variables selected in the previous section. The first, a single classification tree, revealed underlying patterns in the data set, which were then used to examine the role of each predictor variable in controlling dune erosion. Instability in this model, however, ultimately led to the pursuit of a more complex model comprised of many individual classification trees. Details related to this are provided below. This second model had improved performance and thus provided a path forward regarding future applications.

The first model, a single classification tree with a custom misclassification cost matrix ($c = 2$), was grown to its full depth (29 terminal nodes) and then pruned to 11 terminal nodes. The terminal nodes' classes represented the most common damage class associated with the observations sorted into each node. The final depth was selected because it minimized the misclassification rate of the testing set and the 10-fold cross-validation error (Figure 6). The pruned tree is presented in Figure 7 with node statistics listed in Table 4.

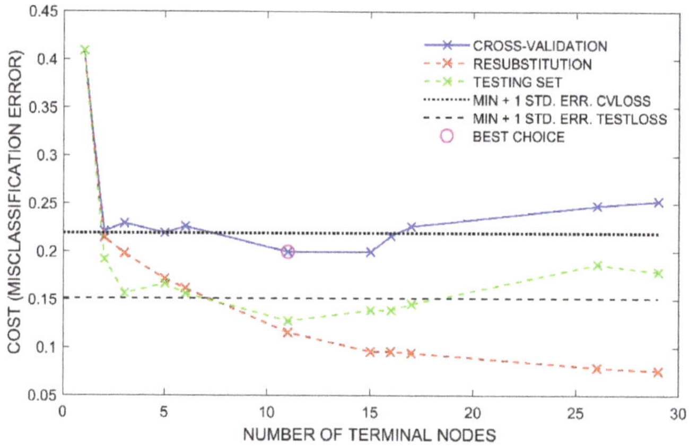

Figure 6. Errors of subtrees of classification tree model with custom misclassification cost matrix ($c = 2$).

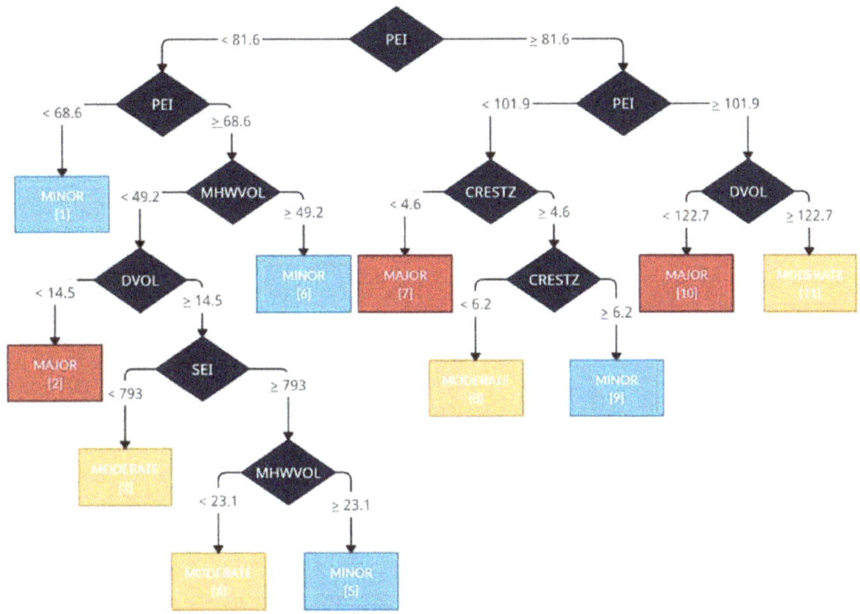

Figure 7. Pruned classification tree with custom misclassification cost matrix ($c = 2$). Predictor parameters and the associated values used to split nodes are defined in the black diamonds and extending arrows. Terminal nodes are represented by blue, yellow, and red rectangles. At each terminal node, both the associated damage class and node ID are defined.

Table 4. Node statistics for pruned classification tree.

Node ID	Assigned Class	Number of Observations in Each Damage Class Based on Known Data			
		Minor	**Moderate**	**Major**	**Total**
1	Minor	394 (96.8%)	12 (2.9%)	1 (0.2%)	407
2	Major	3 (50.0%)	0 (0.0%)	3 (50.0%)	6
3	Moderate	1 (16.7%)	5 (83.3%)	0 (0.0%)	6
4	Moderate	2 (25.0%)	5 (62.5%)	1 (12.5%)	8
5	Minor	29 (87.9%)	4 (12.1%)	0 (0.0%)	33
6	Minor	54 (96.4%)	2 (3.6%)	0 (0.0%)	56
7	Major	2 (20.0%)	3 (30.0%)	5 (50.0%)	10
8	Moderate	9 (31.0%)	19 (65.5%)	1 (3.4%)	29
9	Minor	9 (100.0%)	0 (0.0%)	0 (0.0%)	9
10	Major	2 (7.4%)	4 (14.8%)	21 (77.8%)	27
11	Moderate	0 (0.0%)	6 (100.0%)	0 (0.0%)	6

The tree indicated that both storm intensity and morphological features were important in determining storm-induced dune impacts. The first two splits were based on PEI, indicating that it explained the majority of the variation within the data set. Observations with PEI values less than 69 were typically associated with Minor damage and those with PEI values greater than 102 were typically associated with either Moderate or Major damage, dependent upon dune volume. In between 69 and 102, a full range of classes were expected, dependent upon the pre-storm morphology. The overall accuracy of the model based on the training and testing set combined was 92%, with 91% of the observed Major damages, 55% of the observed Moderate damages, and 96% of the observed Minor damages classified properly (Figure 8, left).

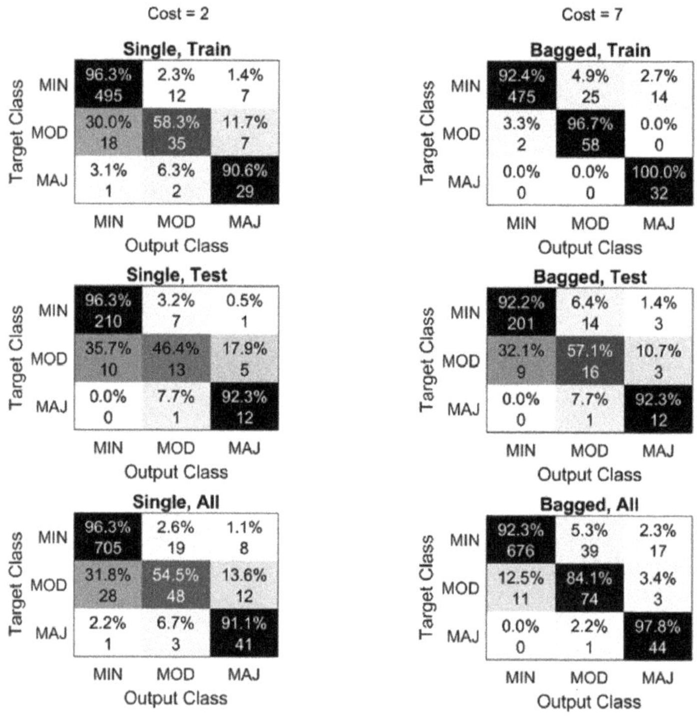

Figure 8. Confusion matrices for the single classification tree (**left**) with custom misclassification cost matrix (c = 2) and for the classification tree ensemble (**right**) with custom misclassification cost matrix (c = 7). Confusion matrices provided based on the training (**top**), testing (**middle**), and combined (**bottom**) data sets.

The presented model was determined to be unstable, meaning that using an alternative random sample of the training set changed both the configuration and performance of the pruned tree. As discussed in the section above, this is a common occurrence with classification trees due to their tendency to optimize the splitting criteria (i.e., variable and value) for a particular training set. For this study, the tendency was likely exacerbated due to the skewness of the data towards Minor damages. Relative to the whole sample there were a small number of observed Moderate and Major damages. The splitting criteria for these damage classes in the model were then sensitive to which samples were selected from the training set. Despite this, one feature which was consistent across the alternative trees was the use of PEI for the initial split (splitting value typically between 70 and 100). This was attributed to the high importance of this variable. However, other properties of the tree including the variables and values used for subsequent splits and the optimal tree size varied. This exercise illustrated the sensitivity of model development to the set of observations included in the training data. Although alternative trees exhibited some similar characteristics, their differences inherently affected the overall accuracies and raised questions of which tree was truly optimal. Therefore, utilizing an ensemble of trees was explored to reduce the variation in model output. Among the methods commonly used to build an ensemble, bootstrap aggregation was found to have the best performance. For brevity, only the results using this method are presented.

A second model, a classification tree ensemble, was grown considering the selected predictor variables. The model included 260 trees, each grown so that the terminal nodes were homogenous (minimum leaf size of 1). This configuration was chosen as it resulted in a smaller misclassification rate of the testing set among the configurations tested. The tested configurations included ensembles composed of up to 500 trees with minimum leaf sizes up to 100 observations. For each minimum leaf size, errors generally stabilized around 250 to 300 trees, if not earlier, indicating that the model would not benefit from the addition of more trees (Figure 9).

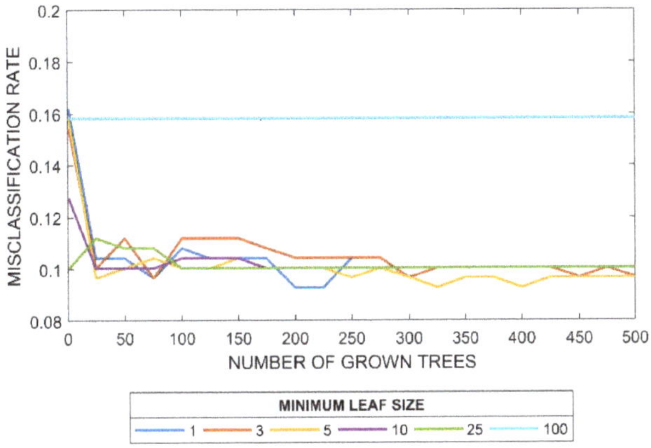

Figure 9. Sensitivity of model performance described by the misclassification rate of the testing set to ensemble size and maximum tree depth.

Of those tested, the poorest performing configuration was that with a minimum leaf size of 100 observations. This model overgeneralized and did not allow for individual sets of observations to be distinguished, no matter how many trees were included. The model was found incapable of predicting either Moderate or Major damages, thus leading to a permanent misclassification rate of the testing set of 16% (Figure 9). With 100 observations in each terminal node, the limited number of Moderate and Major damage observations in the training set were always outnumbered by observations of Minor damage, leading to all terminal nodes being assigned to the Minor damage class.

Across the remaining configurations (i.e., minimum leaf size of 25 or smaller), the misclassification rate of the testing set was similar (approximately 10%). It should be noted, however, that this rate reflected the misclassification across all three classes. As discussed in the previous section, due to the potential implications it could have on pre-storm preparations, correctly classifying Moderate and Major damages was considered more important than correctly classifying Minor damages. With further inspection, it was found that smaller minimum leaf sizes tended to reduce the misclassification rate of Moderate and Major damages. In other words, these configurations created models with higher accuracy at the Moderate and Major damage class. For example, an ensemble of 260 trees with a minimum leaf size of 25 correctly classified 45% of Moderate and Major damages. An ensemble with the same number of trees and a minimum leaf size of 1 correctly classified 72%. Therefore, the selected configuration used a minimum leaf size of 1.

Accuracy at the Moderate and Major damage class was further increased by introducing a custom misclassification cost matrix. Improvements in the accuracy at these classes resulted in a corresponding decrease in the accuracy at the Minor damage class. Generally, this was considered an acceptable tradeoff. After sensitivity testing, a matrix with a cost coefficient of 7 was selected. Accuracy at the Moderate and Major classes increased to 88%, while still maintaining an accuracy of 92% at the Minor damage class (Figure 8, right). Increasing the cost coefficient further showed little benefit to the model. For example, using a coefficient of 15 only increased accuracy at the Moderate and Major damage classes to 92% while decreasing that at the Minor damage class to 77%.

The selected classification tree ensemble configuration was preferable to the single tree for both its stability and accuracy. While the overall accuracies of the two models were similar, the ensemble correctly classified more of the Moderate and Major damages. This improvement was most apparent at the Moderate damage class with accuracy increasing from 55% to 84% (Figure 8). It is believed that this improvement was due to the ability of the ensemble to identify and use patterns that were either not identified or were pruned in the single tree model. As discussed above, the single tree was strategically pruned to 11 terminal nodes to avoid overfitting and improve the model's prediction capability. This tree, therefore, included the most prevalent patterns and removed those which were less important to error reduction. As shown in Table 4 and Figure 8 (left), this led to the misclassification of observed Moderate and Major damages.

The classification tree ensemble, on the other hand, consisted of individual trees where the terminal nodes were homogenous. These trees identified all patterns in the bootstrap samples used to build the trees, separating observations so that no misclassifications occurred within the sample. Individually, these trees would be poor predictors of damage. However, by averaging results over 260 trees, it was anticipated that the correct output class was more likely to be obtained. This prediction took into consideration all underlying patterns, which may have aided in the proper classification of Moderate and Major damage observations originally misclassified by the single tree.

Figure 10 (top) presents predictor importance estimates for the seven predictor variables based on the ensemble model. Predictor importance estimates reflect the difference in resubstitution error if the value of the predictor variable used to split the node were randomly selected versus strategically selected to maximize homogeneity. The higher this value, the more important the variable was considered. In the case of an ensemble, this change in error was averaged across all trees and divided by the standard deviation to reflect the spread. Of the seven variables included, PEI had the highest predictor importance estimate, approximately three times higher than that of the next variable. The value was greater than one, meaning that the average error was greater than the standard deviation. This indicated that across all trees this variable was critical in explaining the variation in dune damages.

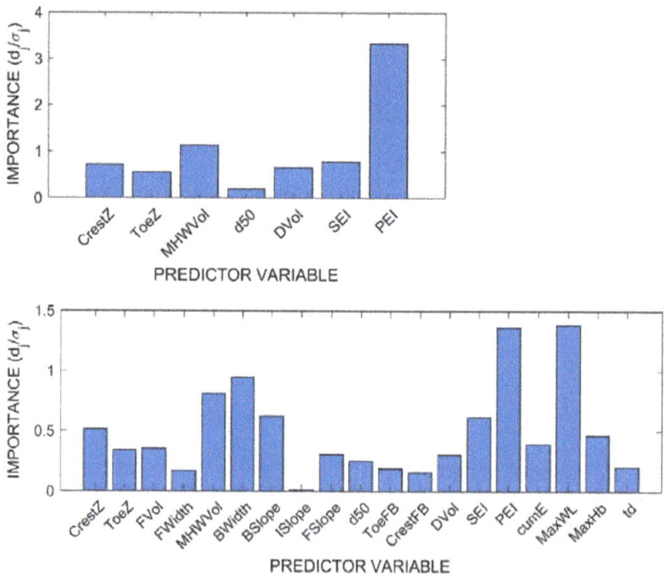

Figure 10. Predictor importance estimates based on classification tree ensemble using selected variables (**top**) and all initially tested variables (**bottom**).

5. Discussion

5.1. Predictors of Dune Impacts

The correlation analysis and classification trees presented in the previous section highlighted the most important parameters in controlling dune impacts and the interaction of those parameters. The predictor variables with the strongest correlations to the response variables were the storm characteristics (Figure 5). This suggests that storm intensity was most important in controlling dune erosion. While increased storm intensities (e.g., PEI) increased the likelihood of more significant impacts, observed spread indicated that secondary factors (i.e., beach and dune morphology) play a role as well.

With regard to storm intensity, both model configurations supported PEI as the most important predictor. The predictor importance estimate for PEI, based on an ensemble of 260 trees, was more than three times larger than the next highest (Figure 10, top). The individual classification tree indicated that it was most important to split the observations by storm intensity (i.e., PEI) before considering site-specific morphological parameters (Figure 7). While not presented, similar patterns were found in the individual trees of the ensemble model as well.

PEI was used by the individual classification tree to initially split the data into four groups. From the training set, 33 observations were sorted into the highest storm intensity group (PEI > 102) where either Moderate or Major damages were typically observed. This implied that under most beach conditions in New Jersey, the wave and water level conditions which combined to create a PEI greater than 102 were large enough to erode the dune. The magnitude of these losses (i.e., Moderate or Major damages) were dependent upon the beach conditions with dunes of larger volume experiencing smaller percent losses. From the training set, 407 observations were sorted into the smallest storm intensity group (PEI < 69), where Minor damages were typically observed. This result implied that under most beach conditions in New Jersey, the wave and water level conditions, which combined to create a PEI of less than 69 were not large enough to erode the dune.

The importance of PEI (versus *SEI*) suggests that dune erosion is controlled by what is happening at a storm's peak. PEI reflects the maximum instantaneous erosion intensity during the storm and is influenced by the breaking wave height, water level, and the timing of the two maxima in relation to one another. At this peak, elevated water levels and wave heights combine in such a way to allow hydrodynamic forces to reach the dune either via wave run-up or direct impact of waves. A separate analysis, where the importance estimates were calculated for all nineteen of the originally considered predictor variables, suggested the importance of PEI was driven by maximum water level (Figure 10, bottom). The predictor importance estimates of PEI and MaxWL were of similar magnitude (1.25–1.5) and were approximately 50% larger than the next highest set (MWHVol and BWidth at 1).

Parameters of secondary importance, reflecting local beach morphology, explained how much erosion a storm of a particular intensity could achieve. Understanding morphological conditions was critical in estimating damages at Moderate storm intensities (PEI 69–102) where there was no consistent morphological response. Specific parameters which were identified as important considering both the individual tree and the ensemble included dune volume, dune crest and toe elevations, and berm volume.

In general, for similar storm intensities, larger dunes both in terms of volume and crest elevation were associated with smaller percent losses. For example, all Major damages in the training set associated with storms of PEI > 102 occurred when dune volumes were less than 123 m^3/m (Figure 7, Table 4). Dune volume controls how much erosion is required to achieve either Moderate or Major damages. A loss of 40 m^3/m for a smaller dune may be catastrophic with the dune being entirely removed (Major damage) while at a location with a much larger pre-storm dune (e.g., >123 m^3/m) the same volumetric loss may be categorized as Moderate. The importance of crest elevation is in part related to dune volume as the two parameters are correlated (Figure 5). Dunes with higher crest elevations tended to have greater total volume than dunes with lower crest elevations. For storms with PEI between 82 and 102, all Moderate and Major damages in the training set occurred when the crest elevation was less than 6.2 m NAVD.

The importance of crest elevation, in addition to toe elevation, suggests that the position of the dune relative to the water level is also important in controlling erosion. Together, the two parameters control how much of the dune face is exposed to hydrodynamic events. These elevations have been previous used by Sallenger [21] to classify expected dune impacts into one of four regimes by comparing the elevations to that of the 2% exceedance wave run-up. Specifically, the toe elevation is important in explaining when hydrodynamic forces begin to directly impact the dune while the crest elevation is important in explaining when the dune is overtopped. While not shown in the presented classification tree, toe elevation was utilized in the fully grown classification tree and the individual trees of the ensemble to split the data. Here, dunes with lower toe elevations were more likely to have Moderate or Major damages. In the training set, the single observed case of Major damage for PEI < 69 occurred at a location where both the toe elevation (1.4 m NAVD) and berm volume (8 m^3/m) were lower than typical.

Finally, the beach berm serves as a buffer for the dune. For observations of similar storm intensities, dunes fronted by larger berms (i.e., higher MHWVol) were typically associated with lower percent losses. For example, for a PEI between 69 and 82, all Major and Moderate damages occurred when the berm volume was less than 49 m^3/m (Figure 7, Table 4). For the ensemble model, berm volume had the second highest predictor importance estimate. Physically, the berm provides protection to the dune through two mechanisms. For storms of lower intensity where the wave run-up does not reach the toe of the dune, the berm volume represents the material available to erode before the dune is directly impacted. However, at higher water levels where the berm is submerged, the berm width (correlated with berm volume, Figure 5) represents the distance over which wave energy is allowed to dissipate before reaching the dune.

While the model explained a majority of the variation in the training and testing sets using seven predictor variables, there remained some misclassified observations. Some of this variation may be explained by inherent inaccuracies in the compiled data set associated with timing of the surveys. Pre-storm surveys typically occurred three months before storm occurrence. While dune volumes are not expected to change significantly over this time by natural mechanisms (unless in the event of a storm), berm volumes and widths are expected to fluctuate. This change can be seasonal, the result of changes in wave climate. Seasonal shoreline changes on the order of tens of meters from the mean position have been reported [42–44]. Therefore, the true berm width and berm volume may be different than that estimated by the available survey. This may introduce errors as berm volume was identified as the second most important control on dune erosion.

The current work focused on parameters believed to be the primary controls of dune erosion and investigated them leveraging existing data sets. A particular emphasis was placed on PEI and *SEI*, parameters used to describe storm intensity, for which a quantitative relationship to observed dune impacts had yet been established. It is recognized that there are parameters outside of the current work that may explain a portion of the remaining variation in the model including proximity to coastal structures [45–47], presence of offshore bars [48], and the cover and type of dune vegetation [49–51].

Of all the parameters considered, this study highlighted the importance of a storm's peak conditions, described through PEI, in controlling dune erosion. Recognizing this, additional parameters describing the length and timing of that peak may become importance to consider. For example, if the peak intensity occurred at the beginning of a storm, the berm width and volume would be close to the pre-storm condition and would provide the greatest level of protection. However, if the peak occurred towards the end, the berm would likely have eroded and would provide a lower level of protection, yielding potentially greater impact to the dune. Other studies have noted improved estimates of erosion using empirical models when considering the time evolution of the profile during the storm [52]. This is expected to be most important for storms of long duration which have the ability to yield significant berm erosion before peaking. It is likely that this is linked to *SEI* which reflects a storm's time-varying erosion intensity and is sensitive to a storm's total duration. Further, storms which maintained peak conditions for a longer duration would likely cause greater dune erosion than more short-lived events as the dune would be exposed to hydrodynamic forces for a greater length of time. Therefore, it may be useful to define a second storm duration capturing this peak. It is suggested that future studies investigate ways to quantify both the timing and duration of a storm's peak using the Storm Erosion Index and explore the role of these parameters in dune erosion.

5.2. Model Performance

With the goal of creating a model that accurately predicted the expected impact to coastal dunes, the performances of two model configurations were compared to determine the optimal one for use in future applications. The two models underwent rigorous optimization and testing to assess sensitivity to training set selection, misclassification cost coefficient (c), tree size, and predictor parameter selection. The single classification tree was pursued first as its simplicity was deemed advantageous. However, because of the instability associated with the configuration, it was superseded by the ensemble over which individual outputs were averaged to determine a final output. Accuracies were improved, particularly at the Moderate damage class. Overall, the ensemble model was most accurate at the two extremes (92% accuracy for Minor damages; 98% for Major damages) and less so for that in between (84% for Moderate damages).

It is important to note that data-driven models, such as those discussed in this study, have inherent uncertainty in the predictions they provide as they do not account for all physical processes that govern a particular outcome such as dune erosion. In the case of the models presented here, predictions were based on seven variables which schematized a given storm and beach profile. While more parameters may be added in the hope of capturing additional aspects of the physical processes, the model is unlikely to capture every consideration included in a process-based model. Despite this limitation, the model developed here was capable of providing reasonable predictions of storm-induced dune erosion and was shown to be a viable alternative, particularly in applications where computational efficiency is required.

When the model is applied, some measures may be taken to limit how much uncertainty there is in the results. For example, model application should be limited to the range of data over which it was trained and tested. Table 5 presents the ranges for the seven predictor variables. It is expected that estimates based on sets of observations far outside of these ranges will have increased errors and will be associated with greater uncertainty. For application of the model on beaches of similar characteristics over typical events, the data were considered adequate as profiles and storms throughout the state over thirty years were captured. The profiles captured conditions ranging from those recently nourished to those due for renourishment. If applying the model in a region where the conditions were very different than that presented, the model would likely need to be retrained.

Table 5. Characteristics of compiled morphological and storm parameters in combined (training and testing) data sets.

Parameter [Units]	Minimum	Mean	Median	Maximum	IQR *
Berm volume [m^3/m]	1.8	75.2	62.5	420.1	56.1
Dune volume [m^3/m]	1.5	49.9	39.2	281.7	46.2
Median grain size [mm]	0.16	0.42	0.38	2.19	0.20
Dune crest elevation [m NAVD]	2.2	5.1	5.1	7.8	1.6
Dune toe elevation [m NAVD]	1.0	2.7	2.6	4.8	1.0
Storm Erosion Index	55	1456	1069	3871	1155
Peak Erosion Intensity	24.3	65.9	60.6	163.3	16.6

* Interquartile Range (IQR) defines the length from the first quartile (25th percentile) to the third quartile (75th percentile).

The model was trained using a range of storm conditions describing events with return periods of at least one year to that of 50 years. Storms less intense than those over which the model was trained are expected to have little impact. An exception to this would be non-typical beach conditions, where the dune and/or beach were very small. Only a single observation such as this appears to exist in the training data set. Storms more intense than those over which the model was trained should be treated with a higher level of uncertainty. For this study, the largest event was Hurricane Sandy, which had a maximum PEI of 163. The developed model determined that for PEI values greater than 102, at least some significant impact (Moderate/Major damages) was likely to occur regardless of the beach conditions. These portions of the trees were trained with a limited number of observations (33 observations). Splits by PEI and other morphological factors may change if considering more intense storms. It is suggested that the model be retrained with a data set supplemented by simulated morphological responses to extreme events to expand the range over which the model is valid.

6. Conclusions

Previous works [14,27] have established qualitative relationships between storm intensity, described by the Storm Erosion Index, and storm-induced impacts on beaches and coastal communities. Prior to this work, however, a direct quantitative relationship did not exist. Previous studies were limited in both the quality and quantity of observed data from which comparisons could be made. This study documented the compilation of a robust data set from which the role of storm intensity, described by *SEI* and PEI, and beach morphology in controlling dune impacts was investigated. Tools including a simple correlation analysis and more involved data-driven models were utilized to achieve the study objectives. Specifically, the role and relative importance of different erosion controls were evaluated while also creating a computational efficient model, which enables predicting dune erosion on a local scale.

Preliminary analysis revealed that dune impacts, described by percent losses, were highly correlated with storm characteristics. However, spread in the data suggested it was likely a combination of parameters that controlled dune impacts. Through the development of two data-driven models, PEI, berm volume, and dune volume were determined to be the most important factors among those tested. Although the single classification tree was ultimately superseded by the ensemble based on performance, the model was valuable in that it provided a visual display on how the different predictors interacted to ultimately predict a damage class. On its own, PEI was successful in distinguishing between storms most likely to result in no impacts (PEI < 69) and those likely to result in some (PEI > 102) regardless of the beach conditions. For intensities in between, where no consistent behavior was observed, beach conditions must be considered in combination with PEI to estimate impacts.

The parameters that were identified as most important in predicting storm-induced impacts were consistent with existing literature. Historically, wave run-up, which is dependent upon water level and wave height, has been linked to dune erosion [17,21]. PEI, which is dependent on similar variables, was identified as the most important parameter in the developed models. PEI reflects the maximum instantaneous erosion intensity during the storm. *SEI*, reflecting the cumulative erosion potential, had a much lower predictor importance factor than PEI. This suggests that the peak conditions (highest water level; largest wave heights) that typically occur over several hours were most critical in predicting dune impacts. This is consistent with that indicated by previous studies where dune erosion was controlled by the conditions under which the dune was directly impacted by waves and the time over which that direct impact occurred, e.g., [52,53]. The importance of berm volume (and width) and dune volume, the second and third most important factors based on this analysis, has been noted in literature as well, e.g., [17,23].

The results presented here demonstrate the usefulness of the Storm Erosion Index (specifically PEI) for evaluating storm intensity on the basis of readily available parameters. While it can help indicate potential impacts to the primary dune, for the better predictions it must be combined with less readily available, and potentially more spatially and temporally variable beach morphology information. The classification tree ensemble developed in pursuit of exploring the relationship between the parameters enables a number of applications where computational efficiency is advantageous such as regional vulnerability assessments, sensitivity testing, and forecasting. These applications are the focus of future work.

Author Contributions: Conceptualization, L.L. and J.K.M.; methodology, L.L. and J.K.M.; formal analysis, L.L.; data curation, L.L.; writing—original draft preparation, L.L.; writing—review and editing, J.K.M.; visualization, L.L.; supervision, J.K.M.; project administration, J.K.M. All authors have read and agreed to the published version of the manuscript.

Funding: This research was funded by the New Jersey Department of Environmental Protection (NJDEP) through the New Jersey Coastal Protection Technical Assistance Service (N.J.S.A. 18A:64L-1).

Institutional Review Board Statement: Not applicable.

Informed Consent Statement: Not applicable.

Data Availability Statement: Publically available datasets were analyzed in this study. Details regarding these sources can be found at The Stockton University Coastal Research Center (https://www.stockton.edu/coastal-research-center/njbpn/index.html, 1 December 2021) and in Lemke and Miller [26].

Acknowledgments: The authors wish to thank the Stockton University Coastal Research Center for providing the survey profile data. The authors would also like to thank Reza Marsooli (Stevens Institute of Technology), Upendra Prasad (Stevens Institute of Technology), and Kristen D. Splinter (UNSW) for their constructive comments regarding this work.

Conflicts of Interest: The authors declare no conflict of interest. The funders had no role in the design of the study; in the collection, analyses, or interpretation of data; in the writing of the manuscript, or in the decision to publish the results.

References

1. Barone, D.; McKenna, K.; Farrell, S. Hurricane Sandy: Beach-dune performance at New Jersey Beach Profile Network sites. *Shore Beach* **2014**, *82*, 13–23.
2. Tomiczek, T.; Kennedy, A.; Zhang, Y.; Owensby, M.; Hope, M.E.; Lin, N.; Flory, A. Hurricane Damage Classification Methodology and Fragility Functions Derived from Hurricane Sandy's Effects in Coastal New Jersey. *J. Waterw. Port Coast. Ocean Eng.* **2017**, *143*, 04017027. [CrossRef]
3. Houser, C.; Wernette, P.; Rentschlar, E.; Jones, H.; Hammond, B.; Trimble, S. Post-storm beach and dune recovery: Implications for barrier island resilience. *Geomorphology* **2015**, *234*, 54–63. [CrossRef]
4. Morton, R.A.; Jeffrey, G.P.; James, C.G. Stages and Durations of Post-Storm Beach Recovery, Southeastern Texas Coast, U.S.A. *J. Coast. Res.* **1994**, *10*, 884–908.
5. Mathew, S.; Davidson-Arnott, R.G.; Ollerhead, J. Evolution of a beach–dune system following a catastrophic storm overwash event: Greenwich Dunes, Prince Edward Island, 1936–2005. *Can. J. Earth Sci.* **2010**, *47*, 273–290. [CrossRef]
6. Schott, T.; Landsea, C.; Hafele, G.; Lorens, J.; Taylor, A.; Thurm, H.; Ward, B.; Willis, M.; Zaleski, W. *The Saffir-Simpson Hurricane Wind Scale*; NOAA/National Weather Service: Silver Spring, MD, USA, 2012.
7. Simpson, R.H.; Saffir, H. The hurricane disaster potential scale. *Weatherwise* **1974**, *27*, 169.
8. Dolan, R.; Davis, R.E. An intensity scale for Atlantic coast northeast storms. *J. Coast. Res.* **1992**, *8*, 840–853.
9. Mendoza, E.T.; Jimenez, J.A.; Mateo, J. A coastal storms intensity scale for the Catalan sea (NW Mediterranean). *Nat. Hazards Earth Syst. Sci.* **2011**, *11*, 2453–2462. [CrossRef]
10. Kriebel, D.; Dalrymple, R.; Pratt, A.; Sakovich, V. Shoreline risk index for northeasters. In Proceedings of the 1996 Conference on Natural Disaster Reduction, Washington, DC, USA, 3–5 December 1996; pp. 251–252.
11. Walker, R.A.; Basco, D.R. Application of Coastal Storm Impulse (Cosi) Parameter to Predict Coastal Erosion. *Coast. Eng. Proc.* **2011**, *1*, 23. [CrossRef]
12. Kraus, N.C.; Wise, R.A. Simulation of January 4, 1992 storm erosion at Ocean City, Maryland. *Shore Beach* **1993**, *61*, 34–41.
13. Zhang, K.Q.; Douglas, B.C.; Leatherman, S.P. Beach erosion potential for severe nor'easters. *J. Coast. Res.* **2001**, *17*, 309–321.
14. Miller, J.K.; Livermont, E. A predictive index for wave and storm surge induced erosion. In Proceedings of the 31st International Conference on Coastal Engineering, Hamburg, Germany, 31 August–5 September 2008; pp. 4143–4153.

15. Morgan, K.L.; Krohn, M.D. Post-Hurricane Sandy coastal oblique aerial photographs collected from Cape Lookout, North Carolina, to Montauk, New York, November 4–6, 2012. *Data Ser.* **2014**, *858*. [CrossRef]
16. The Stockton University Coastal Research Center. *Beach Dune Performance Assessment of New Jersey Beach Profile Network (NJBPN) Sites at Northern Ocean County, New Jersey, after Hurricane Sandy Related to FEMA Disaster DR-NJ-4086*; The Stockton University Coastal Research Center: Port Republic, NJ, USA, 2012.
17. Plant, N.G.; Stockdon, H. Probabilistic prediction of barrier-island response to hurricanes. *J. Geophys. Res. Space Phys.* **2012**, *117*, 117. [CrossRef]
18. Splinter, K.D.; Kearney, E.T.; Turner, I. Drivers of alongshore variable dune erosion during a storm event: Observations and modelling. *Coast. Eng.* **2018**, *131*, 31–41. [CrossRef]
19. Beuzen, T.; Harley, M.D.; Splinter, K.D.; Turner, I.L. Controls of Variability in Berm and Dune Storm Erosion. *J. Geophys. Res. Earth Surf.* **2019**, *124*, 2647–2665. [CrossRef]
20. Palmsten, M.L.; Splinter, K.D.; Plant, N.G.; Stockdon, H.F. Probabilistic estimation of dune retreat on the Gold Coast, Australia. *Shore Beach* **2014**, *82*, 35–43.
21. Sallenger, A.H. Storm impact scale for barrier islands. *J. Coast. Res.* **2000**, *16*, 890–895.
22. Overbeck, J.R.; Long, J.W.; Stockdon, H.F. Testing model parameters for wave-induced dune erosion using observations from Hurricane Sandy. *Geophys. Res. Lett.* **2017**, *44*, 937–945. [CrossRef]
23. Judge, E.K.; Overton, M.F.; Fisher, J.S. Vulnerability Indicators for Coastal Dunes. *J. Waterw. Port Coast. Ocean Eng.* **2003**, *129*, 270–278. [CrossRef]
24. Deierlein, G.G.; Zsarnóczay, A. *State of the Art in Computational Simulation for Natural Hazards Engineering (Version v2)*; Zenodo: Genève, Switzerland, 2021. [CrossRef]
25. Wehof, J.; Miller, J.K.; Engle, J. Application of the Storm Erosion Index (*SEI*) to three unique storms. In Proceedings of the 34th International Conference on Coastal Engineering, Seoul, Korea, 15–20 June 2014.
26. Lemke, L.; Miller, J.K. Evaluation of storms through the lens of erosion potential along the New Jersey, USA coast. *Coast. Eng.* **2020**, *158*, 103699. [CrossRef]
27. Janssen, M.; Lemke, L.; Miller, J. Application of Storm Erosion Index (*SEI*) to parameterize spatial storm intensity and impacts from Hurricane Michael. *Shore Beach* **2019**, *87*, 41–50. [CrossRef]
28. Cheng, J.; Cossu, F.T.; Wang, P. Factors controlling longshore variations of beach changes induced by Tropical Storm Eta (2020) along Pinellas County beaches, west-central Florida. *Shore Beach* **2021**, *89*, 75–85. [CrossRef]
29. Zhang, K.; Leatherman, S. Barrier Island Population along the U.S. Atlantic and Gulf Coasts. *J. Coast. Res.* **2011**, *27*, 356. [CrossRef]
30. Walling, K.; Herrington, T.O.; Miller, J.K. Hurricane Sandy damage comparison: Oceanfront houses protected by a beach and dune system with vs. without a rock seawall. *Shore Beach* **2016**, *84*, 35–41.
31. Janssen, M.S. Risk-Based Assessment of Coastal Defense Projects: Quantifying Hazard, Vulnerability, and Parametric Design Applications. Ph.D. Thesis, Stevens Institute of Technology, Hoboken, NJ, USA, 2021.
32. The Richard Stockton Coastal Research Center. *Beach-Dune System Susceptibility Assessment for the Bourough of Mantoloking*; The Richard Stockton Coastal Research Center: Ocean County, NJ, USA, 2004.
33. Lee, J.M.; Park, J.Y.; Choi, J.Y. Evaluation of Sub-aerial Topographic Surveying Techniques Using Total Station and RTK-GPS for Applications in Macrotidal Sand Beach Environment. *J. Coast. Res.* **2013**, *10065*, 535–540. [CrossRef]
34. Brodie, K.L.; Spore, N.J. Foredune Classification and Storm Response: Automated Analysis of Terrestrial Lidar Dems. In Proceedings of the Coastal Sediments 2015, San Diego, CA, USA, 11–15 May 2015.
35. Breiman, L.; Friedman, J.; Stone, C.J.; Olshen, R.A. *Classification and Regression Trees*; CRC Press: Boca Raton, FL, USA, 1984.
36. De'ath, G.; Fabricius, K.E. Classification and regression trees: A powerful yet simple technique for ecological data analysis. *Ecology* **2000**, *81*, 3178–3192. [CrossRef]
37. Olden, J.D.; Lawler, J.J.; Poff, N.L. Machine Learning Methods Without Tears: A Primer for Ecologists. *Q. Rev. Biol.* **2008**, *83*, 171–193. [CrossRef]
38. Breiman, L. Heuristics of instability and stabilization in model selection. *Ann. Stat.* **1996**, *24*, 2350–2383. [CrossRef]
39. Li, R.-H.; Belford, G.G. Instability of decision tree classification algorithms. In Proceedings of the Eighth ACM SIGKDD International Conference on Knowledge Discovery and Data Mining, Edmonton, AB, Canada, 23–26 July 2002; pp. 570–575.
40. Sutton, C.D. Classification and Regression Trees, Bagging, and Boosting. *Handb. Stat.* **2005**, *24*, 303–329. [CrossRef]
41. Breiman, L. Bagging predictors. *Mach. Learn.* **1996**, *24*, 123–140. [CrossRef]
42. Lemke, L.; Miller, J.K. EOF analysis of shoreline and beach slope variability at a feeder beach constructed within a groin field at Long Branch, New Jersey. *Coast. Eng.* **2017**, *121*, 14–25. [CrossRef]
43. Pearre, N.S.; Puleo, J.A. Quantifying Seasonal Shoreline Variability at Rehoboth Beach, Delaware, Using Automated Imaging Techniques. *J. Coast. Res.* **2009**, *254*, 900–914. [CrossRef]
44. Diez, J.; Cohn, N.; Kaminsky, G.M.; Medina, R.; Ruggiero, P. Spatial and Temporal Variability of Dissipative Dry Beach Profiles in the Pacific Northwest, U.S.A. *J. Coast. Res.* **2018**, *34*, 510–523. [CrossRef]
45. Almarshed, B.; Figlus, J.; Miller, J.; Verhagen, H.J. Innovative Coastal Risk Reduction through Hybrid Design: Combining Sand Cover and Structural Defenses. *J. Coast. Res.* **2019**, *36*, 174–188. [CrossRef]
46. Boers, M.; van Geer, P.; van Gent, M. Dike and dune revetment impact on dune erosion. In Proceedings of the Coastal Sediments 2011: In 3 Volumes, Miami, FL, USA, 2–6 May 2011.

47. Zimmerman, T.; Miller, J.K. UAS-SfM approach to evaluate the performance of notched groins within a groin field and their impact on the morphological evolution of a beach nourishment. *Coast. Eng.* **2021**, *170*, 103997. [CrossRef]
48. Harley, M.D.; Turner, I.L.; Short, A.D.; Ranasinghe, R. An empirical model of beach response to storms-SE Australia. In Proceedings of the Coasts and Ports 2009: In a Dynamic Environment, Wellington, New Zealand, 16–18 September 2009.
49. Feagin, R.; Furman, M.; Salgado, K.; Martinez, M.; Innocenti, R.; Eubanks, K.; Figlus, J.; Huff, T.; Sigren, J.; Silva, R. The role of beach and sand dune vegetation in mediating wave run up erosion. *Estuar. Coast. Shelf Sci.* **2019**, *219*, 97–106. [CrossRef]
50. Sigren, J.M.; Figlus, J.; Highfield, W.; Feagin, R.A.; Armitage, A.R. The Effects of Coastal Dune Volume and Vegetation on Storm-Induced Property Damage: Analysis from Hurricane Ike. *J. Coast. Res.* **2018**, *341*, 164–173. [CrossRef]
51. Charbonneau, B.R.; Wootton, L.S.; Wnek, J.P.; Langley, J.A.; Posner, M.A. A species effect on storm erosion: Invasive sedge stabilized dunes more than native grass during Hurricane Sandy. *J. Appl. Ecol.* **2017**, *54*, 1385–1394. [CrossRef]
52. Palmsten, M.L.; Holman, R.A. Laboratory investigation of dune erosion using stereo video. *Coast. Eng.* **2012**, *60*, 123–135. [CrossRef]
53. Larson, M.; Erikson, L.; Hanson, H. An analytical model to predict dune erosion due to wave impact. *Coast. Eng.* **2004**, *51*, 675–696. [CrossRef]

Journal of
Marine Science and Engineering

MDPI

Article

Effects of Anthropogenic Pressures on Dune Systems—Case Study: Calabria (Italy)

Giandomenico Foti [1,*], Giuseppe Barbaro [1], Giuseppina Chiara Barillà [1] and Ferdinando Frega [2]

1 DICEAM Department, Mediterranea University of Reggio Calabria, Via Graziella Loc. Feo di Vito, 89122 Reggio Calabria, Italy; giuseppe.barbaro@unirc.it (G.B.); chiara.barilla@unirc.it (G.C.B.)
2 Department of Civil Engineering, Calabria University, 87036 Rende, Italy; ferdinando.frega@unical.it
* Correspondence: giandomenico.foti@unirc.it

Abstract: During the second half of the last century, considerable anthropization processes were observed throughout most of the Italian territory. These processes have altered the equilibrium conditions of several river and coastal ecosystems, causing the destruction of numerous dune systems. This issue is particularly important in territories such as Calabria, a region in southern Italy subject to considerable anthropogenic pressures and characterized by over 700 km of coast. The aim of the paper was to evaluate the effects of anthropogenic pressures on the Calabrian dune systems, especially in regard to the triggering of coastal erosion processes. For this purpose, historical and current cartographic data, such as shapefiles, cartography, and satellite imagery, were analyzed using QGIS. This evaluation was carried out through the comparison between the current extension of the dune systems and their extensions after the Second World War, before the anthropogenic pressures. This evaluation was also carried out through the analysis of shoreline changes in coastal areas, where dune systems are currently present, and in coastal areas where dune systems have been partially or totally destroyed by anthropogenic causes, compared to the 1950s, thus excluding coastal areas without dune systems in the 1950s, and analyzing what was built in place of the destroyed dune systems. Two criteria were defined to identify the levels of destruction of the dune systems and to identify the coastal erosion processes. The analysis showed a strong correlation between the destruction of dune systems by anthropogenic causes and the triggering of coastal erosion processes.

Keywords: dune systems; anthropogenic pressure; shoreline changes; coastal erosion; historical cartography; satellite imagery; QGIS; Calabria

Citation: Foti, G.; Barbaro, G.; Barillà, G.C.; Frega, F. Effects of Anthropogenic Pressures on Dune Systems—Case Study: Calabria (Italy). *J. Mar. Sci. Eng.* **2022**, *10*, 10. https://doi.org/10.3390/jmse10010010

Academic Editor: Patrick Hesp

Received: 28 October 2021
Accepted: 21 December 2021
Published: 23 December 2021

1. Introduction

Coastal areas are systems between land and sea, in constant evolution, often directly exposed to wave action. Most coastal areas are particularly important due to the presence of residential settlements, infrastructure, economic activities, archaeological sites, ecological systems, dunes, etc. [1–3]. Among these, the dunes are habitats of important environmental and landscape values. Moreover, dunes form natural coastal defenses because they act both as a reserve of sand and as a physical barrier to protect landward territories [4–7].

The anthropization process that began in the second half of the last century, represented mainly by urban expansion and by industrial and tourism activities, has considerably modified the natural and environmental characteristics of several coastal and river ecosystems, causing intense erosive processes and the destruction of numerous dune systems [8–12]. Indeed, 30% of the world's coasts are currently eroding [13–15] and, in Europe alone, dune systems have decreased by 70% [16]. Furthermore, climate change can increase these processes through sea level rise [17–20].

Generally, dune systems are in flat coastal areas, the seabeds of which have low slopes. These systems consist of sediments of both alluvial and marine origin, and they are continuously subjected to the combined action of multiple physical, chemical, and biological agents, essential for genesis and structuring phases. Dune systems are a highly dynamic environment, whose morphology varies spatially and temporally under the action of both natural and anthropogenic factors [21–24].

From the temporal point of view, there are long-term, middle-term, and short-term variations. Long-term variations are mainly correlated to the Pleistocene deposition of the dune systems. Middle-term variations are of the order of tens of years and depend on alterations in coastal, wind, and river sedimentary balance [25–28]. Often these alterations are caused by anthropogenic factors, such as the construction of buildings, infrastructure, ports, and coastal defense works in coastal areas [29–31] and the construction of hydraulic structures interfering with fluvial dynamics, such as levees, dams, and inert drains from riverbeds [32–35]. On the other hand, short-term variations are mainly related to natural factors, such as single sea storms [36], clusters of sea storms [37,38], extreme flood events [39], or concurrent events [40–42], which cause coastal flooding [43]. A full dune recovery can take several decades [44].

From a spatial point of view, the erosive processes of the dune systems can vary greatly along the coast for various reasons, the main ones being linked to the geomorphological characteristics of the area and to the variability of the wave climate and of the longshore transport [45,46].

Quantitative measurement of the intensity of impact that human activity has on coastal ecosystems can help us better understand the interaction between humans and dune systems, promoting scientific management of the ecological environment and improving the scientific and orderly development of coastal zones [47]. In this regard, the analysis of morphological changes of coastal dunes and the analysis of the related causes is much analyzed in the scientific literature by analyzing case studies [48], with laboratory experiments [49–53] and applying analytical and numerical models [54–56]. Indeed, these variations can be analyzed in probabilistic terms [57,58] to identify critical thresholds of sea storms that cause dune erosion [59], and in terms of the beach response to the action of the most intense sea storms [60]. Furthermore, the analysis of the causes of erosive phenomena [61] is of fundamental importance to quantify and predict the vulnerability of dune systems on different timescales and to correctly plan any interventions on dunes and beaches [62–70].

The aim of the paper was to evaluate the effects of anthropogenic pressures on the Calabrian dune systems, especially in terms of the triggering of coastal erosion processes. This evaluation compares the current and past dune system extension, stretching back to the 1950s, while also analyzing what was built in place of the destroyed dune systems, and in the analysis of the possible correlations between the destruction of dune systems by anthropogenic causes and coastal erosion.

2. Materials and Methods

2.1. Site Description

Calabria is a region in southern Italy with a narrow and elongated morphology, which has been subject to a considerable coastal extension, exceeding 700 km (Figure 1). The coastal areas are characterized by an alternation of beaches, mainly sandy and pebbly, and high coasts, especially in the promontories of Capo Rizzuto, on the Ionian coast, and of Capo Vaticano, on the Tyrrhenian coast. Calabria is bordered by two seas, Tyrrhenian and Ionian, by the Strait of Messina and by the Gulf of Taranto. Tyrrhenian and Ionian coasts are characterized by different climatic conditions and by different fetch extensions. These differences lead to a remarkable variability of meteorological and marine conditions between the different areas of Calabria.

From a morphological point of view, Calabria is mainly characterized by hills and mountains, with a percentage of less than 10% of plains. Much of the Tyrrhenian coast is characterized by reliefs generally located a short distance from the coast and by very few coastal plains. Instead, on the Ionian coast, the reliefs are further away from the coast than the Tyrrhenian coast.

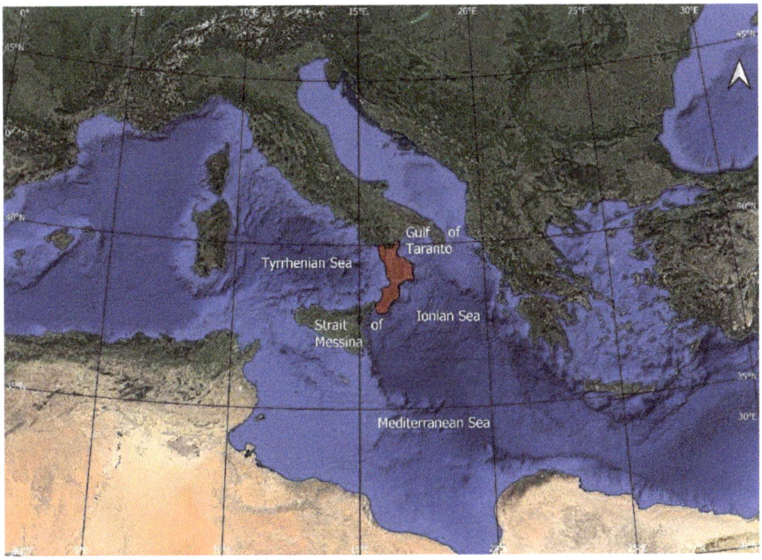

Figure 1. Location of Calabria region (highlighted by the red fill) in Southern Italy in the center of the Mediterranean Sea.

From a hydrological point of view, most of the Calabrian rivers (called *fiumare*) are characterized by high slopes, modest corrivation times, and a torrential regime, so floods occur suddenly [71,72]. For these reasons, river sediment transport is mainly related to soil erosion by water [73,74], which also influences shoreline evolution [75,76].

Furthermore, Calabria is characterized by considerable anthropogenic pressures, concentrated mostly in the second half of the last century after the Second World War. These pressures have encompassed both the construction of towns and infrastructures and the construction of numerous tourist activities and bathing establishments, which make the coastal areas of fundamental importance in the regional economy.

2.2. Methodology

The proposed methodology has two main aims, and it can be divided into five phases, according to the flow chart shown in Figure 2. This methodology was applied at the municipality level. In Calabria, there are 116 coastal municipalities, 71 are on the Ionian coast, 3 on the Strait of Messina, and 42 on the Tyrrhenian coast. Among these municipalities, those with negligible coastal lengths of less than 1 km were excluded. Therefore, 10 municipalities were excluded, 8 of them are in the Ionian coast and the other 2 are in the Tyrrhenian coast. Therefore, the analyzed municipalities are 63 on the Ionian coast, 3 in the Strait of Messina, and 40 on the Tyrrhenian coast.

Regarding the aims, the first of them concerns the comparison between the current extension of the dune systems and their extension after the Second World War, before the considerable anthropic pressures described above. The second aim concerns the analysis of shoreline changes, excluding only the municipalities without dune systems in the 1950s.

The five phases were all developed in QGIS and concerned:

1. Acquisition of historical and current cartographic data available, such as shapefiles, cartography, and satellite imagery;
2. Comparison between past and present dune systems;
3. Analysis of what was built in place of the destroyed dune systems, for example, inhabited centers, promenades etc.;

4. Analysis of the shoreline changes in municipalities where dune systems are currently present and where dune systems have been partially or totally destroyed by anthropogenic causes;
5. Analysis of the possible correlations between the destruction of dune systems by anthropogenic causes and the triggering of coastal erosion processes.

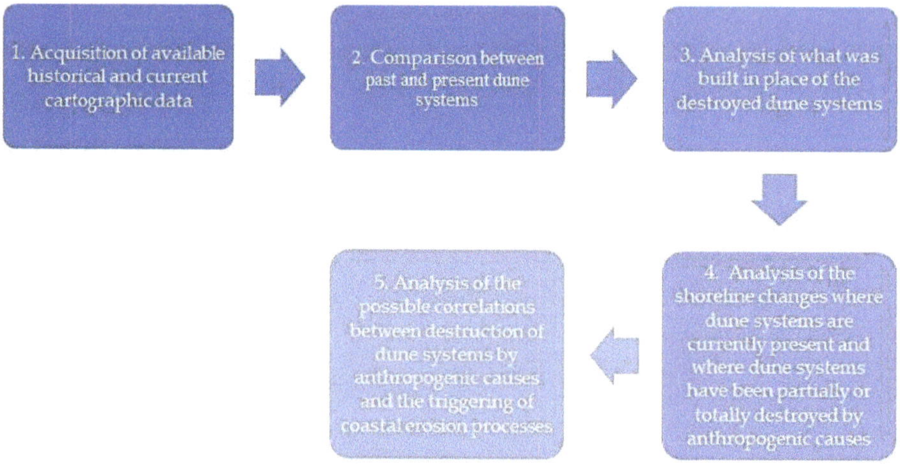

Figure 2. Flow chart of the proposed methodology.

Regarding the first phase, the input data were shapefiles, cartography, and satellite imagery. In detail, the following have been analyzed: the shapefile of the shoreline and the shapefiles of the inhabited centers, both dated 1954, and both digitized based on CASMEZ, "*Cassa del Mezzogiorno*", cartography of 1954, in scale 1:10,000; the shapefile of the dune systems in 1959, digitized starting from Geological cartography of 1959 in scale 1:25,000; the shapefile of the Corine Land Cover fourth level of 2018, where the current dune systems and the inhabited centers are highlighted; shapefiles of the municipal boundaries. These data are available in the Open Data section of the Calabrian Geoportal (http://geoportale.regione.calabria.it/opendata, accessed on 15 December 2021), with the exception of the Corine Land Cover, available in the Open Data section of the Italian Higher Institute for Environmental Protection and Research (https://groupware.sinanet.isprambiente.it/uso-copertura-e-consumo-di-suolo/library/copertura-del-suolo/corine-land-cover/corine-land-cover-2018-iv-livello, accessed on 15 December 2021). Other input data used included the most recent Google satellite imagery. The coverages of these data are not temporally homogeneous, but the images are from between 2019 and 2021, depending on the location examined.

Regarding the resolution of the input data, the Corine Land Cover was obtained through photointerpretation of satellite images in the land thematic area of the Copernicus program and characterized by a minimum cartographic unit for the coverage of 25 hectares and by a minimum width of the linear elements of 100 m. Moreover, the level of detail of both CASMEZ and geological cartographies is not very high, as they have scales of 1:10,000 and 1:25,000, respectively. Therefore, digitization errors would be in the order of meters, especially in geological cartographies. However, this accuracy is consistent with the aims of the paper, which concern the evaluation of the order of magnitude of the extension of past and present dune systems.

In the second phase, the shapefiles of the dune systems of the 1950s were superimposed on the Corine Land Cover fourth level of 2018 to verify which dune systems present in the 1950s are still present, and to quantify and compare the relative extensions.

In the third phase, the shapefiles of the inhabited centers of the 1950s were superimposed on the Corine Land Cover fourth level of 2018, and on the most recent Google satellite imagery, to verify which inhabited centers were built in place of the dune systems of the 1950s, and to analyze what was built in place of the destroyed dune systems, for example, inhabited centers, promenades, ports, etc. It should be noted that the municipalities analyzed are characterized by different morphological and anthropic conditions. We focused on man-made areas with buildings, infrastructure, promenades, ports, and coastal structures, and on areas without them. The criterion adopted to define the level of destruction of the dune systems was the following: intact dune systems if the current extension was the same as that of the 1950s or if the reduction do not exceed 10%; dune systems partially destroyed if this reduction was between 10 and 90%; dune systems totally destroyed if this reduction was greater than 90%.

The analysis of the shoreline changes carried out in the fourth phase involved the comparison between the 1954 shoreline and the shoreline digitized starting from the most recent Google satellite images. The analysis was carried out in coastal areas where dune systems are currently present and in coastal areas where dune systems are partially or totally destroyed by anthropogenic causes compared to the 1950s; thus, excluding coastal areas without dune systems in the 1950s. The results were aggregated at the municipality level. The digitalization of the missing shorelines was carried out using the spatial analysis tools of Google Earth Pro at an eye altitude of 200 m, corresponding to a scale higher than 1:1000. Moreover, transepts were traced, with an average spacing of the order of a hundred meters, and the related baselines were identified for each transept. These lines identify the upper limit of the beach and correspond to promenades, roads, and structures. Shorelines, transepts, and baselines were saved as kml files then saved on QGIS as shapefiles. Baselines were used on QGIS as control points to confirm the accuracy of the procedure.

Generally, the uncertainties in the digitization phase concern georeferencing, the orthorectification process, the resolution of the different imagery sets, the digitizing uncertainty, the uncertainty in the identification of the wet/dry line, and any error caused by a variation in some factors affecting the shoreline change, such as the seasonal cycle of erosion and deposition, the tide excursion, and the impact of storms [77–82]. In this case, the reference line chosen was the wet/dry line. Moreover, the cartography data are all related to the summer period, and no storm conditions were observed in any of the data, so the effects of seasonal variation and individual storms on shoreline change are of limited importance. To estimate the tide excursions, the recordings of the tide gauges of Crotone and Reggio Calabria were analyzed, the Tide Tables of the Italian Marine Hydrographic Institute [83] and the scientific papers were consulted, especially that of Sannino et al. [84]. The maximum-recorded tide height values are about 25 cm in Reggio Calabria and over 80 cm in Crotone; the minimum recorded tide height values are over 50 cm in Reggio Calabria and over 70 cm in Crotone and the average recorded tide height values are less than 30 cm in Reggio Calabria and about 50 cm in Crotone. The maximum tide height value reported in the Tide Tables is 25 cm. Therefore, Calabria is a microtidal environment where the effects on the variation of the shoreline position are of limited importance. The shoreline digitized based on the CASMEZ Cartography of 1954 on a scale of 1:10,000 is affected by a scanning error of less than one meter. On the other hand, regarding the georeferencing error of the Google Earth shorelines, the use of baselines as control points contained the error within a few tens of cm. Finally, the physical component of the error was estimated using the formula of Allan et al. [85]. The error was estimated, starting from the average and maximum values of the tide height and from the beach slope. This last parameter was estimated using the QGIS Profile tool plugin based on the 1 m side square mesh LIDAR DTMs available on the Italian Geoportal (http://www.pcn.minambiente.it/mattm/, accessed on 15 December 2021). The beach slope values of the examined locations varied between 5 and 15% so that the estimated error, assuming maximum tide height conditions, was between 5 and 15 m, and the estimated error (assuming minimum tide height conditions) was between 4 and 14 m. However, these are very precautionary values, as the times of the satellite image was not known and, consequently, it is not possible to know the tide conditions at these times. Furthermore, most of the examined locations were subject to tidal excursions lower than the maximum ones. Therefore, in most of the examined locations, the uncertainties in the shoreline position are in the order of a few meters. This accuracy is compatible with the aims of the paper, which concern the evaluation of the erosion and advancement trends, and not their precise quantification.

In the last phase, a cross-statistical analysis of the results of the previous phases was carried out to verify whether there was a correlation between the destruction of dune systems by anthropogenic causes and the triggering of coastal erosion processes. Through this analysis, the number and the percentage of municipalities where dune systems were present or absent were firstly evaluated. Subsequently, the number of municipalities where dune systems were intact compared to the 1950s, or had been partially (or totally) destroyed by anthropogenic causes, was evaluated. Furthermore, the number of municipalities with shoreline erosion or advancement was evaluated by classifying the erosive processes according to their intensity.

The criterion adopted to identify the coastal erosion processes was the following. For each municipality, the maximum value of the shoreline retreat was considered, not counting retreat of the order of a few meters. Furthermore, the coastal erosion intensity was classified according to a scale with four classes: slight erosion, for maximum retreat of up to 20 m; moderate erosion, for maximum retreat between 20 and 50 m; intense erosions, for maximum retreat between 50 and 100 m; severe erosion, for maximum retreat exceeding 100 m. The same for shoreline retreat, the shoreline advancement was also identified. For each municipality, the maximum value of the shoreline advancement was considered, not counting the advancement of the order of a few meters.

3. Results

The results were aggregated at the municipality level, and they are summarized in Tables 1–4. In detail, Table 1 shows a summary of the dune system condition (intact, partially, or totally destroyed by anthropogenic causes) and of the maximum shoreline retreat of advancement values for each of the 84 analyzed municipality, ordered clockwise from the Ionian coast to the Tyrrhenian coast. Table 2 shows the number of municipalities where dune systems are present or absent both in the 1950s and today. Table 3 shows the comparison between the dune systems in the 1950s and today, highlighting the number of municipalities where these systems are intact and where they were partially (or totally) destroyed by anthropogenic causes. In addition, this table shows the number of municipalities where erosion or shoreline advancement was observed. Finally, Table 4 shows the erosive process classification based on their intensity.

The analysis has shown that, in the 1950s, dune systems were present in most of the Calabrian coastal municipalities. Indeed, most of the inhabited centers were in inland areas, except for the provincial capitals, such as Reggio Calabria and Crotone, and for the larger towns. Dune systems were present in 84 out of 106 municipalities, corresponding to a percentage of 79% and covering an area of about 120 km^2. Consequently, the municipalities without dune systems were just 22 out of 106, corresponding to a percentage of 21%. Of these 22 municipalities without dune systems, 10 are located on the Ionian coast, 3 in the Strait of Messina, and 9 on the Tyrrhenian coast. In contrast, of the 84 municipalities with dune systems, 54 are located on the Ionian coast and the other 30 are located on the Tyrrhenian coast. However, the Ionian coast extends for about 450 km while the Tyrrhenian coast extends for about 250 km. Therefore, in relation to the coastal extension, the dune systems were well divided between the Ionian and Tyrrhenian coasts while most of the municipalities without dune systems lay on the promontory of Capo Vaticano and in the Strait of Messina area.

Table 1. Summary of the dune system condition (intact, partially, or totally destroyed by anthropogenic causes) and of the maximum shoreline retreat of advancement values for each of the 72 analyzed municipality (ordered clockwise from the Ionian coast to the Tyrrhenian coast). Legend: Max Ret. = maximum retreat; Max Ad. = maximum advancement.

Municipality	Dune Condition	Max Ret. (m)	Max Ad. (m)	Municipality	Dune Condition	Max Ret. (m)	Max Ad. (m)
Rocca Imperiale	Intact	-	170	Marina di Gioiosa Ionica	Totally	70	-
Roseto Capo Spulico	Totally	15	-	Siderno	Partially	55	-
Amendolara	Partially	30	-	Locri	Partially	30	-
Trebisacce	Intact	-	25	Sant'Ilario dello Ionio	Partially	60	-
Villapiana	Intact	-	60	Ardore	Partially	30	-
Cassano allo Ionio	Intact	-	80	Bovalino	Partially	40	-
Corigliano Calabro	Partially	60	-	Bianco	Intact	-	70
Rossano	Partially	20	-	Africo	Intact	-	50
Crosia	Partially	60	-	Ferruzzano	Intact	-	10
Calopezzati	Partially	50	-	Brancaleone	Partially	25	-
Pietrapaola	Partially	20	-	Palizzi	Partially	30	-
Mandatoriccio	Partially	70	-	Condofuri	Intact	-	15
Cariati	Partially	50	-	Ricadi	Totally	20	-
Crucoli	Partially	40	-	Parghelia	Partially	75	-
Cirò	Partially	40	-	Zambrone	Totally	75	-
Cirò Marina	Partially	50	-	Curinga	Intact	-	105
Strongoli	Partially	-	70	Lamezia Terme	Intact	-	65
Crotone	Partially	90	-	Gizzeria	Partially	70	-
Isola Capo Rizzuto	Intact	-	5	Falerna	Partially	25	-
Cutro	Partially	-	15	Amantea	Totally	160	-
Botricello	Intact	-	40	Belmonte	Partially	25	-
Cropani	Intact	-	15	Longobardi	Partially	25	-
Sellia Marina	Partially	50	-	Fiumefreddo Bruzio	Partially	60	-
Simeri Crichi	Partially	50	-	Falconara Albanese	Totally	20	-
Borgia	Partially	20	-	San Lucido	Partially	70	-
Squillace	Partially	40	-	Paola	Totally	25	-
Stalettì	Partially	25	-	Fuscaldo	Partially	70	-
Montepaone	Totally	40	-	Guardia Piemontese	Totally	150	-
Soverato	Partially	-	40	Acquappesa	Totally	80	-
Sant'Andrea Apostolo dello Ionio	Intact	-	50	Cetraro	Totally	80	-
Santa Caterina dello Ionio	Intact	-	30	Diamante	Partially	30	-
Guardavalle	Partially	15	-	Grisolia	Intact	-	80
Monasterace	Partially	90	-	Santa Maria del Cedro	Totally	25	-
Camini	Partially	30	-	San Nicola Arcella	Totally	20	-
Riace	Partially	-	30	Praia a Mare	Partially	-	70
Stignano	Intact	-	55	Tortora	Totally	100	-

Table 2. Summary of the municipalities with dune systems present or absent in the 1950s and today, and a summary of the dune system extension in the 1950s and today.

Dune Systems	Municipality		Area	
	1950s	Today	1950s	Today
Present	84 (79%)	70 (66%)	120 km^2	25 km^2
Absent	22 (21%)	36 (34%)	-	-

Table 3. Summary of the comparison between the dune systems in the 1950s and today and a summary of the number of municipalities where shoreline erosion or advancements have been observed, or where shoreline changes are due to other non-anthropogenic causes.

Dune Systems	1950s–Today	Erosion	Advancement	Shoreline Changes Due to Other Non-Anthropogenic Causes
Intact	18	0	17	1
Partially destroyed	52	36	5	11
Totally destroyed	14	14	0	0

Table 4. Summary of the classification of coastal erosion processes according to their intensity.

Dune Systems/Erosion	Slight	Moderate	Intense	Severe
Intact	0	0	0	0
Partially destroyed	1	18	17	0
Totally destroyed	1	6	4	3
Total	2	24	21	3

On the other hand, dune systems are today present in 70 out of 106 municipalities, corresponding to a percentage of 66% and cover an area of about 25 km^2 with a reduction of 14 municipalities and about 100 km^2 compared with the 1950s. Most of these municipalities are in the Tyrrhenian coast, 11 out of 14 to be precise; therefore, dune systems are currently present in 51 municipalities on the Ionian coast and in just 19 municipalities on the Tyrrhenian coast. Consequently, the municipalities without dune systems are 36 out of 106 today, corresponding to a percentage of 34%. Of these 36 municipalities, 13 are located on the Ionian coast, 3 in the Strait of Messina and 20 on the Tyrrhenian coast. Therefore, currently most of the dune systems are located on the Ionian coast.

The comparison between the dune systems between the 1950s and today highlights that, currently, only in 18 out of 84 municipalities are the dune systems intact compared to the 1950s. In contrast, in 52 out of 84 municipalities, the dune systems have been partially destroyed and most of these municipalities are in the Ionian coast, precisely 36 out of 52. On the other hand, in 14 out of 84 municipalities, the dune systems have been totally destroyed and most of these municipalities are in the Tyrrhenian coast, 11 out of 14 to be precise.

Shoreline changes were analyzed in 72 out of 106 municipalities. The 22 municipalities without dune systems in the 1950s, and the 12 municipalities where the destruction of dune systems is not caused by anthropogenic action, but by other non-anthropogenic causes, such as erosion at river mouths and the construction of ports were excluded from the comparison. In the first case, all the dune systems present in the 1950s near the river mouths, in areas subsequently eroded and, therefore, currently covered by the sea, have been included. Indeed, between the 1950s and today, significant retreat of the shoreline has been observed at the mouths of many Calabrian rivers such as Allaro, Petrace, Mesima, Angitola, even of the order of hundreds of meters (Figure 3). On the other hand, the construction of ports, despite being an anthropogenic cause, was considered as a separate cause, as it can create an obstacle to longshore transport by creating both accumulation and erosion areas and, thus, altering the coastal balance. This analysis showed that, in 50 municipalities, there is erosion, while in 22 municipalities there is advancement. Furthermore, in no municipality where the dune systems are intact is there erosion and in almost all, as many as 17 out of 18, there is advancement, while in only 1 municipality the shoreline changes are due to other non-anthropogenic causes. In contrast, in all 14 municipalities, where the dune systems were totally destroyed between the 1950s and today, erosions are observed. Finally, regarding the 52 municipalities where the dunes have been partially destroyed between the 1950s and today, in 36 municipalities there is erosion, and only in 5 municipalities is there advancement, while in 11 municipalities the shoreline changes are due to other non-anthropogenic causes.

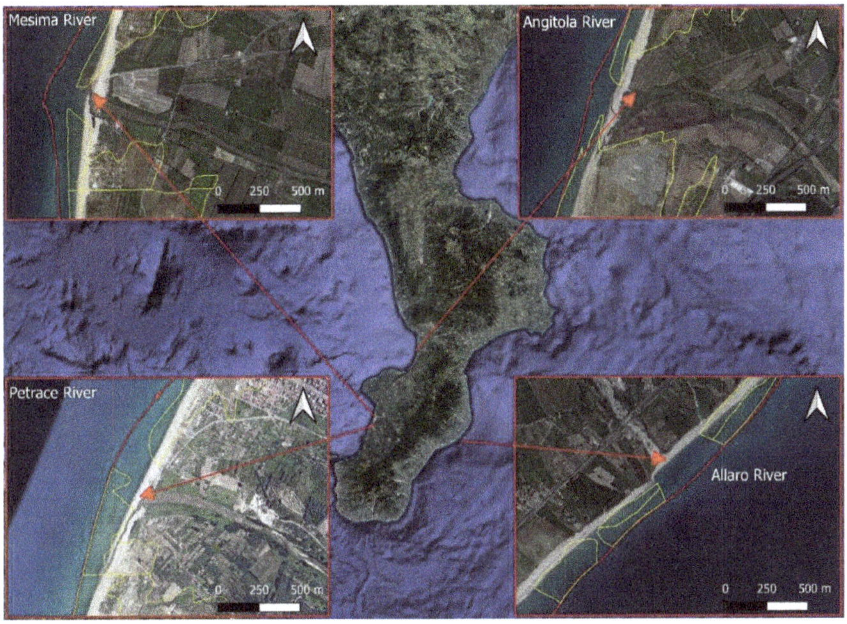

Figure 3. Shoreline erosions near river mouths. Legend: red line = shoreline of 1954; yellow dotted area = dune systems of 1950s; background = most recent Google satellite images.

In the 50 municipalities where erosions were observed, their intensity was also evaluated, according to the scale described above, highlighting that in most of the municipalities, 45 out of 50, there are moderate or intense erosion while only in 2 municipalities are there slight erosions, while in 3 municipalities there are severe erosions. However, severe erosions have been observed only in municipalities where dune systems have been totally destroyed and these municipalities are all on the Tyrrhenian coast. The maximum advance is observed in Rocca Imperiale on the Ionian Sea, with a value of 170 m, and another significant advance is observed in Curinga on the Tyrrhenian Sea, with a value greater than 100 m. The maximum retreat is observed in Amantea, with a value of 160 m in a coastal area not affected by the construction of the port (Figure 4), and another significant retreat is observed in Guardia Piemontese, with a value of 150 m (Figure 5), and in Tortora, with a value of 100 m. Finally, for each of the 65 municipalities with partial or total destruction of the dune systems, what was built in place of the dunes was analyzed. In detail, in almost all municipalities, new settlements have been built or existing ones have been expanded. A total of 42 new towns have been built, while scattered dwellings have been built in another 21 municipalities, the only exceptions being the municipalities of Sant'Ilario dello Ionio, where only a promenade with no inhabited center was built, and Bruzzano, where a railway was built. Finally, in 44 out of 65 municipalities, a promenade has also been built in the new inhabited centers.

Figure 4. Amantea: overlap between CASMEZ cartography and most recent Google satellite image. Legend: red line = shoreline of 1954; yellow dotted area = dune systems of 1950s.

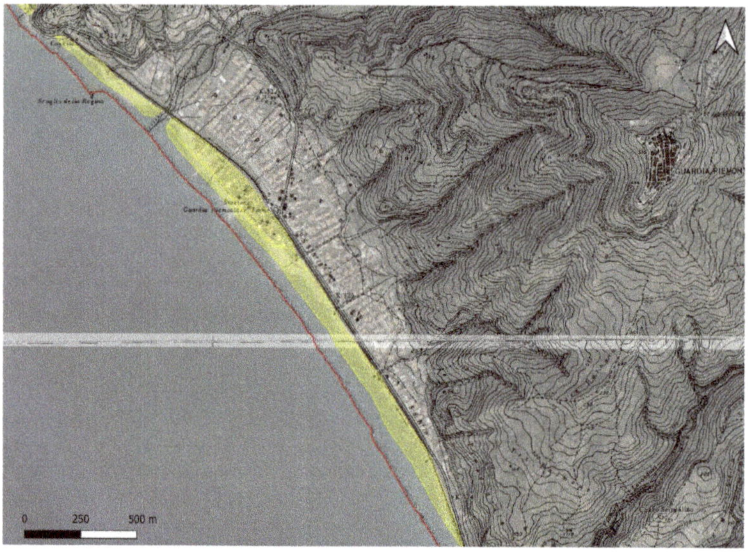

Figure 5. Guardia Piemontese: overlap between CASMEZ cartography and most recent Google satellite image. Legend: red line = shoreline of 1954; yellow dotted area = dune systems of 1950s.

4. Discussion

The paper analyzes the effects of anthropogenic pressures on dune systems, especially in terms of the triggering of coastal erosion processes, using Calabria, a region in southern Italy, as a case study. This issue is particularly topical and important because, during the second half of the last century, considerable anthropization processes were observed in many territories around the world [10–12]. On the other hand, Calabria is an interesting case study for the considerable anthropogenic pressures and for its geomorphological peculiarities. Among these peculiarities, two are highlighted: the border with two different seas, the Ionian and the Tyrrhenian, each of them exposed to very different meteorological and marine conditions from the other, and the presence of numerous coastal areas where sea and mountains are very close, especially along the Tyrrhenian Sea.

The analysis showed that these anthropogenic pressures had strong impacts on the Calabrian dune systems, which are currently greatly reduced from 120 km^2 in the 1950s to just 25 km^2 today. This significant reduction is also visible by analyzing on a municipalities scale. Indeed, 79% of the municipalities where dune systems were present in the 1950s are currently characterized by partially (62%) or totally (17%) destroyed dune systems, while in the remaining 21% the current dune systems are intact compared with those of the 1950s. In place of the destroyed dune systems, entire inhabited centers or small hamlets have been built.

No municipality where the current dune systems are intact compared to the 1950s show shoreline retreats. Indeed, in almost all these municipalities, the shoreline is advanced. In contrast, all the municipalities where the dune systems were totally destroyed between the 1950s and today show shoreline retreats. It should be noted that all three locations with severe erosions are located on the Tyrrhenian Sea. These locations are Amantea, Guardia Piemontese, and Tortora, with maximum retreat equal to 160, 150, and 100 m, respectively. Finally, in the municipalities where the dune systems were partially destroyed between the 1950s and today, the shoreline retreats prevail, and only in five municipalities did the shoreline advanced.

A total of 12 municipalities were excluded from this analysis because the shoreline changes were not due to the destruction of the dune systems, but to other non-anthropogenic causes. These localities include Catanzaro, Isca sullo Ionio, Badolato, Caulonia, Roccella Ionica, Bruzzano Zeffirio, Palmi, Gioia Tauro, San Ferdinando, Nicotera, Pizzo, and Nocera Terinese. Among these localities, those with erosion near the river mouths are Caulonia near the Allaro River, Bruzzano Zeffirio near the Bruzzano River, Palmi, and Gioia Tauro near the Petrace River, San Ferdinando and Nicotera near the Mesima River and Pizzo near the Angitola River and, in the latter, there is a dam. On the other hand, the other locations are close to ports. In detail, Isca sullo Ionio is located near the port of Badolato, which was built in a straight coastal area causing considerable shoreline variations on the two sides of the port. Specifically, there are shoreline advancements exceeding 150 m from the Badolato side and shoreline retreats of about 50 m from the Isca side. Moreover, Nocera Terinese is located near the port of Amantea. Like the port of Bodolato, the construction of the port of Amantea caused shoreline retreats in Nocera Terinese and shoreline advancements in Amantea. Only in San Ferdinando was a shoreline advancement observed, while in the other 10 municipalities, there was shoreline retreat. San Ferdinando is located near the port of Gioia Tauro, which is an inland port that has two breakwaters extending over 200 m from the shoreline, thereby causing a groin-like effect on the adjacent coast.

The anthropization process, with consequent destruction of the dune systems, is more present on the Tyrrhenian coast than on the Ionian one and can be correlated to the morphology of the territory. Indeed, the short distance between the reliefs and the coast may have facilitated the anthropization of the few flat coastal areas, often destroying the existing dune systems, as in the case of Tortora (Figure 6). In the 1950s, in fact, the town of Tortora was located only in the hills, while on the coast, there was an extensive dune system of over 650,000 m^2, with only a few sporadic buildings. Currently, instead of the dune system, the Tortora Marina town has been built, with a promenade and several buildings built not far from the shoreline. Moreover, the beach width is between 30 and a few meters, decreasing towards the north, and a maximum erosion of about 100 m is observed compared to 1954. Furthermore, on the Ionian coast, there is generally a greater distance between the coast and the reliefs, so several inhabited centers have been built behind the existing dunes, as in the case of Villapiana in the Ionian coast (Figure 7). In this case, the dune system has a length of about 5 km and an area of over 700,000 m^2. The inhabited center was built almost entirely in the second half of the last century, behind the dune system, reducing it only partially within a strip of a few hundred meters where some buildings were built. The beach has a width of between 50 and 100 m and is advancing compared to 1954, with a maximum advancement of the order of 50 m.

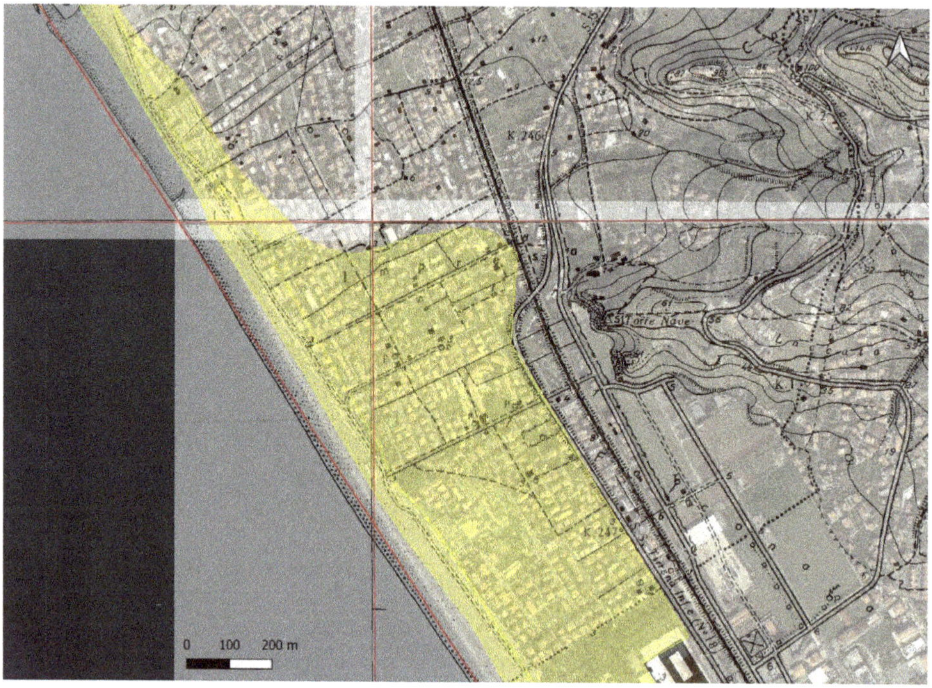

Figure 6. Guardia Piemontese: overlap between CASMEZ cartography and the most recent Google satellite image. Legend: red line = shoreline of 1954; yellow dotted area = dune systems of 1950s.

Finally, it is useful to consider the effects of climate change through an analysis of sea level rise. In Calabria, Barbaro et al. [41] analyzed this parameter based mainly on the study of Nerem et al. [86], which estimated that the average sea level has increased by 7 cm over the past 25 years, previously it was about 3 mm/year, and show that the growth rate is non-linear, but is accelerated by 0.084 mm/year. These results were obtained using data from 1993 to date, from various satellites, such as TOPEX/Poseidon, Jason-1, Jason-2, and Jason-3, and they are in accordance with the forecasts described in the fifth IPCC Report [87]. Therefore, the average value is expected to be about 10 cm, in the next 20 years, and about 80 cm in the next 100 years. Considering the beach slope values described above, in most of the examined localities this sea level rise would cause linear retreats of the shorelines of 15 m in the next 100 years. Therefore, this retreat would have a limited impact on Calabrian dune systems.

Figure 7. Villapiana: overlap between CASMEZ cartography and the most recent Google satellite image. Legend: red line = shoreline of 1954; yellow dotted area = dune systems of 1950s.

5. Conclusions

The aim of this paper was to evaluate the effects of anthropogenic pressures on the Calabrian dune systems, especially in terms of the triggering of coastal erosion processes. The analysis highlights a strong correlation between the destruction of dune systems by anthropogenic causes and the triggering of coastal erosion processes. This correlation is highlighted above all by the result that in all the municipalities where the dune systems are intact, compared to the 1950s, there are no shoreline retreats, while in all the municipalities where the dune systems have been totally destroyed by anthropogenic causes, compared to the 1950s, there are shoreline retreats.

The methodology proposed in this paper is easily applicable as it is based on open-source software, such as QGIS. Furthermore, it can be easily replicated, as it is based on freely accessible cartographic data, so it is sufficient to find similar data to carry out the same analysis in any other location.

Finally, this analysis is also of interest in the field of planning and management of coastal areas, to limit the anthropogenic impacts on coasts in the future, favoring the restoration of dune systems.

Author Contributions: Conceptualization, G.F.; methodology, G.F., G.B., G.C.B., and F.F.; software, G.F. and G.C.B.; validation, G.F., G.B., G.C.B., and F.F.; formal analysis, G.F. and G.C.B.; investigation, G.F. and G.C.B.; resources, G.F. and G.C.B.; data curation, G.F. and G.C.B.; writing—original draft preparation, G.F.; writing—review and editing, G.F., G.B., G.C.B., and F.F.; visualization, G.F.; supervision, G.B.; project administration, G.B. All authors have read and agreed to the published version of the manuscript.

Funding: This research received no external funding.

Institutional Review Board Statement: Not applicable.

Informed Consent Statement: Not applicable.

Conflicts of Interest: The authors declare no conflict of interest.

References

1. Cadenasso, M.L.; Pickett, S.T.A.; Schwarz, K. Spatial heterogeneity in urban ecosystems: Reconceptualizing land cover and a framework for classification. *Front. Ecol. Environ.* **2007**, *5*, 80–88. [CrossRef]
2. European Committee of the Regions. European Union Sustainable Tourism in the Mediterranean. Report. 2012. Available online: https://op.europa.eu/en/publication-detail/-/publication/30fa94a5-ce3f-4d5a-ad97-d0fdbca0ad8f/language-en (accessed on 20 October 2021).
3. Yanes, A.; Botero, C.M.; Arrizabalaga, M.; Vásquez, J.G. Methodological proposal for ecological risk assessment of the coastal zone of Antioquia, Colombia. *Ecol. Eng.* **2018**, *130*, 242–251. [CrossRef]
4. Fryberger, S.G.; Dean, G. Dune forms and wind regime. In *A Study of Global Sand Seas*; McKee, E.D., Ed.; United States Geological Survey Professional Paper 1052; US Government Printing Office: Washington, DC, USA, 1979; pp. 137–169.
5. Jay, H. Beach–Dune Sediment Exchange and Morphodynamic Responses: Implications for Shoreline Management, The Sefton Coast, NWEngland. Ph.D. Thesis, University of Reading, Reading, UK, 1998.
6. Sabatier, F.; Anthony, E.J.; Héquette, A.; Suanez, S.; Musereau, J.; Ruz, M.-H.; Regnauld, H. Morphodynamics of beach/dune systems: Examples from the coast of France. *Géomorphol. Relief Process. Environ.* **2009**, *15*, 3–22. [CrossRef]
7. Harley, M.; Ciavola, P. Managing local coastal inundation risk using real-time forecasts and artificial dune placements. *Coast. Eng.* **2013**, *77*, 77–90. [CrossRef]
8. Syvitski, J.P.M.; Vörösmarty, C.J.; Kettner, A.J.; Green, P. Impact of Humans on the Flux of Terrestrial Sediment to the Global Coastal Ocean. *Science* **2005**, *308*, 376–380. [CrossRef] [PubMed]
9. Yi, L.; Chen, J.; Jin, Z.; Quan, Y.; Han, P.; Guan, S.; Jiang, X. Impacts of human activities on coastal ecological environment during the rapid urbanization process in Shenzhen, China. *Ocean Coast. Manag.* **2018**, *154*, 121–132. [CrossRef]
10. Wang, J.; Zhou, W.; Pickett, S.T.; Yu, W.; Li, W. A multiscale analysis of urbanization effects on ecosystem services supply in an urban megaregion. *Sci. Total Environ.* **2019**, *662*, 824–833. [CrossRef] [PubMed]
11. Aguilera, M.A.; Tapia, J.; Gallardo, C.; Núñez, P.; Varas-Belemmi, K. Loss of coastal ecosystem spatial connectivity and services by urbanization: Natural-to-urban integration for bay management. *J. Environ. Manag.* **2020**, *276*, 111297. [CrossRef]
12. Zhai, T.; Wang, J.; Fang, Y.; Qin, Y.; Huang, L.; Chen, Y. Assessing ecological risks caused by human activities in rapid urbanization coastal areas: Towards an integrated approach to determining key areas of terrestrial-oceanic ecosystems preservation and restoration. *Sci. Total Environ.* **2020**, *708*, 135153. [CrossRef]
13. Luijendijk, A.; Hagenaars, G.; Ranasinghe, R.; Baart, F.; Donchyts, G.; Aarninkhof, S. The State of the World's Beaches. *Sci. Rep.* **2018**, *8*, 6641. [CrossRef]
14. Mentaschi, L.; Vousdoukas, M.I.; Pekel, J.-F.; Voukouvalas, E.; Feyen, L. Global long-term observations of coastal erosion and accretion. *Sci. Rep.* **2018**, *8*, 12876. [CrossRef] [PubMed]
15. Vousdoukas, M.I.; Ranasinghe, R.; Mentaschi, L.; Plomaritis, T.A.; Athanasiou, P.; Luijendijk, A.; Feyen, L. Sandy coastlines under threat of erosion. *Nat. Clim. Chang.* **2020**, *10*, 260–263. [CrossRef]
16. European Commission. Article 17 Technical Report 2001–2006. European Topic Centre on Biological Diversity. 2008. Available online: https://ec.europa.eu/environment/nature/knowledge/rep_habitats/index_en.htm#heading2001/06 (accessed on 20 October 2021).
17. FitzGerald, D.M.; Fenster, M.S.; Argow, B.A.; Buynevich, I.V. Coastal Impacts Due to Sea-Level Rise. *Annu. Rev. Earth Planet. Sci.* **2008**, *36*, 601–647. [CrossRef]
18. De Figueiredo, S.A.; Calliari, L.J.; Machado, A.A. Modelling the effects of sea-level rise and sediment budget in coastal retreat at Hermenegildo Beach, Southern Brazil. *Braz. J. Oceanogr.* **2018**, *66*, 210–219. [CrossRef]
19. Forgiarini, A.P.P.; de Figueiredo, S.A.; Calliari, L.J.; Goulart, E.S.; Marques, W.; Trombetta, T.B.; Oleinik, P.H.; Guimarães, R.C.; Arigony-Neto, J.; Salame, C.C. Quantifying the geomorphologic and urbanization influence on coastal retreat under sea level rise. *Estuar. Coast. Shelf Sci.* **2019**, *230*, 106437. [CrossRef]
20. Reguero, B.G.; Losada, I.J.; Méndez, F.J. A recent increase in global wave power as a consequence of oceanic warming. *Nat. Commun.* **2019**, *10*, 205. [CrossRef]
21. Pye, K. Physical and human influences on coastal dune development between the Ribble and Mersey estuaries, Northwest England. In *Coastal Dunes: Processes and Morphology*; Nordstrom, K.F., Psuty, N.P., Carter, R.W.G., Eds.; Wiley: Chichester, UK, 1990; pp. 339–359.
22. Sancho, F.; Abreu, T.; D'Alessandro, F.; Tomasicchio, G.R.; Silva, P.A. Surf hydrodynamics under collapsing coastal dunes. *J. Coast. Res.* **2011**, *64*, 144–148.
23. Anthony, E.J. Storms, shoreface morphodynamics, sand supply, and the accretion and erosion of coastal dune barriers in the southern North Sea. *Geomorphology* **2013**, *199*, 8–21. [CrossRef]
24. Feagin, R.; Furman, M.; Salgado, K.; Martinez, M.; Innocenti, R.; Eubanks, K.; Figlus, J.; Huff, T.; Sigren, J.; Silva, R. The role of beach and sand dune vegetation in mediating wave run up erosion. *Estuar. Coast. Shelf Sci.* **2019**, *219*, 97–106. [CrossRef]

25. Pye, K.; Blott, S. Decadal-scale variation in dune erosion and accretion rates: An investigation of the significance of changing storm tide frequency and magnitude on the Sefton coast. *Geomorphology* **2008**, *102*, 652–666. [CrossRef]
26. Swann, C.; Bride, K.; Spore, N. *Coast Foredunes: Identifying Coastal, Aeolian and Management Interactions Driving Morphological State Change*; ERDC/CHL TR-17873; U.S. Army Corps of Engineers: Washington, DC, USA, 2014.
27. Walker, I.J.; Davidson-Arnott, R.G.D.; Bauer, B.O.; Hesp, P.A.; Delgado-Fernandez, I.; Ollerhead, J.; Smyth, T.A.G. Scale-dependent perspectives on the geomorphology and evolution of beach-dune systems. *Earth-Sci. Rev.* **2017**, *171*, 220–253. [CrossRef]
28. Tomasicchio, G.R.; Francone, A.; Simmonds, D.J.; D'Alessandro, F.; Frega, F. Prediction of Shoreline Evolution. Reliability of a General Model for the Mixed Beach Case. *J. Mar. Sci. Eng.* **2020**, *8*, 361. [CrossRef]
29. Pagán, J.; Aragonés, L.; Tenza-Abril, A.; Pallarés, P. The influence of anthropic actions on the evolution of an urban beach: Case study of Marineta Cassiana beach, Spain. *Sci. Total Environ.* **2016**, *559*, 242–255. [CrossRef] [PubMed]
30. Miduri, M.; Foti, G.; Puntorieri, P. Impact generated by Marina of Badolato on adjacent coasts. In Proceedings of the 13th International Congress on Coastal and Marine Sciences, Engineering, Management & Conservation (MEDCOAST), Mellieha, Malta, 31 October–4 November 2017; Volume 2, pp. 935–945.
31. Pagán, J.; López, I.; Aragonés, L.; Garcia-Barba, J. The effects of the anthropic actions on the sandy beaches of Guardamar del Segura, Spain. *Sci. Total Environ.* **2017**, *601–602*, 1364–1377. [CrossRef] [PubMed]
32. Zema, D.A.; Bombino, G.; Boix-Fayos, C.; Tamburino, V.; Zimbone, S.M.; Fortugno, D. Evaluation and modeling of scouring and sedimentation around check dams in a Mediterranean torrent in Calabria, Italy. *J. Soil Water Conserv.* **2014**, *69*, 316–329. [CrossRef]
33. Aragonés, L.; Pagán, J.; López, M.; García-Barba, J. The impacts of Segura River (Spain) channelization on the coastal seabed. *Sci. Total Environ.* **2016**, *543*, 493–504. [CrossRef]
34. Zhu, L.; He, Q.; Shen, J.; Wang, Y. The influence of human activities on morphodynamics and alteration of sediment source and sink in the Changjiang Estuary. *Geomorphology* **2016**, *273*, 52–62. [CrossRef]
35. Foti, G.; Barbaro, G.; Manti, A.; Foti, P.; La Torre, A.; Geria, P.F.; Puntorieri, P.; Tramontana, N. A methodology to evaluate the effects of river sediment withdrawal: The case study of the Amendolea River in southern Italy. *Aquat. Ecosyst. Health Manag.* **2020**, *23*, 465–473. [CrossRef]
36. Dissanayake, P.; Brown, J.; Karunarathna, H. Impacts of storm chronology on the morphological changes of the Formby beach and dune system, UK. *Nat. Hazards Earth Syst. Sci.* **2015**, *15*, 1533–1543. [CrossRef]
37. Splinter, K.D.; Carley, J.T.; Golshani, A.; Tomlinson, R. A relationship to describe the cumulative impact of storm clusters on beach erosion. *Coast. Eng.* **2014**, *83*, 49–55. [CrossRef]
38. Dissanayake, P.; Brown, J.; Wisse, P.; Karunarathna, H. Comparison of storm cluster vs isolated event impacts on beach/dune morphodynamics. *Estuar. Coast. Shelf Sci.* **2015**, *164*, 301–312. [CrossRef]
39. Wernette, P.; Houser, C.; Lehner, J.; Evans, A.; Weymer, B. Investigating the Impact of Hurricane Harvey and Driving on Beach-Dune Morphology. *Geomorphology* **2020**, *358*, 107119. [CrossRef]
40. Barbaro, G.; Petrucci, O.; Canale, C.; Foti, G.; Mancuso, P.; Puntorieri, P. Contemporaneity of Floods and Storms. A Case Study of Metropolitan Area of Reggio Calabria in Southern Italy. In Proceedings of the 3rd International Symposium New Metropolitan Perspectives (ISTH2020), Reggio Calabria, Italy, 22–25 May 2018; pp. 614–620. [CrossRef]
41. Canale, C.; Barbaro, G.; Foti, G.; Petrucci, O.; Besio, G.; Barillà, G.C. Bruzzano river mouth damage due to meteorological events. *Int. J. River Basin Manag.* **2021**, 1–17. [CrossRef]
42. Canale, C.; Barbaro, G.; Petrucci, O.; Fiamma, V.; Foti, G.; Barillà, G.C.; Puntorieri, P.; Minniti, F.; Bruzzaniti, L. Analysis of floods and storms: Concurrent conditions. *Ital. J. Eng. Geol. Environ.* **2020**, *1*, 23–29. [CrossRef]
43. Barbaro, G.; Foti, G.; Nucera, A.; Barillà, G.C.; Canale, C.; Puntorieri, P.; Minniti, F. Risk mapping of coastal flooding areas. Case studies: Scilla and Monasterace (Italy). *Int. J. Saf. Secur. Eng.* **2020**, *10*, 59–67. [CrossRef]
44. Castelle, B.; Marieu, V.; Bujan, S.; Splinter, K.D.; Robinet, A.; Sénéchal, N.; Ferreira, S. Impact of the winter 2013–2014 series of severe Western Europe storms on a double-barred sandy coast: Beach and dune erosion and megacusp embayments. *Geomorphology* **2015**, *238*, 135–148. [CrossRef]
45. Houser, C. Alongshore variation in beach–dune morphology: Implications for barrier island response. *Geomorphology* **2013**, *199*, 48–61. [CrossRef]
46. de Winter, R.; Gongriep, F.; Ruessink, B. Observations and modeling of alongshore variability in dune erosion at Egmond aan Zee, the Netherlands. *Coast. Eng.* **2015**, *99*, 167–175. [CrossRef]
47. Yi, L.; Yu, Z.; Qian, J.; Kobuliev, M.; Chen, C.; Xing, X. Evaluation of the heterogeneity in the intensity of human interference on urbanized coastal ecosystems: Shenzhen (China) as a case study. *Ecol. Indic.* **2021**, *122*, 107243. [CrossRef]
48. Pye, K.; Neal, A. Coastal dune erosion at Formby Point, north Merseyside, England: Causes and Mechanisms. *Mar. Geol.* **1994**, *119*, 39–56. [CrossRef]
49. van Thiel de Vries, J.S.M.; Van Gent, M.R.A.; Walstra, D.J.R.; Reniers, A.J.H.M. Analysis of dune erosion processes in large-scale flume experiments. *Coast. Eng.* **2008**, *55*, 1028–1040. [CrossRef]
50. Tomasicchio, G.R.; D'Alessandro, F.; Barbaro, G. Composite modelling for large-scale experiments on wave–dune interaction. *J. Hydraul. Res.* **2011**, *49*, 15–19. [CrossRef]
51. Tomasicchio, G.R.; Sánchez-Arcilla, A.; D'Alessandro, F.; Ilic, S.; James, M.R.; Sancho, F.; Fortes, C.J.; Schüttrumpf, H. Large-scale experiments on dune erosion processes. *J. Hydraul. Res.* **2011**, *49*, 20–30. [CrossRef]

52. D'Alessandro, F.; Tomasicchio, G.R. Wave–dune interaction and beach resilience in large-scale physical model tests. *Coast. Eng.* **2016**, *116*, 15–25. [CrossRef]
53. Palmsten, M.; Splinter, K.D. Observations and simulations of wave runup during a laboratory dune erosion experiment. *Coast. Eng.* **2016**, *115*, 58–66. [CrossRef]
54. Larson, M.; Erikson, L.; Hanson, H. An analytical model to predict dune erosion due to wave impact. *Coast. Eng.* **2004**, *51*, 675–696. [CrossRef]
55. Roelvink, D.; Reniers, A.; Van Dongeren, A.; De Vries, J.V.T.; McCall, R.; Lescinski, J. Modelling storm impacts on beaches, dunes and barrier islands. *Coast. Eng.* **2009**, *56*, 1133–1152. [CrossRef]
56. D'Alessandro, F.; Tomasicchio, G.R.; Musci, F.; Ricca, A. Dune Erosion Physical, Analytical and Numerical Modelling. In Proceedings of the 33rd International Conference on Coastal Engineering, Santander, Spain, 1–6 July 2012; Volume 1. [CrossRef]
57. Esteves, L.S.; Brown, J.M.; Williams, J.J.; Lymbery, G. Quantifying thresholds for significant dune erosion along the Sefton Coast, Northwest England. *Geomorphology* **2012**, *143–144*, 52–61. [CrossRef]
58. Li, F.; van Gelder, P.; Vrijling, J.; Callaghan, D.; Jongejan, R.; Ranasinghe, R. Probabilistic estimation of coastal dune erosion and recession by statistical simulation of storm events. *Appl. Ocean Res.* **2014**, *47*, 53–62. [CrossRef]
59. Furmańczyk, K.; Dudzińska-Nowak, J.; Paplińska-Swerpel, B.; Brzezowska, N.; Furmańczyk, K. Critical storm thresholds for the generation of significant dune erosion at Dziwnow Spit, Poland. *Geomorphology* **2012**, *143–144*, 62–68. [CrossRef]
60. Tătui, F.; Vespremeanu-Stroe, A.; Preoteasa, L. Alongshore variations in beach-dune system response to major storm events on the Danube Delta coast. *J. Coast. Res.* **2014**, *70*, 693–699. [CrossRef]
61. Cohn, N.; Ruggiero, P.; García-Medina, G.; Anderson, D.; Serafin, K.A.; Biel, R. Environmental and morphologic controls on wave-induced dune response. *Geomorphology* **2019**, *329*, 108–128. [CrossRef]
62. D'Alessandro, F.; Tomasicchio, G.R.; Frega, F.; Carbone, M. Design and management aspects of a coastal protection system. A case history in the South of Italy. *J. Coast. Res.* **2011**, *64*, 492–495.
63. Suanez, S.; Cariolet, J.-M.; Cancouët, R.; Ardhuin, F.; Delacourt, C. Dune recovery after storm erosion on a high-energy beach: Vougot Beach, Brittany (France). *Geomorphology* **2012**, *139–140*, 16–33. [CrossRef]
64. Scott, T.; Masselink, G.; O'Hare, T.; Saulter, A.; Poate, T.; Russell, P.; Davidson, M.; Conley, D. The extreme 2013/2014 winter storms: Beach recovery along the southwest coast of England. *Mar. Geol.* **2016**, *382*, 224–241. [CrossRef]
65. Castelle, B.; Bujan, S.; Ferreira, S.; Dodet, G. Foredune morphological changes and beach recovery from the extreme 2013/2014 winter at a high-energy sandy coast. *Mar. Geol.* **2017**, *385*, 41–55. [CrossRef]
66. Smith, E.R.; D'Alessandro, F.; Tomasicchio, G.R.; Gailani, J.Z. Nearshore placement of a sand dredged mound. *Coast. Eng.* **2017**, *126*, 1–10. [CrossRef]
67. D'Alessandro, F.; Tomasicchio, G.R.; Francone, A.; Leone, E.; Frega, F.; Chiaia, G.; Saponieri, A.; Damiani, L. Coastal sand dune restoration with an eco-friendly technique. *Aquat. Ecosyst. Health Manag.* **2020**, *23*, 417–426. [CrossRef]
68. Fernández-Montblanc, T.; Duo, E.; Ciavola, P. Dune reconstruction and revegetation as a potential measure to decrease coastal erosion and flooding under extreme storm conditions. *Ocean Coast. Manag.* **2020**, *188*, 105075. [CrossRef]
69. Leone, E.; Kobayashi, N.; Francone, A.; Bartolo, S.; Strafella, D.; D'Alessandro, F.; Tomasicchio, G. Use of Nanosilica for Increasing Dune Erosion Resistance during a Sea Storm. *J. Mar. Sci. Eng.* **2021**, *9*, 620. [CrossRef]
70. Sanromualdo-Collado, A.; García-Romero, L.; Peña-Alonso, C.; Hernández-Cordero, A.I.; Ferrer-Valero, N.; Hernández-Calvento, L. Spatiotemporal analysis of the impact of artificial beach structures on biogeomorphological processes in an arid beach-dune system. *J. Environ. Manag.* **2021**, *282*, 111953. [CrossRef]
71. Sabato, L.; Tropeano, M. Fiumara: A kind of high hazard river. *Phys. Chem. Earth Parts A/B/C* **2004**, *29*, 707–715. [CrossRef]
72. Sorriso-Valvo, M.; Terranova, O. The Calabrian fiumara streams. *Z. Geomorphol.* **2006**, *143*, 109–125.
73. Barbaro, G.; Foti, G.; Mandaglio, G.; Mandaglio, M.; Sicilia, C.L. Estimate of sediment transport capacity in the basin of the Fiumara Annunziata (RC). Atti del 8600B0 Congresso Nazionale della Società Geologica Italiana, Arcavacata di Rende (CS), 18-20 settembre. *Rend. Online Soc. Geol. Ital.* **2012**, *21*, 696–697.
74. Terranova, O.; Antronico, L.; Coscarelli, R.; Iaquinta, P. Soil erosion risk scenarios in the Mediterranean environment using RUSLE and GIS: An application model for Calabria (Southern Italy). *Geomorphology* **2009**, *112*, 228–245. [CrossRef]
75. Barbaro, G.; Bombino, G.; Foti, G.; Borrello, M.M.; Puntorieri, P. Shoreline evolution near river mouth: Case study of Petrace River (Calabria, Italy). *Reg. Stud. Mar. Sci.* **2019**, *29*, 100619. [CrossRef]
76. Foti, G.; Barbaro, G.; Bombino, G.; Fiamma, V.; Puntorieri, P.; Minniti, F.; Pezzimenti, C. Shoreline changes near river mouth: Case study of Sant'Agata River (Reggio Calabria, Italy). *Eur. J. Remote Sens.* **2019**, *52*, 102–112. [CrossRef]
77. Masselink, G.; Pattiaratchi, C. Seasonal changes in beach morphology along the sheltered coastline of Perth, Western Australia. *Mar. Geol.* **2001**, *172*, 243–263. [CrossRef]
78. Cooper, A.; Jackson, D.; Navas, F.; McKenna, J.; Malvarez, G. Identifying storm impacts on an embayed, high-energy coastline: Examples from western Ireland. *Mar. Geol.* **2004**, *210*, 261–280. [CrossRef]
79. Boak, E.H.; Turner, I. Shoreline Definition and Detection: A Review. *J. Coast. Res.* **2005**, *214*, 688–703. [CrossRef]
80. Hapke, C.J.; Himmelstoss, E.A.; Kratzmann, M.G.; List, J.H.; Thieler, E.R. *National Assessment of Shoreline Change: Historical Shoreline Change along the New England and Mid-Atlantic Coasts*; US Geological Survey 2010, Open-File Report 1118; U.S. Geological Survey: Reston, VA, USA, 2010. [CrossRef]

81. Del Río, L.; Gracia, F.J. Error determination in the photogrammetric assessment of shoreline changes. *Nat. Hazards* **2012**, *65*, 2385–2397. [CrossRef]
82. Puig, M.; Del Río, L.; Plomaritis, T.A.; Benavente, J. Contribution of storms to shoreline changes in mesotidal dissipative beaches: Case study in the Gulf of Cádiz (SW Spain). *Nat. Hazards Earth Syst. Sci.* **2016**, *16*, 2543–2557. [CrossRef]
83. Istituto Idrografico della Marina. *Tavole di Marea e Delle Correnti di Marea*; Istituto Idrografico della Marina Italiana: Genova, Italy, 2020; 144p, ISBN 97888II3133. (In Italian)
84. Sannino, G.; Carillo, A.; Pisacane, G.; Naranjo, C. On the relevance of tidal forcing in modelling the Mediterranean thermohaline circulation. *Prog. Oceanogr.* **2015**, *134*, 304–329. [CrossRef]
85. Allan, J.C.; Komar, P.D.; Priest, G.R. Shoreline variability on the high-energy Oregon coast and its usefulness in erosion-hazard assessments. *J. Coast. Res.* **2003**, *38*, 83–105.
86. Nerem, R.S.; Beckley, B.D.; Fasullo, J.T.; Hamlington, B.D.; Masters, D.; Mitchum, G.T. Climate-change–driven accelerated sea-level rise detected in the altimeter era. *Proc. Natl. Acad. Sci. USA* **2018**, *115*, 2022–2025. [CrossRef] [PubMed]
87. IPCC. Climate Change 2013: The Physical Science Basis. In *Contribution of Working Group I to the Fifth Assessment Report of the Intergovernmental Panel on Climate Change*; Stocker, T.F., Qin, D., Plattner, G.-K., Tignor, M., Allen, S.K., Boschung, J., Nauels, A., Xia, Y., Bex, V., Midgley, P.M., Eds.; Cambridge University Press: Cambridge, UK; New York, NY, USA, 2013.

Journal of
*Marine Science
and Engineering*

MDPI

Article

A Numerical Study on the Impact of Building Dimensions on Airflow Patterns and Bed Morphology around Buildings at the Beach

Paran Pourteimouri *, Geert H. P. Campmans, Kathelijne M. Wijnberg and Suzanne J. M. H. Hulscher

Water Engineering and Management Group, Faculty of Engineering Technology, University of Twente, 7500 AE Enschede, The Netherlands; g.h.p.campmans@utwente.nl (G.H.P.C.); k.m.wijnberg@utwente.nl (K.M.W.); s.j.m.h.hulscher@utwente.nl (S.J.M.H.H.)

* Correspondence: p.pourteimouri@utwente.nl

Abstract: The attractiveness of beaches to people has led, in many places, to the construction of buildings at the beach–dune interface. Buildings change the local airflow patterns which, in turn, alter the sediment transport pathways and magnitudes. This induces erosion and deposition patterns around the structures. In this study, a numerical model is developed using the open-source computational fluid dynamics solver OpenFOAM. First, the model is used to predict the airflow patterns around a single rectangular building. The model predictions are validated with wind-tunnel data, which show good agreements. Second, a reference beach building is introduced and then the building dimensions are increased in length, width and height, each up to three times the reference building dimension. The impact of each dimensional extent on the near-surface airflow patterns is investigated. The results show that the near-surface airflow patterns are least dependent on the length of the building in the wind direction and they depend most on the width of the building perpendicular to the wind direction. Third, the convergence of the third-order horizontal near-surface velocity field is calculated to interpret the impact of changes in airflow patterns on potential erosion and deposition patterns around the building. The numerical predictions are compared with the observed erosion and sedimentation patterns around scale models in the field. The comparisons show satisfactory agreements between numerical results and field measurements.

Keywords: airflow around buildings; aeolian sediment transport; erosion and deposition patterns; beach structures; computational fluid dynamics (CFD)

Citation: Pourteimouri, P.; Campmans, G.H.P.; Wijnberg, K.M.; Hulscher, S.J.M.H. A Numerical Study on the Impact of Building Dimensions on Airflow Patterns and Bed Morphology around Buildings at the Beach. *J. Mar. Sci. Eng.* **2022**, *10*, 13. https://doi.org/10.3390/jmse10010013

Academic Editors: Giuseppe Roberto Tomasicchio, Ferdinando Frega and Felice D'Alessandro

Received: 4 November 2021
Accepted: 22 December 2021
Published: 24 December 2021

Publisher's Note: MDPI stays neutral with regard to jurisdictional claims in published maps and institutional affiliations.

1. Introduction

Coastal zones worldwide have always been attractive to humans, since they provide a wide variety of valuable resources and recreational activities. Population growth near coastlines leads to an increased demand for construction of restaurants, sailing clubs, holiday cottages and pavilions at the beach–dune interface. Figure 1 shows some typical examples of these structures.

A considerable number of studies indicated that the coastlines worldwide have been modified over millennia by human interventions, and this development is continuously growing [1–4]. There are complex interactions between airflow patterns, sediment transport and bed morphology on the beach. These interactions vary over a wide range of spatial and temporal scales and determine the shape, size, spacing and alignment of beaches and aeolian sand dunes [5]. The impact of buildings at the beach can be schematized by a loop as in Figure 2.

Buildings at the beach affect local airflow patterns and as a result aeolian sediment transport. These airflow patterns depend on building dimension, geometry, orientation, elevation from ground level, surface roughness and the positioning and distance in a row of buildings on the beach [6]. According to Jackson and Nordstrom [4], the dimensions of a building affect the degree to which a structure acts as an obstacle against wind flow and sediment migration. This affects the ability of airflow or sediment particles to move across the top of the structure or around the lateral sides of the structure. Therefore, buildings at the beach–dune interface locally alter wind flow patterns and change the location of erosion and accretion on the beach [7,8]. At longer time scales, the buildings could potentially change the dynamic state of the adjacent dune system, as they may modify the amount and spatial distribution of aeolian sand supply from the beach to the dune. Dunes provide natural flood protections against storm surges. Therefore, the coastal safety might be affected as dunes become more mobile and variability in height increases. Furthermore, a building could locally increase deposition or cause intensive erosion around the structure. These morphological changes affect the buildings at the beach. They might result in the need for sediment removal or even cause the tilting of the structure that affects the building functionality. Therefore, people move their houses elsewhere due to the excessive erosion and deposition, or change the shape of their houses by constructing on poles, for example, to prevent the buildings' dysfunctions. The impacts of buildings on wind flow and impacts of wind flow on buildings have been well addressed in the literature, focusing on applications such as pedestrian wind comfort, air pollutant dispersion, heat transfer, natural ventilation and wind-driven snow or rain around buildings. However, only a few studies have been conducted on the effects of building characteristics, specifically the impacts of building dimensions, on near-surface wind flow patterns and bed morphology at the beach. Fackrell [9] and Beranek [10] conducted wind-tunnel studies to investigate wind flow around buildings with various dimensions. Fackrell [9] found that the length of the recirculation region behind the building, which was defined as the distance between the leeward face of the structure and the reattachment point of the separated flow, increases with increasing building width normal to the flow direction as well as with decreasing building length parallel to the flow direction. Martinuzzi and Tropea [11] performed experiments to study the impact of width-to-height aspect ratio, W/H, of surface mounted obstacles on the flow patterns and parameters including windward separation and leeward reattachment lengths. They found that the separation length in front of the obstacle increases with increasing width up to about $W/H \approx 6$ and then decreases slightly for higher ratios. The indications showed that the reattachment length behind the obstacle increases linearly with increasing width up to about $W/H \approx 4$ and then asymptotically approaches a constant value.

Considering the impact of roughness elements, similar to buildings, on bed topography, Iversen et al. [12,13] conducted wind-tunnel tests on the sand bed to determine the impacts of obstacles with different dimensions on local aeolian erosion and deposition patterns. They found that the flow patterns and therefore the sand transport, depend considerably on the obstacle aspect ratio which was defined as the ratio of obstacle height to lateral width. Their studies showed that the observed erosion on the windward side of the rectangular object was caused by the formation of a horseshoe-type vortex. In a more detailed study performed by Tominaga et al. [14], sand erosion and deposition patterns around a surface-mounted cube was investigated using a wind-tunnel experiment. The results showed a considerable erosion at the upwind edges of the cube extended downwind along the lateral faces, and a small amount of sand accumulation at the leeward face of the cube. They found that the largest amount of erosion in the streamwise direction, x, occurs at $x/H = -0.75$ in front of the windward face of the cube, while the largest amount of erosion in the spanwise direction, z, occurs at $z/H = 0.85$ from the lateral sides of the cube, where H was defined as the cube height. Luo et al. [15] performed wind-tunnel tests to improve the understanding of the airflow patterns downwind of cuboid obstacles and to interpret the formation of the sand shadows observed behind obstacles in arid regions. In their studies, they investigated the impact of obstacle shape ratio on both horizontal and vertical airflow patterns around the structures. The shape ratio was defined as the ratio of the top area of the obstacle to its frontal area normal to the wind direction. Considering H as the height of the obstacle, the measurements showed that the flow begins to reattach and move along the bed surface at some distance between $2.5H$ to $3H$ from the separation point. They concluded that the formation of the low-velocity bubble downwind the obstacle causes sediment deposition behind the leeward face. Sutton and Neuman [16] studied the impact of vortical structures formed in the vicinity and in the wake of the cylindrical objects on the initiation of sediment transport. Their results show that the two counter-rotating vortices in the lee of the cylindrical objects allow the sediment entrainment to occur at lower wind speeds than that of required far away from the objects and in their wakes. The spacing between the cylindrical objects influences the strength of the two counter-rotating vortices, therefore may cause an increase in the sediment activity around cylindrical objects.

(**a**) (**b**)

Figure 1. Buildings at beach–dune interface on the (**a**) Egmond beach, and (**b**) Zandvoort beach (https://www.hollandluchtfoto.nl (accessed on 1 December 2021)), The Netherlands.

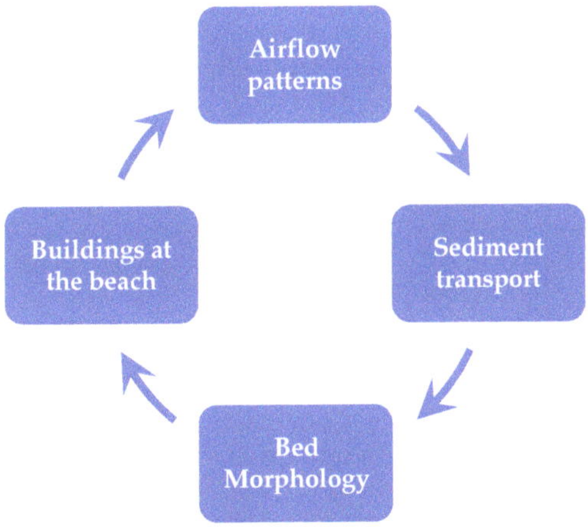

Figure 2. Morphological loop indicating the interactions between buildings at the beach, airflow patterns, sediment transport and bed morphology.

Beyers and Waechter [17] developed a CFD model to investigate the development of wind-driven snowdrifts around buildings. As they noted in their study, it is necessary to take the impacts of flow divergence into account in order to predict the development of snowdrifts realistically, while the commonly used models only rely on the threshold wind shear velocities to derive the snow erosion and deposition patterns around buildings.

The aforementioned studies show that the previous research mainly focused on the general airflow patterns around the buildings. However, the detailed quantitative impacts of building dimensions on near-surface airflow patterns have remained poorly understood, despite their important role in near-surface aeolian sediment transport. In addition, first attempts to find a relation between near-surface airflow patterns and near-surface erosion and deposition patterns around the buildings go back to the experimental work by Luo et al. [15]. However, their study was limited to the airflow patterns behind the obstacles that cause the evolution of sand shadows in arid regions over time. Furthermore, Poppema et al. [18] studied the size of deposition patterns around single buildings of different dimensions. However, their study does lack the detailed information on airflow patterns inducing those patterns.

Therefore, in the present study we systematically investigate the impact of building dimensions on the nature and extent of near-surface airflow patterns and the potential morphological changes induced by those flow fields when buildings are placed at a sand surface. We consider a wide area around the buildings to also capture the deposition patterns like those observed by Poppema et al. [18]. The building dimensions considered are length, width and height. The systematic study means that the building dimension is increased in each direction, while the other two dimensions remain unchanged. The two main research questions this study addresses are: (Q1) What are the detailed quantitative impacts of building length, width and height on near-surface airflow patterns which drive wind-driven sediment transport around buildings?; (Q2) What are the qualitative impacts of building length, width and height, on initial morphologic changes driven by wind around buildings at the beach?

In this paper, first a general description of airflow patterns and complex flow structures around an isolated building or a cube are presented in Section 1.1. In Section 2, the numerical modelling approach is explained. The detailed explanation of the numerical model itself and the validation of the model are provided in Appendices A and B, respectively. Results related to Question 1, on near-surface airflow patterns around buildings with different length, width and height, are presented in Section 3.1. Results related to Question 2, on the impact of building dimensions on wind-driven erosion and deposition patterns, are presented in Section 3.2. The paper ends with the discussion in Section 4, and conclusions that are presented in Section 5.

1.1. Wind Flow around an Isolated Building

The wind flow pattern in the vicinity of an isolated building is highly complex. The intrusion of a building, that acts as an impermeable obstacle, into the atmospheric boundary layer causes strong perturbations and complex flow structures in its vicinity. This perturbation is characterized by converting mean kinetic energy to turbulent kinetic energy due to the formation of eddies that are rotating faster or slower than the eddies in the mean flow [19]. Figure 3 shows flow features around an isolated cubical building with an orientation normal to the incident wind flow.

As wind approaches a building, the flow streamlines are deflected over and around the structure which is due to the formation of high-pressure gradients on the windward face. A stagnation point with the highest pressure is formed on the windward face of the building at an elevation approximately two thirds of the building height [20,21]. The location of the stagnation point depends on the building frontal aspect ratio, the height of the building in comparison with the height of the atmospheric boundary layer and the surface roughness upwind of the building [20]. The approaching flow diverges from the stagnation point to the zones with lower pressure including up over the roof, around the lateral sides and down the windward face towards the surface. When the upward and sideward flows encounter the windward edges of the building, they are detached from the surface and flow separation takes place. The separation bubbles on both the roof and the lateral sides of the building are characterized by the reverse flows, low velocity distributions and relatively high turbulence intensities [21]. This happens due to the air suction induced by low-pressure zones on the roof and lateral sides of the building. The detached flow might reattach to the roof or side walls of the building depending on the top and lateral aspect ratios, and upstream surface roughness that determines the turbulence intensity of the incidence flow [20,22]. As mentioned earlier, some of the flow approaching the windward face of the building is deflected downwards to the ground and moves in the reverse direction compared to the incident wind direction. The reversed flow undercuts the incident wind flow and causes it to be detached from the ground level and creates a standing vortex near the bed surface just upstream the windward face of the building [19]. This primary roll-like vortex induces formation of additional vortices that are smaller in size and weaker than the main vortex structure and are eventually connected to the primary vortex around the lateral sides of the building. This vortex is then stretched around the side walls and is extended downwind the building creating a so-called horseshoe-shape vortex, shaded blue in Figure 3 [20].

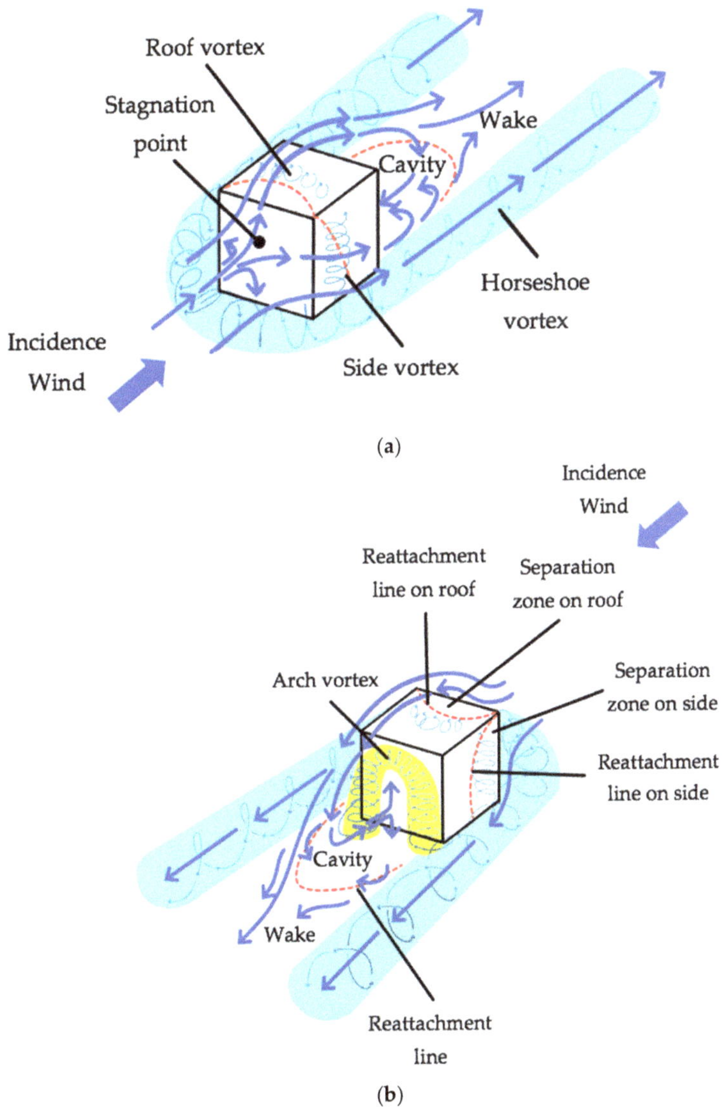

Figure 3. Schematic airflow patterns around an isolated cubical building with wind incidence angle perpendicular to the building (**a**) windward view, and (**b**) leeward view (modified from Oke et al. [19] and Blocken et al. [21]).

The flow structures formed behind the building are very complex. The low-pressure zone at the leeward face of the building creates air suction in a so-called cavity region. In this region, the along-wind flow passing over the roof of the building and two horizontally-oriented flows around the lateral sides of the building move in the reverse direction compared to the incidence wind flow, creating a recirculating zone just downstream of the leeward face of the building. The dashed line downstream of the building in Figure 3b shows the end of the cavity region where the streamlines are reattached to the ground surface [19,21]. For a wind incidence angle perpendicular to the upwind face of a cubical building, the height of the cavity region is about 1.5H, where H is the building height and the length of the cavity region extends to about 2.5 to 3H, measured from the upstream face of the building. The flow interference increases with increasing building width normal to the wind direction, therefore the cavity region height increases to some extent and its length reaches 12H for wide buildings with small height-to-width aspect ratios [23]. The horizontal flow patterns behind the rear face of the building show the formation of two counter-rotating vortices that join their extensions at the vertical symmetry plane (yellow shaded vortex in Figure 3b). These spiral vortices entrain some air from the horseshoe-shape vortex, created near the ground level, and whirl it upwards to create a vertically-oriented arch-shape vortex just downstream of the building [11,20,24]. Beyond the cavity region, the reattached flow moving in the direction of approaching flow requires some distance to recover the features of incidence wind flow and release all perturbations, separation impacts and secondary flow structures induced by the presence of building. This occurs in the so-called wake region that is characterized by velocity deficits, higher turbulence intensities and smaller scale eddies compared to the eddies in the incidence wind flow [20,23]. The wake region typically persists to about 5 to 30H downwind of the building and its height reaches to about 3 to 4H at a distance of 10H downstream of the building [19,23].

2. Methods

2.1. Computational Fluid Dynamics (CFD)

In the past few decades, the advances in computing power have led to a significant progress in the application of two and three-dimensional computational fluid dynamics models in wind engineering and aeolian geomorphology [21,25,26]. In CFD models, the flow motion is solved numerically using the Navier–Stokes equations that are a set of partial differential equations including the conservation of mass, conservation of momentum in three dimensions and the conservation of energy. Considering the finite volume method, the computational domain is discretized into a finite number of control volumes and using numerical algorithms, the governing Navier–Stokes equations are integrated over all control volumes. This results in the conversion of partial differential equations into a set of algebraic equations before solving them [27].

For a systematic study of the effect of building dimensions on airflow, application of computational fluid dynamics offers considerable advantages over field measurements and wind-tunnel experiments. The main advantage is that the geometrical design and boundary conditions such as wind speed, incidence angle and shear velocity as well as surface roughness can be changed relatively quickly to systematically analyze the influence of an individual parameter on results. Moreover, the flow field can be solved in very small control volumes that enables the observation of detailed flow features. In addition, CFD simulations avoid scaling issues that might happen in wind-tunnel experiments as the geometrical design can be modelled exactly at the dimensions of interest. This facilitates the validation procedure, since the flow features at the same spatial scales can be compared in both numerical model and experimental results. Furthermore, CFD models avoid the impacts of walls in wind-tunnel experiments, using appropriate boundary conditions. This permits wind flow to leave the computational domain from the lateral sides and the outlet of the domain, and avoids the reflective impacts of walls [21,26,27]. A main disadvantage of CFD is that it can be computationally expensive when increasing the resolution of the computational mesh and/or the size of the computational domain. In practice, the required level of detail and the minimum required three-dimensional space to be simulated, put a limit to the number of cases that are feasible to simulate in a given study.

In this study, a numerical model is developed using OpenFOAM, which is an open-source CFD software. The details of the numerical model including governing equations, turbulence modelling, boundary conditions and initial internal fields are presented in Appendix A of this paper. The model validation is presented in Appendix B of this paper, which shows the capability of the numerical model in predicting the airflow patterns around an isolated building.

2.2. Computational Domain

A three-dimensional rectangular computational domain, shown in Figure 4, is considered for modelling airflow patterns around an isolated building. The definition of the geometric parameters shown in Figure 4, are given in Table 1. Essentially, the scale models of buildings in a numerical wind-tunnel without side wall effects are simulated. We study the impact of relative increases in each dimension, using scale model sized buildings. The dimensions of both the computational domain and the building are selected based on the wind-tunnel experiments performed by Leitl and Schatzmann [28], and their measurements are used in Appendix B of this paper for the model validation. It should be noted that using such a small-scale model in comparison with real buildings at the beach shows the capability of CFD in simulating scaled models. The computational domain inlet is located at $x = 0$ m, and the domain length, width and height are $\left(L''_u + L''_d\right) \times 2W'' \times H''$, respectively. A rectangular surface-mounted building with the length of l'', width of w'' and height of h'' is specified within the computational domain, where the building center is located at $x = 1$ m. The computational domain and the building dimensions are symmetric in the spanwise direction, z. The so-called blockMesh and refineMesh utilities in OpenFOAM are used to create structured hexahedral mesh over the computational domain.

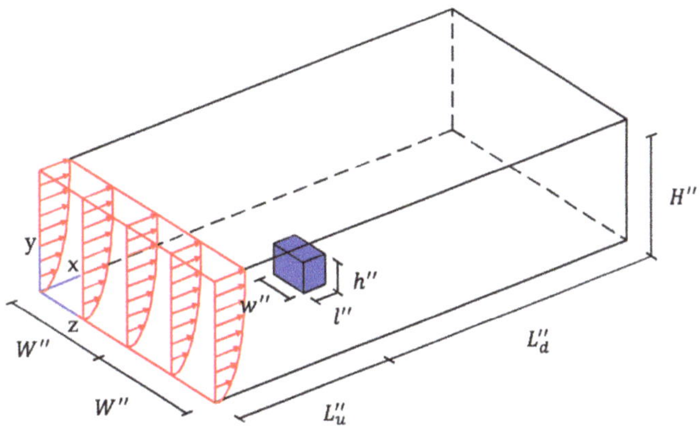

Figure 4. Schematic configuration of the computational domain and the surface-mounted building.

In this study, a computational domain with the length of 3 m, width of 2 m and height of 1.5 m is used. The reference building with the length of l''_0, width of w''_0, and height of h''_0, is considered within the domain. The atmospheric boundary layer parameters are chosen based on the wind-tunnel experiments performed by Leitl and Schatzmann [28] that are presented in Appendix A.3. Considering computational grids with a size almost equal to 0.0125 m and a height of the ground adjacent cells of 0.03 m, the total number of cells in the mesh is approximately 4.57 million. In order to systemati-cally study the impacts of building length, width and height, the reference building is increased in each direction up to three times the reference quantity, while the other two dimensions remain constant. This results in thirteen different simulations as specified in Table 2. It should be noted that the building center in all simulations is located at $x = 1$ m.

Table 1. Definition and values of the geometric parameters of the computational domain and the surface-mounted building.

Parameter	Definition
L''_u	Upstream distance between the domain inlet and the building centerline
L''_d	Downstream distance between the domain outlet and the building centerline
W''	Lateral distance between the lateral sides of the domain and the building centerline
H''	Height of the domain
l''	Length of the building
w''	Width of the building
h''	Height of the building

Table 2. An overview of the conducted simulations and the building dimensions in each case. The length, width and height of the computational domain is 3 m, 2 m and 1.5 m, respectively.

Simulation ID	Building Length (l'') [m]	Building Width (w'') [m]	Building Height (h'') [m]
	Reference building		
$l''_0 \times w''_0 \times h''_0$	0.1000	0.1500	0.1250
	Impact of building length		
$1.5l''_0 \times w''_0 \times h''_0$	0.1500	0.1500	0.1250
$2l''_0 \times w''_0 \times h''_0$	0.2000	0.1500	0.1250
$2.5l''_0 \times w''_0 \times h''_0$	0.2500	0.1500	0.1250
$3l''_0 \times w''_0 \times h''_0$	0.3000	0.1500	0.1250
	Impact of building width		
$l''_0 \times 1.5w''_0 \times h''_0$	0.1000	0.2250	0.1250
$l''_0 \times 2w''_0 \times h''_0$	0.1000	0.3000	0.1250
$l''_0 \times 2.5w''_0 \times h''_0$	0.1000	0.3750	0.1250
$l''_0 \times 3w''_0 \times h''_0$	0.1000	0.4500	0.1250
	Impact of building height		
$l''_0 \times w''_0 \times 1.5h''_0$	0.1000	0.1500	0.1875
$l''_0 \times w''_0 \times 2h''_0$	0.1000	0.1500	0.2500
$l''_0 \times w''_0 \times 2.5h''_0$	0.1000	0.1500	0.3125
$l''_0 \times w''_0 \times 3h''_0$	0.1000	0.1500	0.3750

2.3. Methodology for Deriving Bed Level Change from Airflow Patterns

In this study, we are interested in predicting the potential impact of airflow patterns around buildings on the bed level changes of the surrounding area when that area consists of moveable substrate. Commonly used sediment transport models show that the sediment transport rate, q, is proportional to the third-order velocity field ($q \propto \vec{U}^3$) [29–38]. In this study, it is assumed that the sediment will transport at the near-bed wind speed, and it will stay close to the bed. Therefore, as a first step, the vertical component of the velocity field can be neglected, and the sediment transport rate can be written in the following form:

$$q \propto \left(\left| \vec{U_H} \right|^2 \vec{U_H} \right) \tag{1}$$

where the index H denotes the horizontal near-surface velocity field. The Exner equation states that the temporal rate of change in bed elevation is proportional to the convergence of sediment transport rate:

$$\frac{\partial z_b}{\partial t} \propto -\nabla \cdot q \tag{2}$$

where z_b [m] is the bed elevation, t [s] is the time, and q [kg/m ∗ s] is the sediment transport rate. Substituting Equation (1) into Equation (2) gives:

$$\frac{\partial z_b}{\partial t} \propto -\nabla \cdot \left(\left| \vec{U_H} \right|^2 \vec{U_H} \right) \quad (3)$$

Considering Equation (3), a positive convergence of the third-order horizontal wind velocity field in a near-surface plane implies a decrease in sand transport rate hence deposition. Similarly, a negative convergence of the third-order horizontal wind velocity field in a near-surface plane implies an increase in sand transport rate hence erosion:

$$\begin{cases} -\nabla \cdot \left(\left| \vec{U_H} \right|^2 \vec{U_H} \right) > 0 & \rightarrow \quad \textit{Deposition} \\ -\nabla \cdot \left(\left| \vec{U_H} \right|^2 \vec{U_H} \right) < 0 & \rightarrow \quad \textit{Erosion} \end{cases} \quad (4)$$

3. Results

3.1. Near-Surface Airflow Pattern

The impacts of building dimensions on airflow patterns near the beach surface and the potential implications for bed morphology are investigated. We focus on investigating the impacts of building length, width and height on near-surface airflow patterns, as our main motivation for this work is its implication for sediment transport. Therefore, we show the results of wind velocity magnitude at a horizontal plane close to the bed, i.e., y = 0.0125 m, which is located at an elevation equal to ten percent of the reference building height.

The impact of building length parallel to the incidence wind direction on near-surface wind velocity magnitude is presented in Figure 5a–e. The first glance into the results shows that the building length does not have significant impact on the near-surface airflow patterns adjacent to the building. In order to take a deeper look into results, the effect of building length on the length of the downwind recirculation region just behind the building, L_d, which is defined as the distance between the reattachment point of the separated flow and the leeward face of the building, is shown in Figure 5f. It should be noted that the reattachment point at a near-surface plane, y = 0.0125 m, is located on the centerline of the computational domain, where the horizontal component of the velocity, u, changes in sign.

As shown in Figure 5f, the length of the recirculation zone just downstream of the building rear face decreases with increasing building length. This happened due to the reattachment of the detached flow on the roof of the longer building. In order to understand to what extent the velocity deficits due to the presence of the buildings with different lengths continue downwind of the reattachment point, Figure 5g shows the changes in wind velocity magnitude along the domain centerline with respect to the distance from the reattachment point. According to Figure 5g, the wind velocity magnitude increases gradually with increasing distance from the reattachment point until it eventually reaches the undisturbed wind velocity magnitude. Furthermore, the results show that it takes a bit longer distance for the wind to reach a certain speed for the longer building.

Figure 5h shows how streamwise and vertical velocity components change along the domain centerline as the wind approaches the windward face of the building. Figure 5h shows that the near-surface streamwise wind velocity generally decreases with decreasing distance from the windward face of the building, where it changes more rapidly when the distance from the building front face is smaller. The vertical wind velocity is approximately zero until the wind is close the building's windward face, where a local peak occurs due to the small recirculation region that forms in front of the building and close to the surface. This recirculation develops due to downward deflection of the flow along the front face of the building to the ground, where it deflects again leading to near-bed flow in the opposite direction of the incident wind direction. The results further show that building length has no influence on the length of the upstream area with reduced streamwise wind velocities.

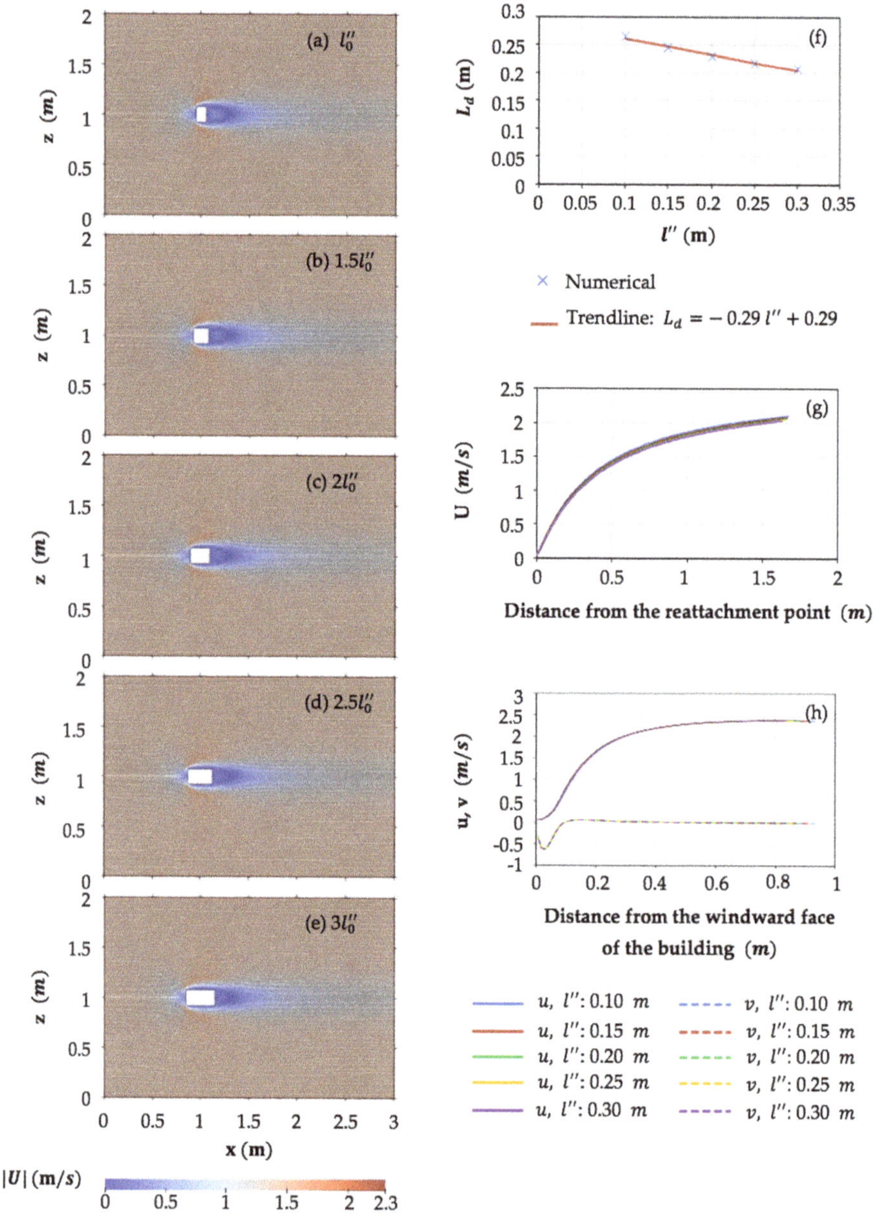

Figure 5. The impact of building length on (**a**–**e**) wind velocity magnitude (color) and direction (streamlines), (**f**) length of the downwind recirculation region, L_d, just behind the leeward face of the building along the centerline, (**g**) wind velocity magnitude downstream of the building after flow reattachment along the centerline, and (**h**) streamwise, u, and vertical wind velocity components, v, upstream of the building along the centerline. Results are derived for a horizontal near-surface plane, $y = 0.0125$ m. The building width and height are w''_0 and h''_0, respectively. The building length, l'', is varied as l''_0, $1.5l''_0$, $2l''_0$, $2.5l''_0$, and $3l''_0$.

The impact of building width on near-surface wind velocity magnitude is presented in Figure 6a–e. The general flow patterns show that the wider building disturbs a longer and wider area both in front of the windward face and behind the leeward face of the building. Furthermore, the wind velocity magnitude of the flow passing around the windward edges and the lateral sides of the building increases with increasing building width. The reason is that the wind flow approaching the front face of the building is separated into two flow branches in the horizontal plane, passing around the sidewalls of the building. The pressure gradient between the point of separation, in the middle of the building width, and the upwind edges of the wider building is greater, causing the higher wind velocity magnitude values around the windward edges and the lateral sides of the building.

Figure 6f shows the effect of building width on the length of the recirculation region just behind the leeward face of the building, L_d. The comparisons between the five different building widths show that the wider building causes the formation of a longer recirculation region just downstream of the building. It can be seen that there is a linear relation between the width of the building and the length of the downwind recirculation zone. The steep slope of the trendline shows that the length of the recirculation zone is highly sensitive to the building width. The results of wind velocity magnitude downstream of the flow reattachment point presented in Figure 6g show that behind a wider building the near surface wind velocity magnitude recovers more slowly over distance from the velocity deficit at the flow reattachment point, where near surface wind velocity magnitude is almost zero.

The changes in streamwise and vertical wind velocities upstream of the windward face of the building are presented in Figure 6h. The results show that the wider the building, the further upwind of the building the minimums of the wind velocity components occur, meaning that the size of the near-bed recirculation region in front of the building increases with increasing building width. Furthermore, it can be concluded from the figure that the streamwise wind velocity deficit for the wider building continue for a longer distance upstream of the building. However, the rate of change decreases with increasing building width. The streamwise wind velocity reaches the undisturbed wind field far away from the building. The negative streamwise wind velocity shows the reversed flow, which depends on the size of the recirculation region in front of the building, and the elevation at which the results were plotted ($y = 0.0125$ m in this study).

The impact of building height on near-surface wind velocity magnitude is presented in Figure 7a–e. The overall flow patterns show more substantial disturbance downstream of the building than upstream of the building. Furthermore, the wind velocity magnitude around the upwind edges and the lateral sides of the building increase considerably with increasing building height. It is obvious that the increase in the near-surface wind velocity magnitude is greater when the building is getting higher in comparison to getting wider. This can be explained by both the pressure gradient and the friction effects that dissipate higher amounts of kinetic energy of the wind flow when passing around the wider buildings. For the wider building, the near-surface flow approaching the building and deflecting towards the lateral sides, travels a longer distance towards the flow detachment at the windward corners of the building compared to the higher building. Therefore, frictional effects act over a longer distance and dissipate higher amounts of kinetic energy of the airflow which, in turn, result in lower wind velocity magnitude around the lateral sides of the wider building. Figure 7f shows that the taller building creates two longer counter-rotating vortices, therefore a longer recirculation region downstream of the building, L_d.

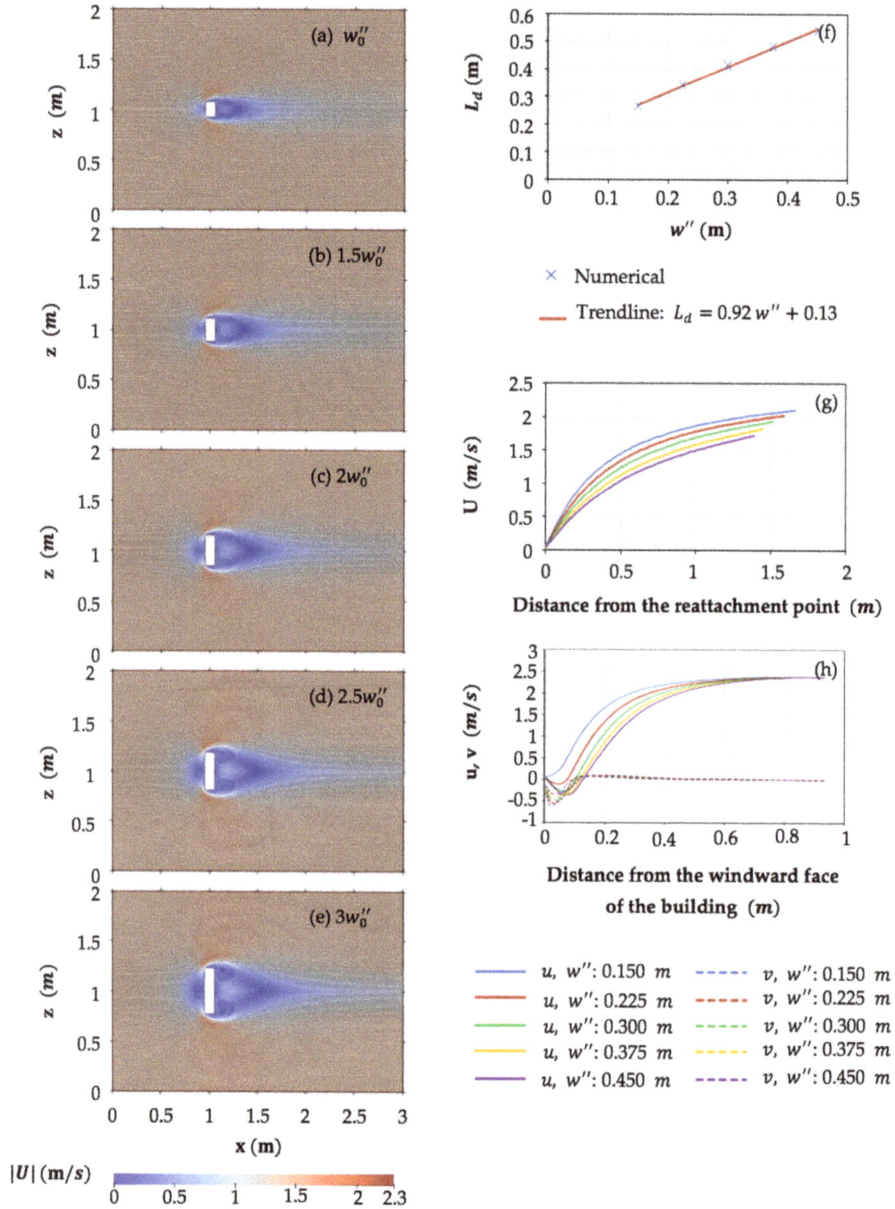

Figure 6. The impact of building width on (**a–e**) wind velocity magnitude (color) and direction (streamlines), (**f**) length of the downwind recirculation region, L_d, just behind the leeward face of the building along the centerline, (**g**) wind velocity magnitude downstream of the building after flow reattachment along the centerline, and (**h**) streamwise, u, and vertical wind velocity components, v, upstream of the building along the centerline. Results are derived for a horizontal near-surface plane, $y = 0.0125$ m. The building length and height are l_0'' and h_0'', respectively. The building width, w'', is varied as w_0'', $1.5w_0''$, $2w_0''$, $2.5w_0''$, and $3w_0''$.

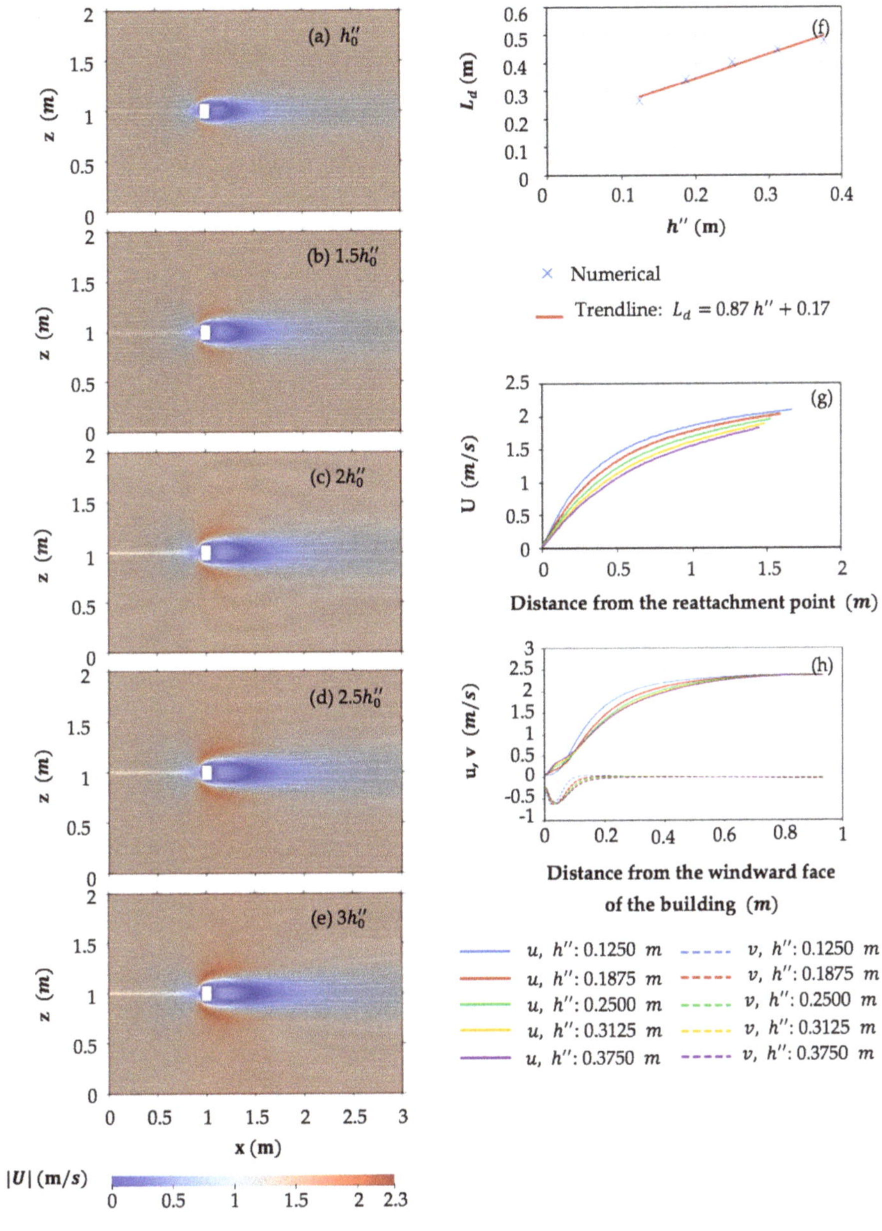

Figure 7. The impact of building height on (**a–e**) wind velocity magnitude (color) and direction (streamlines), (**f**) length of the downwind recirculation region, L_d, just behind the leeward face of the building along the centerline, (**g**) wind velocity magnitude downstream of the building after flow reattachment along the centerline, and (**h**) wind velocity magnitude, streamwise, u, and vertical velocity components, v, upstream of the building along the centerline. Results are derived for a horizontal near-surface plane, $y = 0.0125$ m. The building length and width are l_0'' and w_0'', respectively. The building height, h'', is varied as h_0'', $1.5h_0''$, $2h_0''$, $2.5h_0''$, and $3h_0''$.

The comparisons between five buildings with different heights (Figure 7f) show the high sensitivity of the length of the downwind recirculation zone to building height, however the slope of the trendline shows that the impact of building height on the length of the downwind recirculation zone is smaller than that of building width. As shown in Figure 7g, the influence of building height on the extension of the velocity deficits downwind of the flow reattachment point follows the same pattern as the building width, meaning that the velocity deficits continue for a longer distance from the reattachment point for the taller building. However, the rate of wind velocity magnitude increase over distance is smaller than that for buildings with different widths.

An effect of building height on the streamwise and vertical wind velocities upwind of the building is that a taller building creates the local wind velocity magnitude peak at a larger distance from the windward face of the building (Figure 7h). In addition, Figure 7h shows that the wind velocity magnitude at the center of the recirculation region formed in front of the windward face of the taller building is highest. For the taller building, the wind flow can be less easily released over the building. Therefore, a larger portion of the wind flow moves downwards along the windward face of the building and towards the lateral sides of the building. This implies winds at higher elevation, with higher amount of kinetic energy, move downward. The high-speed downward directed airflow causes the formation of the recirculation region with high-speed magnitudes in front of the taller building. It can be understood from the figure that the length of the low-speed region in front of the building increases with increasing building height, however the rate of change decreases when the building height increases. It is also noteworthy that the magnitude of the local minimum in near-surface windspeed increases with building height to such an extent that there is hardly a difference for the tallest building. The reason is that the flow can move over the smaller building, while more flow blockage happens when the building height increases. This might increase the size of the circulation region in front of the building, and the local minimum in near-surface windspeed.

3.2. Impacts of Building Dimensions on Initial Bed Level Change

3.2.1. Convergence of the Third-Order Horizontal Near-Surface Flow Field as a Proxy for Initial Bed Level Change

In order to validate the assumption that the convergence of the third-order horizontal near-surface flow field is a good proxy for initial bed level change, and to study the impact of scaling on erosion and deposition patterns around buildings, two new airflow simulations were made. The first simulation is a full-scale building with the length, width and height of 12 m, 2.5 m and 2.5 m, respectively, that is exactly with the same dimension as the full-scale building tested at the Noordwijk beach. It should be noted that he inflow boundary condition for the large-scale simulation are the same as the smaller scale simulations (Appendix A.3). The second simulation is a small-scale building with the length, width and height of 0.72 m, 0.15 m and 0.15 m, respectively. It should be noted that these dimensions were selected in a way to reproduce the same frontal and lateral aspect ratios as the full-scale building tested at the Noordwijk beach. Both full-scale and small-scale numerical model results for the bed level rate of change derived from the convergence of the third-order horizontal wind velocity field at near-surface planes are compared qualitatively to field observations of erosion and deposition patterns around experimental models at the beach (Figures 8 and 9). The full-scale model at the beach near Noordwijk in the Netherlands consists of two shipping containers that were placed besides each other with the total size of 12 m × 2.5 m × 2.5 m in length, width and height, respectively. The containers were placed alongshore and parallel to the dunes with a distance of 20 m from the dune foot. The dominant wind direction was parallel or in a small angle with model's centerline. A more extensive description of the experimental models at the beach can be found in Poppema et al. [18].

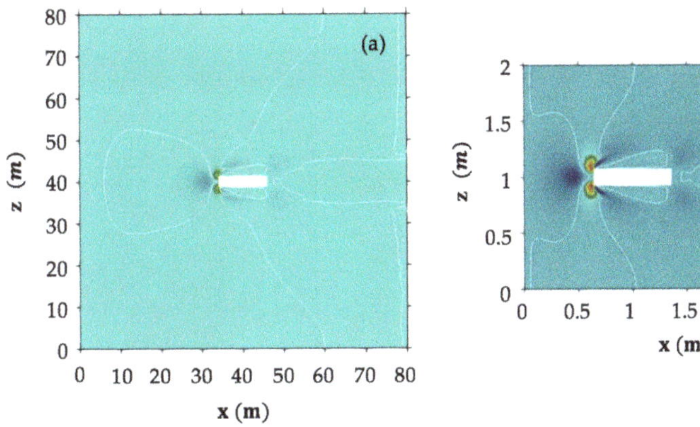

Figure 8. Numerical model predictions of the erosion and deposition patterns inferred from the convergence of the third-order horizontal wind velocity field at a horizontal near-surface plane, (**a**) $y = 0.25$ m for the full-scale building, and (**b**) $y = 0.015$ m for the small-scale building. The wind direction is from the left and perpendicular to the front face of the buildings. The white lines are zero contours.

As shown by the yellow to red shaded colors in Figure 8 (both small-scale and full-scale buildings), the highest negative values of the convergence of the third-order horizontal near-surface velocity field occur in a small area around the upwind corners of the building, indicating this location is prone to erosion in case of a moveable bed and the most intensive amount of erosion is expected to happen there. Figure 9a–d show erosion undercutting around the upwind edges of the full-scale container at beach, which is the same pattern as what is predicted based on the numerical model results. Note that the upwind scour zone in front of the building is more strongly developed in the field observations than in the calculated patterns. This might be an effect of topographic feedback, which is absent in the calculations of initial bed level change. The numerical results predict a deposition region upstream of the building (blue shaded colors). Furthermore, two deposition tails starting from some distance away from the lateral sides of the building to downstream of the building form. The field observations given in Figure 9c–f show the same deposition region with lighter colored sand upwind of buildings, which is accompanied with two tails that are formed at some distance from the sidewalls of the building and extended to some extent downstream of the building. It is also seen in both the numerical model results and field observations that erosion happens directly along the lateral sides of the building, bounded on the outside by the inner edges of the deposition tails. The above comparisons show that there is a quite good qualitative agreement between observed and modeled erosion and deposition patterns. This provides support for our assumption that the convergence of the third-order horizontal wind velocity field at a near-surface plane is a suitable proxy for predicting bed level change. Furthermore, the small-scale and large-scale numerical simulations qualitatively show the same patterns of bed morphology, indicating that the overall erosion and deposition patterns are not affected by the scale of the simulation. Therefore, the results of this study are applicable for full-scale buildings at the beach. It is also noteworthy that the erosion and deposition patterns both in the model as well as in the field develop slower for the larger scale buildings. Furthermore, the actual rates of erosion and deposition around buildings depend on the proportionality coefficient in Equation (1). It should be noted that the shape and the dimension of the erosion and deposition patterns around buildings do not depend on the proportionality coefficient.

Figure 9. Field observations of erosion and deposition patterns around a (**a**–**d**) full-scale model at Noordwijk beach, and (**e**,**f**) small-scale models at De Zandmotor beach, the Netherlands. The arrows show the wind directions. (**c**,**e**,**f** were obtained from Poppema et al. [18]).

3.2.2. Relation between Building Dimensions and Patterns of Wind-Driven Bed Level Change

The results of the convergence of the third-order horizontal wind velocity field at a near-surface plane are derived for the previously mentioned simulations specified in Table 2. The impact of building length on bed morphology is investigated using Figure 10. The two deposition tails that form at some distance from the lateral sides of the building to downstream increase in length, while the deposition rate decreases with increasing building length (see the blue-shaded colors with positive convergence). Furthermore, the eroding regions that occur between the lateral sides of the building and the inner boundary of the deposition tails increase in size when the building length increases (see the zero contour lines). The overall results show that the impact of building length on the convergence pattern of the third-order horizontal near-surface velocity field around the building is small. Therefore, it is expected that the magnitude of the building length does not have a large effect on the bed morphology that develops around the building.

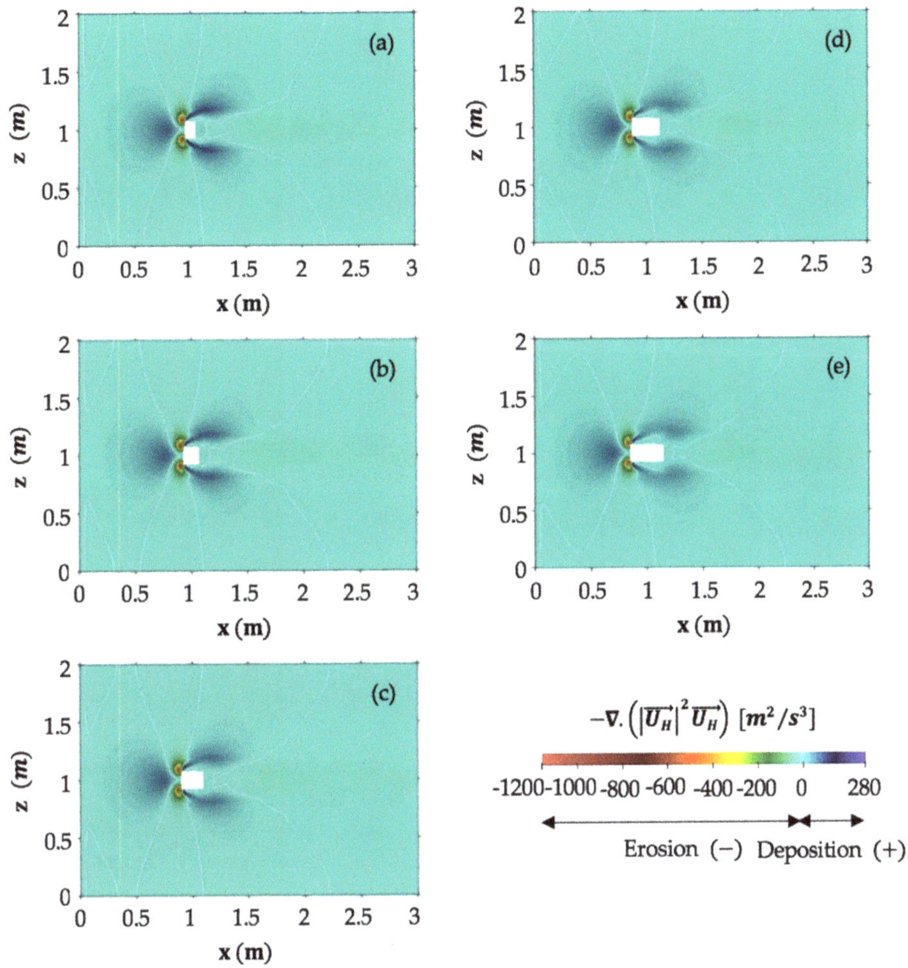

Figure 10. The impact of building length on erosion and deposition patterns inferred from the convergence of the third-order horizontal wind velocity field at a near-surface plane, $y = 0.0125$ m. The building width and height are w''_0 and h''_0, respectively. The building length, l'', is varied as (**a**) l''_0, (**b**) $1.5l''_0$, (**c**) $2l''_0$, (**d**) $2.5l''_0$, and (**e**) $3l''_0$.

Figure 11 shows the impact of building width on bed morphology. The figure shows that for all five tested building widths, the most intensive erosion is expected to happen around the upwind edges of the building. The results show that when the building width increases, the spatial extent of the deposition region in front of the upwind face of the building increases too, and the two deposition tails become slightly longer and wider. The rate of deposition however slightly decreases as the building becomes broader, both in front of the building and in the deposition tails. In addition, the deposition rate just behind the leeward face of the building slightly increases as the building width increases (see the small areas with darker blue-shaded colors just behind the building). It is also noteworthy that initially the erosion is maximum at the centerline behind the building. However, as the building gets wider, the maximum erosion splits and the erosion rate at the centerline reduces in magnitude and regions of deposition start to form.

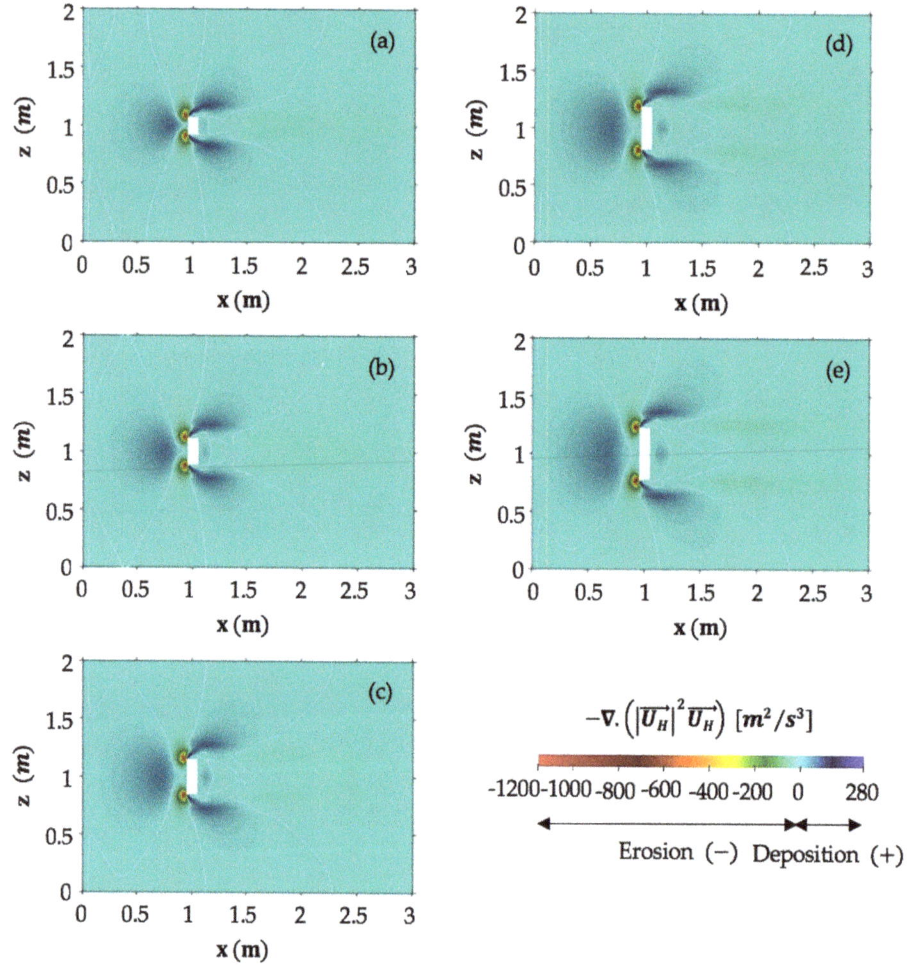

Figure 11. The Impact of building width on erosion and deposition patterns inferred from the convergence of the third-order horizontal wind velocity field at a near-surface plane, $y = 0.0125$ m. The building length and height are l''_0 and h''_0, respectively. The building width, w'', is varied as (**a**) w''_0, (**b**) $1.5w''_0$, (**c**) $2w''_0$, (**d**) $2.5w''_0$, and (**e**) $3w''_0$.

The impact of building height on bed morphology is investigated using Figure 12. Results show that the areal extent of the upwind erosion around the windward edges of the building increases with increasing building height. Furthermore, it is revealed that an increase in building height leads to a substantial reduction in the deposition rate upwind of the building. However, the area with deposition (darker blue-shaded colors) in front of the building increases in spatial extent with increasing building height. The downstream and lateral extension of the two deposition tails increases when the building becomes taller. These implications can be explained by Figure 7a–e, where the higher wind velocity magnitudes occur around the upwind edges and lateral sides of the taller building. This higher-speed wind flow erodes more sediment particles upwind the building and carries them for a longer distance downstream of the building. Therefore, it is probable that when the building increases in height, a shift occurs from mainly sedimentation in front of the building to mainly sedimentation in the tails. In addition, Figure 12 shows that the rate of sand accumulation in a small area just behind the leeward face of the building slightly increases when the building height increases.

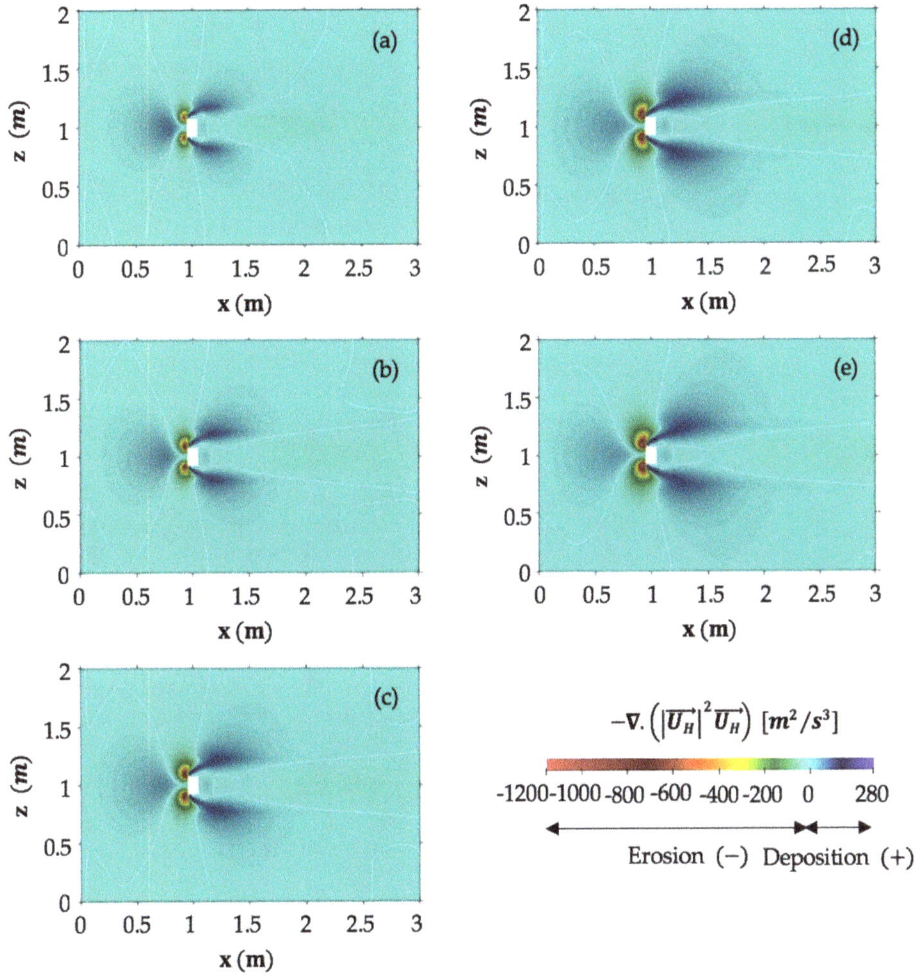

Figure 12. The Impact of building height on erosion and deposition patterns inferred from the convergence of the third-order horizontal wind velocity field at a near-surface plane, $y = 0.0125$ m. The building length and width are l''_0 and w''_0, respectively. The building height, h'', is varied as h''_0, $1.5h''_0$, $2h''_0$, $2.5h''_0$, and $3h''_0$.

4. Discussion

In the present study, a CFD model using OpenFOAM was developed to investigate the impacts of building dimensions, specifically building length, width and height on near-surface airflow patterns and bed morphology. The numerical results of the flow field around the building were consistent with the observed flow patterns by Martinuzzi and Tropea [11] and Leitl and Schatzmann [28]. Similar to the observations by Martinuzzi and Tropea [11], we found that the size of the near-bed recirculation region in front of the building increases with increasing building width. This can be realized by comparing the length of the local peaks shown in Figure 6h. Furthermore, the results of the present study shown in Figure 6f, indicates that the downstream reattachment length increases linearly with building width, which is consistent with the findings by Fackrell [9] and Martinuzzi and Tropea [11]. It should be noted that these findings are based on tested simulations with certain range of width-to-height aspect ratio up to about four, $w''/h'' \approx 4$.

For the bed surface of the numerical domain, representing a sandy bed, a uniform aerodynamic roughness length, y_0, was assumed based on the wind-tunnel experiments performed by Leitl and Schatzmann [28]. This assumption leads to some inconsistencies with the real condition on the beach that can be considered as the model limitations. On dry enough parts of the beach, sand ripples can form over time under windy conditions, changing the bed roughness and therefore the near-surface airflow changes. However, the good agreements between the model predictions of the erosion and deposition patterns with field observations indicate that ripples do not affect the overall patterns, but they might only modify the spatial extents.

The convergence of the third-order horizontal near-surface wind velocity field was used in this study as a proxy for initial bed level change, because in commonly used sediment transport models the sediment transport rate is assumed to be proportional to the third-order velocity field. Strictly speaking, this assumption is only valid for a situation with transport limited conditions; however, on the beach supply limited conditions also occur due to the effects of moisture [39–41] which may affect the rate at which bed level changes develop. Moisture may affect the amount of sediment in transport approaching the building. This may specifically affect the rate at which deposition patterns around a building develop. In the extreme condition, if no sediment is in transport at all due to very wet conditions, no deposition patterns can develop because there is no sediment in the airflow. Furthermore, moist beach surface around the building itself may affect the rate at which building induced erosion occurs, hence it takes more time for erosive features to develop. In addition, in this study, the threshold wind speed was not taken into account. It should be noted that if the wind speed becomes less than the threshold wind speed at which the sediment particles start moving, no sediment transport will happen. Figure 13 shows the results of the convergence of the third-order horizontal wind velocity field at a near-surface plane when the wind approaches the building at higher wind velocity magnitude, $u_{ref} = 17$ m/s (compared to $u_{ref} = 6$ m/s in Figure 8a). The results show that at higher wind speeds where the majority of the domain is well above the threshold wind speed, the similar erosion and deposition patterns develop. It should be noted that the rate of development of patterns increases with increasing the wind velocity magnitude (compared to Figure 8a).

McKenna Neuman and Bédard [42] showed that the fluid perturbation and the system of vortices that develop around buildings depend on the integration between both buildings and the bed surface. In the present study, we used steady airflow patterns around buildings to infer initial effects on bed morphology, hence morphological feedback was not taken into account in this approach. Nevertheless, the erosion and deposition patterns predicted by the numerical model showed good agreement with field observations around the full-scale model and the small-scale model at the beach. The model results showed a considerable erosion around the windward edges of the building, extending less intensively than at the edges along the lateral sides. In addition, the deposition region in front of the upwind face of the building and two deposition tails starting away from the lateral sides of the building and extending downstream of the building occur in all simulations. These findings are similar to the observations by Iversen et al. [12,13], Tominaga et al. [14] and Poppema et al. [18].

To consider the impact of scaling on numerical results shown in this study, the erosion and deposition patterns between a full-scale model and a small-scale model at the beach were compared qualitatively. The comparisons showed that the erosion and deposition patterns that develop around the buildings are not influenced when the scale of the simulation increases. However, the rate of growth of the patterns decreases with increasing the scale of the simulation. This is also valid for small-scale and full-scale experimental models at the beach.

The initial morphological changes predicted by the model show a central downwind deposition just behind the leeward face of the building starts to appear when the building width increases (see Figure 11). The reason is that two horizontal counter-rotating vortices become stronger as the building width increases (see Figure 6a–e). These two vortices push the air towards the centerline behind the building. The air gets pushed upward resulting in an upward component of the wind. On the other hand, the vertical recirculation just behind the building generates a downward motion at the position above the deposition region. For wider buildings, the upward effect of the two horizontal vortices becomes stronger than the downward effect of the vertical vortex which leads to a net upward flow. The net upward flow results in positive convergence of third-order horizontal velocity field at a near-surface plane. Therefore, the erosion and deposition patterns show the formation of a deposition region just behind the leeward face of the wider building.

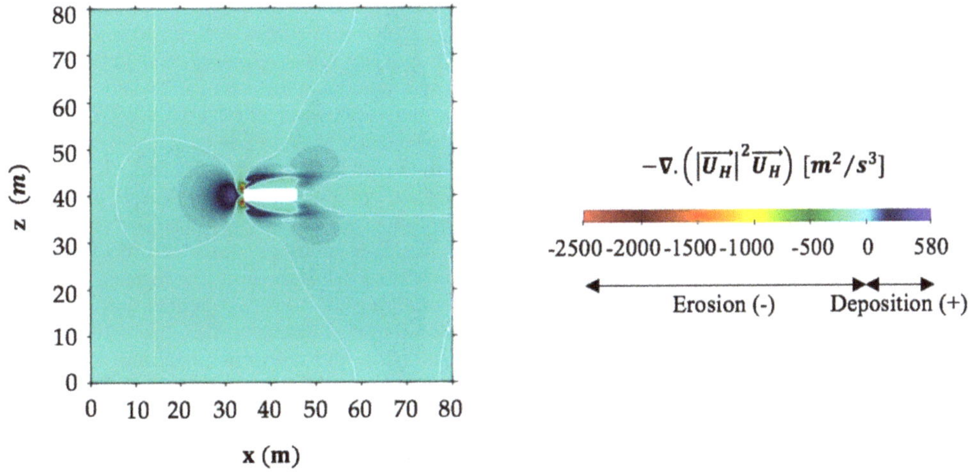

Figure 13. Numerical model predictions of the erosion and deposition patterns inferred from the convergence of the third-order horizontal wind velocity field at a horizontal near-surface plane, $y = 0.25$ m, around the full-scale building. The reference wind velocity magnitude is 17 m/s at a reference height of 1.8 m. The wind direction is from the left and perpendicular to the front face of the building. The white lines are zero contours.

In this paper we studied buildings on a flat sand surface, whereas on an actual beach buildings are often built in front of a dune. Although dune topography was not included in the simulations, some preliminary implications of building dimensions for sand supply from the beach to the dunes may be formulated from the presented results. The convergence patterns in the near-surface flow field induced by the building demonstrate that ambient sand transport will be captured by the building, both upwind and in two downwind tails. The rate of deposition in the two tails varies with building dimension. If we assume these trends prevail also in the presence of dune topography downwind of the building, such higher rate of deposition would mean an increasingly higher deposition on the dune front at the tail location. From our results of the deposition patterns downstream of the buildings, it can then be derived that the highest building is expected to give the strongest local increase in deposition at the dune front and the longest building the least. Note that this holds for situations with predominantly onshore wind. The ridge-like deposition in front of the wider building increases in size as the building becomes broader. In addition, the eroding region around the windward edges of the building increases substantially with increasing the building height. These might lead to buildings' mis-functioning and probably tilting that forces the owners of beach buildings to consider measures to smooth the surface or prevent the development of these patterns.

5. Conclusions

In this study, the impacts of building dimensions on near-surface airflow patterns were investigated as well as the implications for bed morphology in the case of a surrounding sandy bed. Specifically, three building characteristics were studied: the building length parallel to the incident wind direction, building width perpendicular to the incidence wind direction and building height. For this purpose, a CFD model using OpenFOAM was developed. The numerical model predictions showed satisfactory agreement with wind-tunnel data of vertical and horizontal wind velocity profiles in the vicinity of the scale model of buildings, providing confidence in the capability of the model to predict the detailed airflow patterns around an isolated building at the beach.

Using this model, a systematic investigation revealed the effects of building length, width and height on airflow patterns at a horizontal plane close to the bed. The results on the relation between building dimensions and near-surface airflow patterns were consistent with those found in Martinuzzi and Tropea [11] as well as Iversen et al. [12,13], where width and height of the building were most influential in the extent and nature of near-surface airflow patterns around the building and downwind length least. Only the length of the recirculation region just behind the leeward face of the building decreases slightly when the building increases in length. This was consistent with findings in Fackrell [9]. By focusing on near-surface flows both in front of the building and downstream of the building, our simulations further revealed that a wider building disturbs a longer and wider region both in front of the building and downstream of the building. The distance at which the upwind deceleration of the airflow starts as it approaches the building increases with building width. The numerical results presented in this study highlighted that with increasing building height, the length of the two counter-rotating vortices just behind the leeward face of the building increases considerably, and with it the length of the recirculation region. Furthermore, with taller buildings it takes a longer distance for the near-surface wind leeward of the building to increase its speed back to the undisturbed wind velocity magnitude.

The convergence of the third-order horizontal near-surface wind velocity field was used as a proxy for sediment transport rate. Although inferred erosion and deposition patterns technically only relate to initial bed level changes, it was found that these compared well to those observed around a full-scale model and a small-scale model at Noordwijk beach and De Zandmotor beach in the Netherlands (Poppema et al. [18]).

As for the near-surface airflow patterns, the initial bed morphology was most dependent on the building width normal to the incidence wind direction as well as the building height, and least on the length of the building parallel to the incidence wind direction. In addition, it was found that for all studied building dimensions, the most intensive erosion is expected to happen around the upwind edges of the building, where the undercutting was observed in the field experiments. The initial bed morphology revealed that the areal extent of these eroding regions directly adjacent to the windward edges of the building increases when the building becomes taller. By focusing on the initial deposition patterns around buildings, the numerical simulations highlighted that the deposition tails downstream of the buildings develop more slowly as the building length increases. Furthermore, sediment deposition that occurs in front of the upwind face of the building becomes larger in spatial extent, and the length of the two deposition tails downstream of the building slightly increases as the building becomes broader. The deposition rate both in the ridge-like deposition in front of the building and in the deposition tails slightly decreases with increasing building width. This implies that the wider the building, the broader and shallower the deposition region in front of the windward face of the building. A small area of sand accumulation appears just behind the leeward face of the building as the building width increases. The numerical results further revealed that with increasing building height the sedimentation rate further upwind of the building decreases, while it increases in the downwind tails.

Author Contributions: Conceptualization, P.P., G.H.P.C. and S.J.M.H.H.; methodology, P.P., G.H.P.C. and S.J.M.H.H.; software, P.P.; validation, P.P.; formal analysis, P.P.; investigation, P.P.; writing—original draft preparation, P.P.; writing—review and editing P.P., G.H.P.C., K.M.W. and S.J.M.H.H.; visualization, P.P.; supervision, G.H.P.C., K.M.W. and S.J.M.H.H.; project administration, K.M.W. and S.J.M.H.H.; funding acquisition, K.M.W. All authors have read and agreed to the published version of the manuscript.

Funding: This research was funded by Netherlands Organisation for Scientific Research (NWO) (contract number ALWTW.2016.036) and co-funded by Rijkswaterstaat (RWS) and Hoogheemraadscap Hollands Noorderkwartier (HHNK).

Institutional Review Board Statement: Not applicable.

Informed Consent Statement: Not applicable.

Data Availability Statement: The study did not report any data.

Acknowledgments: This research is part of the ShoreScape project, which is a joint research project of the University of Twente and Delft University of Technology. ShoreScape focuses on sustainable co-evolution of the natural and built environment along sandy shores. We would like to thank Sander Vos for inviting us in the Noordwijk experiments and Ir. Daan Poppema for discussions and the photo of his scale model experiments at the Zandmotor.

Conflicts of Interest: The authors declare no conflict of interest.

Appendix A. Development of the Numerical Model

Appendix A.1. Governing Equations

In this study, a three-dimensional numerical model is developed using OpenFOAM. The wind flow in this model is considered as incompressible, since for the flow with Mach numbers less than 0.3, the change in flow density is negligible and the divergence of flow velocity can be considered zero [43]. The Mach number is a dimensionless parameter, defined as the ratio of the speed of flow to the speed of sound in the surrounding flow [41]. The Mach number of 0.3 corresponds to the airflow with the speed of about 100 m/s in its normal condition, which is typically greater than the wind speed [44]. In addition, considering the Reynolds number criteria, the wind flow in this study is assumed fully turbulent. Therefore, among all available OpenFOAM solvers for incompressible flows, the so-called simpleFoam solver is selected, which is recommended for steady state simulation of turbulent flows.

The simpleFoam algorithm solves the Reynolds-averaged Navier–Stokes (RANS) equations for turbulent flows using the finite volume method (FVM). The RANS method employs the Reynolds decomposition of flow variables into mean and fluctuating parts, and solves the additional Reynolds stresses that appear in the momentum equations of flow using a turbulence model [27,43,45]. Considering that the conservation of energy is only applied for compressible flows, the governing equations of the three-dimensional flow field can be expressed by continuity and momentum equations that can be written in their steady states as follows:

$$\vec{\nabla} \cdot \vec{U} = 0 \tag{A1}$$

$$\vec{\nabla} \cdot \left(\vec{U} \otimes \vec{U}\right) = -\vec{\nabla} p_k + \vec{\nabla} \cdot \left(v_{eff} \vec{\nabla} \vec{U}\right) \tag{A2}$$

where $\vec{U} = (u,\ v,\ w)$ $[\mathrm{m/s}]$ is the three-dimensional flow velocity vector; p_k $[\mathrm{m^2/s^2}]$ is the kinematic pressure, defined as the ratio of the static pressure, p_s $[\mathrm{Pa}]$, to the flow density, ρ $[\mathrm{kg/m^3}]$; v_{eff} $[\mathrm{m^2/s}]$ is the effective kinematic viscosity, defined as the sum of the kinematic viscosity of the flow, v $[\mathrm{m^2/s}]$, and the turbulent (eddy) kinematic viscosity, v_t $[\mathrm{m^2/s}]$, which is calculated from the turbulence model. In the present study, the kinematic viscosity of the air, v, is considered as 1.4×10^{-5} $\mathrm{m^2/s}$.

Although Equations (A1) and (A2) introduce 4 equations and 4 unknowns $(u,\ v,\ w,\ p_k)$, they are difficult to be solved numerically. The main reason is that there is an equation for each component of the flow velocity, but there is no equation for the pressure. It should be noted that the continuity equation plays a role of restriction for velocity components, meaning that the computed velocity components from Equation (A2) must satisfy the Equation (A1). In addition, the wind flow in this study is considered as incompressible, therefore the equation of state cannot be used to compute the pressure, as the flow density and temperature are assumed constant. The next reason is that the convective term in the momentum equation, $\vec{\nabla} \cdot \left(\vec{U} \otimes \vec{U}\right)$, is non-linear which makes the solution of abovementioned equations more complex. The simpleFoam solver uses the SIMPLE (semi-implicit method for pressure-linked equations) algorithm to solve the continuity and momentum equations.

SIMPLE Algorithm

The SIMPLE algorithm uses an iterative procedure, where the intermediate velocity field is first calculated by solving the momentum equation for an initial guessed pressure field. The intermediate velocity field does not satisfy the continuity restriction. Deriving and solving the pressure equation, the so-called Poisson equation, from the continuity and momentum equations result in a pressure corrector that is used for adjusting the intermediate velocity field and obtaining a new velocity field, that satisfies the continuity restriction after doing several iterations [27,45,46].

The first step to initiate the SIMPLE approach is to express the momentum equation, Equation (A2), in the matrix form as follows:

$$A\vec{U} = -\vec{\nabla} p_k \tag{A3}$$

where A is the known matrix of coefficients that its components are calculated using the discretization of the partial differential terms in the momentum equation. Note that the nonlinearity of the advection terms will involve some kind of linearization method. The above equation includes $3n$ equations, where n is the total number of control volumes and each equation is written for the centroid of each control volume. The second step is to decompose the matrix of coefficients, A, into diagonal and non-diagonal components:

$$A\vec{U} = B\vec{U} - C \tag{A4}$$

where B and C are the known diagonal and non-diagonal components of matrix A, respectively. Substituting Equation (A4) into Equation (A3) and rearranging the momentum equation in terms of the flow velocity yields the following equation:

$$\vec{U} = B^{-1}C - B^{-1}\vec{\nabla} p_k \tag{A5}$$

The third step is to derive the pressure equation, Poisson equation, by substituting the above equation into the continuity equation, Equation (A1), that can be written in the following form:

$$\vec{\nabla} \cdot \left(B^{-1} \nabla p_k \right) = \vec{\nabla} \cdot \left(B^{-1} C \right) \tag{A6}$$

In the next step, the iterative procedure starts with an initial guess for the pressure field, p_k^*. Using the guessed pressure field, the momentum equation is solved to compute the intermediate velocity field, $\vec{U^*} = (u^*, v^*, w^*)$. This intermediate velocity field does not satisfy the continuity restriction until the solution has been converged. The correct pressure field, p_k^n, is then obtained by solving the Poisson equation. Subsequently, the pressure corrector, p'_k, is calculated by subtracting the guessed pressure field from the correct pressure field, $p'_k = p_k^n - p_k^*$. Then, the velocity field is updated, $\vec{U^n} = (u^n, v^n, w^n)$, using the momentum equation and is checked for the continuity restriction. If the updated velocity field does not satisfy the continuity restriction, a new iteration cycle will start, using p'_k as the initial guess and $\vec{U^n}$ for the calculation of the matrix of coefficient. The iteration continues until convergence occurs. In this case, the continuity restriction is satisfied in each control volume of the computational domain and the pressure corrector becomes zero [46].

Appendix A.2. Turbulence Modelling

Using the RANS method to solve the Navier–Stokes equations introduces additional unknowns, so-called Reynolds stresses, into the main equations. To close the set of equations, a turbulence closure model must be applied. In this study, the well-known standard $k - \varepsilon$ model proposed by Launder and Spalding [47] is used. The standard $k - \varepsilon$ model is a two-equation model based on the Boussinesq approximation. It assumes that the impacts of turbulence on flow can be expressed by an increased kinematic viscosity, and the additional Reynolds stresses can be related to the mean velocity gradients of flow by the turbulent (eddy) kinematic viscosity, v_t [43]. This turbulent kinematic viscosity can be formulated by the turbulence kinetic energy, $k\ [\mathrm{m^2/s^2}]$, and its rate of dissipation, $\varepsilon\ [\mathrm{m^2/s^3}]$, as follows [48]:

$$v_t = C_\mu \frac{k^2}{\varepsilon} \tag{A7}$$

where $C_\mu\ [-]$ is a dimensionless constant.

In the standard $k - \varepsilon$ model, the k and ε are computed from their transport equations that can be written in their steady states as follows [27,45]:

$$\vec{\nabla} \cdot \left(k\vec{U} \right) = \vec{\nabla} \cdot \left(v_{eff,k} \vec{\nabla} k \right) + \frac{P_k}{\rho} - \varepsilon \tag{A8}$$

$$\vec{\nabla} \cdot \left(\varepsilon\vec{U} \right) = \vec{\nabla} \cdot \left(v_{eff,\varepsilon} \vec{\nabla} \varepsilon \right) + C_{\varepsilon 1} P_k \frac{\varepsilon}{\rho k} - C_{\varepsilon 2} \frac{\varepsilon^2}{k} \tag{A9}$$

where $C_{\varepsilon 1}\ [-]$ and $C_{\varepsilon 2}\ [-]$ are dimensionless constants; $P_k\ [\mathrm{kg/m{\cdot}s^3}]$ is the production of turbulent kinetic energy; and $v_{eff,k}\ [\mathrm{m^2/s}]$ and $v_{eff,\varepsilon}\ [\mathrm{m^2/s}]$ are defined as below:

$$v_{eff,k} = v + \frac{v_t}{\sigma_k} \tag{A10}$$

$$v_{eff,\varepsilon} = v + \frac{v_t}{\sigma_\varepsilon} \tag{A11}$$

where $\sigma_k\ [-]$ and $\sigma_\varepsilon\ [-]$ are dimensionless constants. The first and second terms in Equations (A8) and (A9) represent the transport of k or ε by advection and diffusion, respectively; while the last two terms show the rate of production and the destruction of k or ε, respectively. The empirical model coefficients for the standard $k - \varepsilon$ model are given in

Table A1 [47,49]. It should be noted, in the SIMPLE algorithm, the transport equations of k and ε are solved just after computing the updated velocities and the check for the continuity restriction. The turbulent (eddy) kinematic viscosity, v_t, is then updated and be used in the momentum equation for the next iteration cycle.

Table A1. Values of the empirical constants in the standard $k-\varepsilon$ model [47,49].

Parameter	Value
C_μ $[-]$	0.09
$C_{\varepsilon 1}$ $[-]$	1.44
$C_{\varepsilon 2}$ $[-]$	1.92
σ_k $[-]$	1.00
σ_ε $[-]$	1.30

Appendix A.3. Boundary Conditions and Initial Internal Fields

Considering neutral stratification conditions, fully-developed profiles of mean wind speed, U, and turbulence quantities including turbulence kinetic energy, k, and turbulence dissipation rate, ε, are applied at the inlet of the computational domain shown in Figure A1, using the following equations proposed by Richards and Hoxey [48]:

$$U(x=0,y,z)=\frac{u^*}{\kappa}\ln\left(\frac{y-y_g+y_0}{y_0}\right) \tag{A12}$$

$$k(x=0,y,z)=\frac{u^{*2}}{\sqrt{C_\mu}} \tag{A13}$$

$$\varepsilon(x=0,y,z)=\frac{u^{*3}}{\kappa(y-y_g+y_0)} \tag{A14}$$

where u^* $[\mathrm{m/s}]$ is the friction velocity, κ $[-]$ is the von Karman constant defined as 0.41 in OpenFOAM, y $[\mathrm{m}]$ is the vertical coordinate, y_g $[\mathrm{m}]$ is the minimum y-coordinate or the ground level, y_0 $[\mathrm{m}]$ is the aerodynamic roughness length, and C_μ is the dimensionless constant in the standard $k-\varepsilon$ model defined as 0.09.

The height of the computational domain is significantly smaller than the atmospheric boundary layer (ABL) height, therefore the friction velocity in Equations (A12)–(A14) can be assumed constant with height [50], and is calculated using the following equation proposed by Richards and Hoxey [48]:

$$u^*=\frac{\kappa\, u_{ref}}{\ln\left(\frac{y_{ref}+y_0}{y_0}\right)} \tag{A15}$$

where u_{ref} $[\mathrm{m/s}]$ is the reference velocity at a reference height, y_{ref} $[\mathrm{m}]$.

In the present study, the values of the ABL parameters used in inlet profiles are selected based on the wind-tunnel experiments performed by Leitl and Schatzmann [28]. These values are given in Table A2.

Table A2. Values of the atmospheric boundary layer parameters used in inlet profiles [28].

Parameter	Value
y_g $[m]$	0.0000
y_0 $[m]$	0.0007
u_{ref} $[m/s]$	6.0000
y_{ref} $[m]$	0.5000

The analytical inlet profiles of mean wind speed, U, turbulence kinetic energy, k, and turbulence dissipation rate, ε, proposed by Richards and Hoxey [48] are shown in Figure A1.

A zero-gauge pressure boundary condition is applied at the outlet of the computational domain. The no-slip boundary condition for the velocity is used for the bottom of the computational domain and the building walls. It assumes that the speed of the wind flow in direct contact with the bottom of the domain and the walls of the building is identical to the speed of movement of these boundaries, which is equal to zero. Furthermore, the free-slip boundary condition is used for the top and lateral boundary surfaces of the computational domain for all scalar and vector quantities.

The initial internal fields for pressure and velocity are chosen zero, while the initial internal fields for turbulence kinetic energy and turbulence dissipation rate are calculated using the following equations [27,47,48]:

$$k = \frac{1}{2}\left(u'^2 + v'^2 + w'^2\right) \tag{A16}$$

$$\varepsilon = \frac{C_\mu^{0.75}\, k^{1.5}}{l} \tag{A17}$$

where u' [m/s], v' [m/s] and w' [m/s] are fluctuating components of velocity in the x, y and z directions, respectively; C_μ is the dimensionless constant in the standard $k-\varepsilon$ model defined as 0.09; and l [m] is the turbulence length scale or the characteristic length for the macroscale of turbulence.

Assuming an isotropic inlet turbulence, $u' = v' = w'$, and estimating the fluctuations to be 5 percent of the reference velocity at the inlet of the computational domain, and selecting $l = 0.32$ m for the turbulence length scale based on the wind-tunnel experiments performed by Leitl and Schatzmann [28], the initial internal values of the turbulence kinetic energy, k, and the turbulence dissipation rate, ε, are calculated as 0.135 m^2/s^2 and 0.0255 m^2/s^3, respectively.

Appendix A.4. Wall Functions

As mentioned in Appendix A.3., the vertical wind velocity profile changes from zero at the bottom of the computational domain, due to the no-slip boundary condition, to its free stream value far away from the wall (see Figure A2a). The largest gradients of the velocity occur in the near-wall region, where the velocity profile becomes quite steep. To accurately simulate the flow behavior close to the wall, it is important to precisely capture the velocity variations in the near-wall region. Using a second-order accurate finite volume discretization method in the OpenFOAM model, the flow features are calculated for the cell centers and the variation of variables is linear between the cell centroids. Therefore, the standard method to accurately simulate the flow in the near-wall region is to gradually decrease the mesh size when approaching the wall boundary [51,52] (see Figure A2b). Applying a very fine mesh close to the wall significantly increases the number of cells, that increases the computational time considerably. Furthermore, increasing the mesh resolution normal to the wall results in the formation of cells with high aspect ratios. The skewness of the cells causes poor cell qualities that increases the instabilities of the CFD solution.

An alternative way to accurately resolve the high velocity gradients close to the wall is to replace a single large cell instead of those thin cells in the near-wall region and introduce a function, so-called wall function, to reproduce the high velocity variations between the cell centroid and the wall boundary [51] (see Figure A2c). Using the wall function approach, the number of cells and therefore the computational time decreases considerably. In addition, the stability of the CFD solution increases significantly.

Considering the universal law of the wall based on the experimental measurements of fully developed turbulent channel flows, the variation of dimensionless tangential velocity close to the wall, $U^+ = U/u_\tau$, with the dimensionless distance normal to the wall, $y^+ = yu_\tau/v$, is derived. The u_τ [m/s] is the wall friction velocity or the characteristic velocity based on the wall shear stress as the velocity at wall is zero. It can be different from u^*, and is calculated using the following equation [50,53]:

$$u_\tau = \sqrt{\frac{\tau_w}{\rho}} \tag{A18}$$

where τ_w $\left[N/m^2\right]$ is the wall shear stress; and ρ is the flow density. The near-wall region is then subdivided into three distinct layers as follows [54,55]:

- Viscous layer for $0 < y^+ < 5$
- Buffer layer for $5 < y^+ < 30$
- Inertial layer for $30 < y^+ < 200$

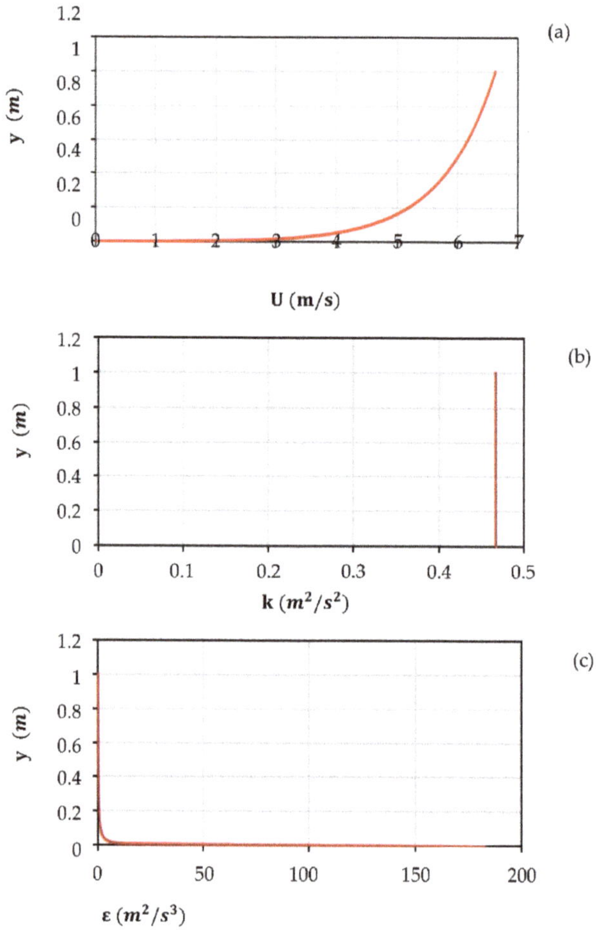

Figure A1. The analytical inflow conditions of (**a**) mean wind speed, (**b**) turbulence kinetic energy, k, and (**c**) turbulence dissipation rate, ε.

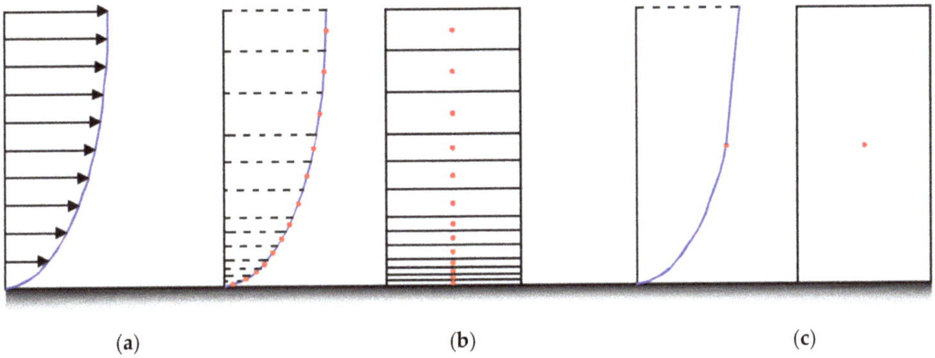

Figure A2. Methods of simulating wind flow velocity in the near-wall region (**a**) The typical vertical wind velocity profile with high velocity gradients close to the wall, (**b**) Standard linear method, and (**c**) Wall function approach.

Wall functions are empirical functions that are best fitted to the observed flow behavior close to the wall. The standard wall functions are a linear and a logarithmic function that precisely reproduce the flow behavior in the viscous layer and the inertial layer, respectively. These wall functions can be expressed by the following equations [52,56]:

$$U^{+} = y^{+} \tag{A19}$$

$$U^{+} = \frac{1}{\kappa}\ln\left(Ey^{+}\right) \tag{A20}$$

where κ is the von Karman constant defined as 0.41 in OpenFOAM; and E $[-]$ is an empirical wall roughness coefficient defined as 9.8 in OpenFOAM. The abovementioned wall functions for viscous and inertial layers intersect in the buffer layer at approximately $y^{+} = 11.225$, so-called y^{+}_{lam} in OpenFOAM. The following conditional statement is then applied in the OpenFOAM model to predict the flow behavior in the near-wall region:

$$U^{+} = \begin{cases} y^{+} & if\ y^{+} \leq y^{+}_{lam} \\ \frac{1}{\kappa}ln\left(Ey^{+}\right) & if\ y^{+} > y^{+}_{lam} \end{cases} \tag{A21}$$

It should be noted, the Equation (A21) cannot precisely predict the flow behavior in the buffer layer, therefore it is not recommended to placing cells in this region.

Appendix B. Model Validation

In order to validate the numerical model, the wind-tunnel measurements performed by Leitl and Schatzmann [28] in the meteorological institute of Hamburg university are compared with the numerical model predictions of the vertical and horizontal wind speed profiles in the vicinity of the building. The atmospheric boundary layer parameters are selected based on the values presented in Table A2, respectively. The total number of cells in the mesh is approximately 1.64 million, consisting of cells with the length, width and height almost equal to 0.0125 m. The results are then derived from sixteen different sections from which eight are located on the symmetry plane, z =0.325 m, and the other eight a near-surface plane, y = 0.035 m, as shown in Figure A3. The comparisons and RMSE values presented in Figures A4 and A5 show satisfactory agreements between the experimental data and the numerical model results. This shows the capability of the numerical model to predict both vertical and horizontal airflow patterns around the building.

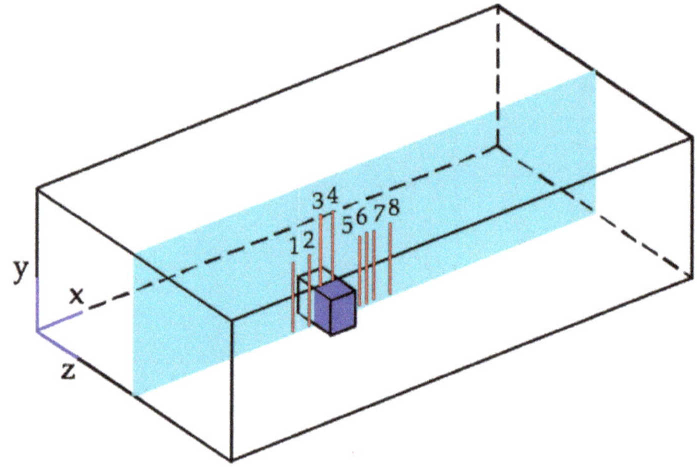

(a)

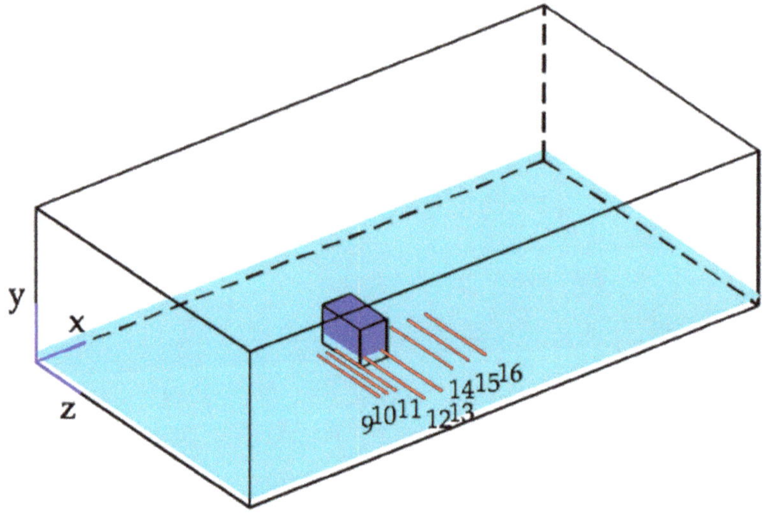

(b)

Figure A3. Schematic representation of the vertical and horizontal measurement sections locating on the (**a**) symmetry plane, $z = 0.325$ m, and (**b**) a near-surface plane, $y = 0.035$ m.

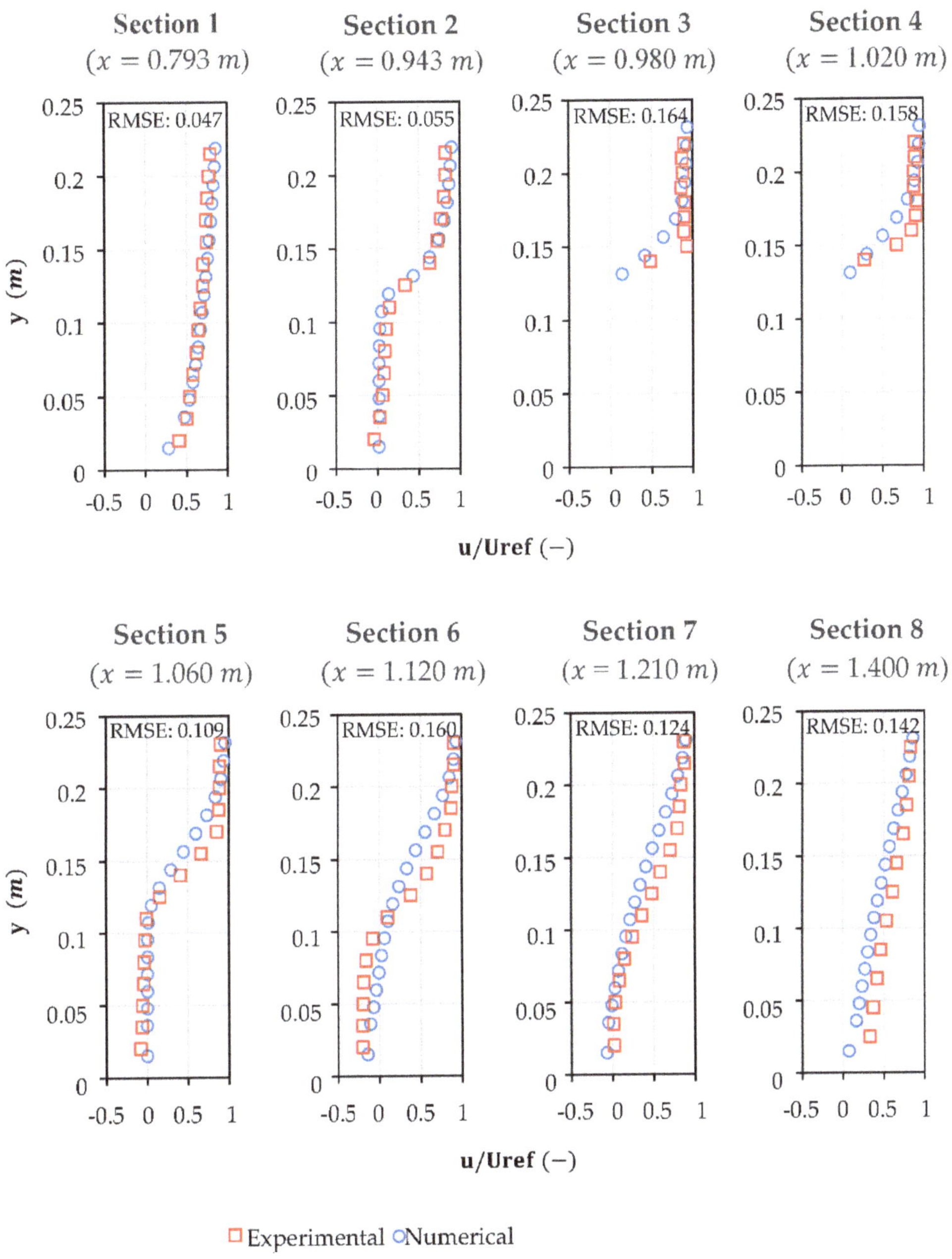

Figure A4. Comparisons between the wind-tunnel measurements performed by Leitl and Schatzmann [28] and the numerical model predictions of the vertical velocity profiles derived from eight different sections locating on the symmetry plane, $z = 0.325$ m.

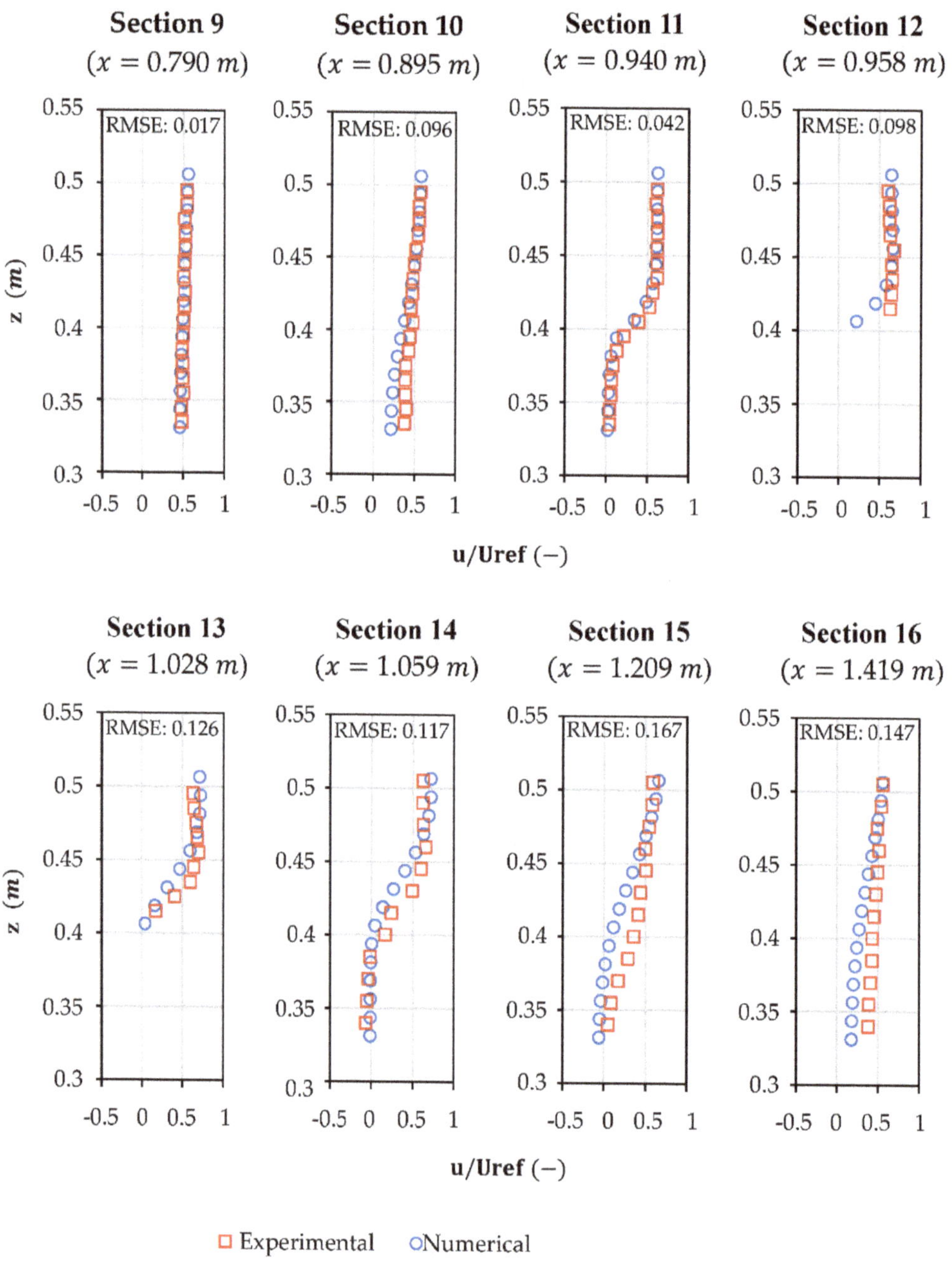

Figure A5. Comparisons between the wind-tunnel measurements performed by Leitl and Schatzmann [28] and the numerical model predictions of the horizontal velocity profiles derived from eight different sections locating on a near-surface plane, $y = 0.035$ m. The comparisons are performed for the right half side of the domain, where the building centreline and the lateral face of the building are located at z = 0.325 m and z = 0.400 m, respectively.

References

1. Marsh, G.P. *The Earth as Modified by Human Action [A New Edition of Manand Nature]*; Sampson Low, Marston, Low, and Searle: London, UK, 1874.
2. Walker, H.J. Man's impact on shorelines and nearshore environments: A geomorphological perspective. *Geoforum* **1984**, *15*, 395–417. [CrossRef]
3. Nordstrom, K.F. Beaches and dunes of human-altered coasts. *Prog. Phys. Geogr.* **1994**, *18*, 497–516. [CrossRef]
4. Jackson, N.L.; Nordstrom, K.F. Aeolian sediment transport and landforms in managed coastal systems: A review. *Aeolian Res.* **2011**, *3*, 181–196. [CrossRef]
5. Walker, I.J.; Nickling, W.G. Dynamics of secondary airflow and sediment transport over and in the lee of transverse dunes. *Prog. Phys. Geogr.* **2002**, *26*, 47–75. [CrossRef]
6. Nordstrom, K.F. *Beaches and Dunes of Developed Coasts*; Cambridge University Press: Cambridge, UK, 2004.
7. Nordstrom, K.F.; McCluskey, J.M. The effects of houses and sand fences on the eolian sediment budget at Fire Island, New York. *J. Coast. Res.* **1985**, *1*, 39–46.
8. Nordstrom, K.F.; McCluskey, J.M.; Rosen, P.S. Aeolian processes and dune characteristics of a developed shoreline. In *Aeolian Geomorphology*; Nickling, W.G., Ed.; Allen and Unwin: Boston, MA, USA, 1986; pp. 131–147.
9. Fackrell, J.E. Parameters characterising dispersion in the near wake of buildings. *J. Wind. Eng. Ind. Aerodyn.* **1984**, *16*, 97–118. [CrossRef]
10. Beranek, W.J. *Wind Environment Around Single Buildings of Rectangular Shape; And, Wind Environment Around Building Configurations. In Stevin-Laboratory of the Department of Civil Engineering*; Delft University of Technology: Delft, The Netherlands, 1984.
11. Martinuzzi, R.; Tropea, C. The flow around surface-mounted, prismatic obstacles placed in a fully developed channel flow (data bank contribution). *J. Fluids Eng.* **1993**, *115*, 85–92. [CrossRef]
12. Iversen, J.D.; Wang, W.P.; Rasmussen, K.R.; Mikkelsen, H.E.; Hasiuk, J.F.; Leach, R.N. The effect of a roughness element on local saltation transport. *J. Wind. Eng. Ind. Aerodyn.* **1990**, *36*, 845–854. [CrossRef]
13. Iversen, J.D.; Wang, W.P.; Rasmussen, K.R.; Mikkelsen, H.E.; Leach, R.N. Roughness element effect on local and universal saltation transport. In *Aeolian Grain Transport*; Springer: Vienna, Austria, 1991; pp. 65–75.
14. Tominaga, Y.; Okaze, T.; Mochida, A. Wind tunnel experiment and CFD analysis of sand erosion/deposition due to wind around an obstacle. *J. Wind. Eng. Ind. Aerodyn.* **2018**, *182*, 262–271. [CrossRef]
15. Luo, W.; Dong, Z.; Qian, G.; Lu, J. Wind tunnel simulation of the three-dimensional airflow patterns behind cuboid obstacles at different angles of wind incidence, and their significance for the formation of sand shadows. *Geomorphology* **2012**, *139*, 258–270. [CrossRef]
16. Sutton, S.L.F.; Neuman, C.M. Sediment entrainment to the lee of roughness elements: Effects of vortical structures. *J. Geophys. Res. Earth Surf.* **2008**, *113*, F2. [CrossRef]
17. Beyers, M.; Waechter, B. Modeling transient snowdrift development around complex three-dimensional structures. *J. Wind. Eng. Ind. Aerodyn.* **2008**, *96*, 1603–1615. [CrossRef]
18. Poppema, D.W.; Wijnberg, K.M.; Mulder, J.P.; Vos, S.E.; Hulscher, S.J. The effect of building geometry on the size of aeolian deposition patterns: Scale model experiments at the beach. *Coast. Eng.* **2021**, *168*, 103866. [CrossRef]
19. Oke, T.R.; Mills, G.; Christen, A.; Voogt, J.A. *Urban Climates*; Cambridge University Press: Cambridge, UK, 2017.
20. Peterka, J.A.; Meroney, R.N.; Kothari, K.M. Wind flow patterns about buildings. *J. Wind. Eng. Ind. Aerodyn.* **1985**, *21*, 21–38. [CrossRef]
21. Blocken, B.; Stathopoulos, T.; Carmeliet, J.; Hensen, J.L. Application of computational fluid dynamics in building performance simulation for the outdoor environment: An overview. *J. Build. Perform. Simul.* **2011**, *4*, 157–184. [CrossRef]
22. Hunt, J.C.R. The effect of single buildings and structures. *Philos. Trans. R. Soc. Lond. Ser. A Math. Phys. Sci.* **1971**, *269*, 457–467.
23. Meroney, R.N. *Turbulent Diffusion Near Buildings, Engineering Meteorology*; CEP78-79RNM23; Colorado State University: Fort Collins, CO, USA, 1982.
24. Meinders, E.R.; Hanjalic, K.; Martinuzzi, R.J. Experimental study of the local convection heat transfer from a wall-mounted cube in turbulent channel flow. *J. Heat Transf. Trans ASME* **1999**, *121*, 564–573. [CrossRef]
25. Bitsuamlak, G.T.; Stathopoulos, T.; Bedard, C. Numerical evaluation of wind flow over complex terrain. *J. Aerosp. Eng.* **2004**, *17*, 135–145. [CrossRef]
26. Smyth, T.A. A review of Computational Fluid Dynamics (CFD) airflow modelling over aeolian landforms. *Aeolian Res.* **2016**, *22*, 153–164. [CrossRef]
27. Versteeg, H.K.; Malalasekera, W. *An Introduction to Computational Fluid Dynamics: The Finite Volume Method*; Pearson Education: London, UK, 2007.
28. Leitl, B.; Schatzmann, M. Cedval at Hamburg University. 2010. Available online: http://www.mi.uni-hamburg.de/cedval (accessed on 1 December 2021).
29. Bagnold, R.A. The movement of desert sand. *Proc. R. Soc. Lond. Ser. A-Math. Phys. Sci.* **1936**, *157*, 594–620. [CrossRef]
30. O'brien, M.P.; Rindlaub, B.D. The transportation of sand by wind. *Civ. Eng.* **1936**, *6*, 325–327.
31. Kawamura, R. Study on sand movement by wind. *Rept. Inst. Sci. Technol.* **1951**, *5*, 95–112.
32. Zingg, A.W. Wind tunnel studies of the movement of sedimentary material. In Proceedings of the 5th Hydraulic Conference Bulletin; Inst. of Hydraulics Iowa City: Iowa City, IA, USA, 1953; Volume 34, pp. 111–135.

33. Owen, P.R. Saltation or uniform grains in air. *J. Fluid Mech.* **1964**, *20*, 225–242. [CrossRef]
34. Hsu, S.A. Wind stress criteria in eolian sand transport. *J. Geophys. Res.* **1971**, *76*, 8684–8686. [CrossRef]
35. Iversen, J.D.; Greeley, R.; White, B.R.; Pollack, J.B. The effect of vertical distortion in the modeling of sedimentation phenomena: Martian crater wake streaks. *J. Geophys. Res.* **1976**, *81*, 4846–4856. [CrossRef]
36. Maegley, W.J. Saltation and Martian sandstorms. *Rev. Geophys.* **1976**, *14*, 135–142. [CrossRef]
37. Lettau, H.; Lettau, H.H. Experimental and micro-meteorological field studies of dune migration. In *Exploring the World's Driest Climate*; IES Report; Leattau, H.H., Lettau, K., Eds.; University of Wisconsin: Madison, WI, USA, 1978; Volume 101, pp. 110–147.
38. White, B.R. Soil transport by winds on Mars. *J. Geophys. Res. Solid Earth* **1979**, *84*, 4643–4651. [CrossRef]
39. Delgado-Fernandez, I. A review of the application of the fetch effect to modelling sand supply to coastal foredunes. *Aeolian Res.* **2010**, *2*, 61–70. [CrossRef]
40. Nolet, C.; Poortinga, A.; Roosjen, P.; Bartholomeus, H.; Ruessink, G. Measuring and modeling the effect of surface moisture on the spectral reflectance of coastal beach sand. *PLoS ONE* **2014**, *9*, e112151. [CrossRef]
41. Hoonhout, B.; de Vries, S. Simulating spatiotemporal aeolian sediment supply at a mega nourishment. *Coast. Eng.* **2019**, *145*, 21–35. [CrossRef]
42. McKenna Neuman, C.; Bédard, O. A wind tunnel study of flow structure adjustment on deformable sand beds containing a surface-mounted obstacle. *J. Geophys. Res. Earth Surf.* **2015**, *120*, 1824–1840. [CrossRef]
43. Ferziger, J.H.; Perić, M.; Street, R.L. *Computational Methods for Fluid Dynamics*; Springer: Berlin/Heidelberg, Germany, 2002; Volume 3, pp. 196–200.
44. Shapiro, A.H. *Compressible Fluid Flow*; Ronald Press: New York, NY, USA, 1953; p. 409.
45. Moukalled, F.; Mangani, L.; Darwish, M. *The Finite Volume Method in Computational Fluid Dynamics*; Springer: Berlin/Heidelberg, Germany, 2016; Volume 6.
46. Caretto, L.S.; Gosman, A.D.; Patankar, S.V.; Spalding, D.B. Two calculation procedures for steady, three-dimensional flows with recirculation. In Proceedings of the Third International Conference on Numerical Methods in Fluid Mechanics, Paris, France, 3–7 July 1972; pp. 60–68.
47. Launder, B.E.; Spalding, D.B. The Numerical Computation of Turbulent Flows. *Comput. Methods Appl. Mech. Eng.* **1974**, *3*, 269–289. [CrossRef]
48. Richards, P.J.; Hoxey, R.P. Appropriate boundary conditions for computational wind engineering models using the k-ε turbulence model. In *Computational Wind Engineering 1*; Elsevier: Amsterdam, The Netherlands, 1993; pp. 145–153.
49. Launder, B.E.; Sharma, B.I. Application of the energy-dissipation model of turbulence to the calculation of flow near a spinning disc. *Lett. Heat Mass Transf.* **1974**, *1*, 131–137. [CrossRef]
50. Blocken, B.; Stathopoulos, T.; Carmeliet, J. CFD simulation of the atmospheric boundary layer: Wall function problems. *Atmos. Environ.* **2007**, *41*, 238–252. [CrossRef]
51. Bredberg, J. *On the Wall Boundary Condition for Turbulence Models*; Internal Report 00/4; Chalmers University of Technology, Department of Thermo and Fluid Dynamics: Goteborg, Sweden, 2000; pp. 8–16.
52. Fluent, A.N.S.Y.S. *ANSYS Fluent Theory Guide 15.0*; ANSYS: Canonsburg, PA, USA, 2013; p. 33.
53. Schlichting, H. *Boundary Layer Theory*; McGraw-Hill: New York, NY, USA, 1968; Volume 960.
54. Tennekes, H.; Lumley, J.L. *A First Course in Turbulence*; The Massachusetts Institute of Technology Press: Cambridge, MA, USA, 1972.
55. White, F.M. *Viscous Fluid Flow*; Mac Graw-Hill International Editions: New York, NY, USA, 1991.
56. Liu, F. *A Thorough Description of How Wall Functions Are Implemented in OpenFOAM*; Nilsson, H., Ed.; Chalmers University of Technology: Göteborg, Sweden, 2016; Available online: http://www.tfd.chalmers.se/~hani/kurser/OS_CFD (accessed on 1 December 2021).

Journal of
Marine Science and Engineering

MDPI

Article

Beach and Dune Erosion: Causes and Interventions, Case Study: Kaulon Archaeological Site

Giuseppe Barbaro [1], Giandomenico Foti [1,*], Giuseppina Chiara Barillà [1] and Ferdinando Frega [2]

1 DICEAM Department, Mediterranea University of Reggio Calabria, Via Graziella loc. Feo di Vito, 89122 Reggio Calabria, Italy; giuseppe.barbaro@unirc.it (G.B.); chiara.barilla@unirc.it (G.C.B.)
2 Department of Civil Engineering, Calabria University, 87036 Arcavacata di Rende, Italy; ferdinando.frega@unical.it
* Correspondence: giandomenico.foti@unirc.it

Abstract: The dune systems are very important from an environmental, landscape, and coastal defense point of view within coastal areas. Currently, dune systems are significantly reduced compared to a few decades ago and, in Europe alone, dune systems have decreased by 70%. During the same period, intense beach erosion processes have often been observed, and, currently, 30% of the world's coasts are eroding. These processes have various causes, both natural and anthropogenic, and the knowledge of the causes of the erosive processes are very important for an effective planning and management of coastal areas and to correctly plan any interventions on dunes and beaches. The paper, through a case study, analyzes the beach and dune erosive processes, their causes, and the possible interventions. The case study concerns the archaeological site of Kaulon, located on a dune in the Ionian coast of Calabria (Italy). The beach near the site was affected by erosive processes and during the winter of 2013–2014, the site was damaged by two sea storms. To identify the causes of these processes, three erosive factors were analyzed. These factors are anthropogenic pressure, wave climate and sea storms, and river transport. The effects produced by these factors were assessed in terms of shoreline changes and of damage to the beach–dune system, also evaluating the effectiveness of the defense interventions. The main causes of the erosive processes were identified through the cross analysis of erosive factors and their effects. This analysis highlighted that in the second half of the last century the erosive processes are mainly correlated to anthropogenic pressure while, recently, natural factors prevail, especially sea storms. Regarding the interventions, the effects produced by two interventions carried out during the winter of 2013–2014, one built in urgency between the first and second sea storm and the other built a few years after the second sea storm were analyzed. This analysis highlighted that the latter intervention was more effective in defending the site.

Keywords: dune; erosion; wave climate; sea storms; longshore transport; river transport

Citation: Barbaro, G.; Foti, G.; Barillà, G.C.; Frega, F. Beach and Dune Erosion: Causes and Interventions, Case Study: Kaulon Archaeological Site. *J. Mar. Sci. Eng.* **2022**, *10*, 14. https://doi.org/10.3390/jmse10010014

Academic Editor: Rodger Tomlinson

Received: 9 October 2021
Accepted: 22 December 2021
Published: 24 December 2021

1. Introduction

Coastal dunes are a habitat of important environmental and landscape value. Additionally, dunes are a natural costal defense because they act both as a reserve of sand and as a physical barrier to protect the territories behind [1–4]. This issue is particularly topical and important because of the significant anthropization processes that occurred from the second half of the last century. Indeed, these processes have totally or partially destroyed numerous dune systems, have altered the coastal equilibrium conditions, and have triggered beach and dune erosion. Currently, 30% of the world's coasts are eroding [5–7] and, in Europe alone, dune systems have decreased by 70% [8].

Dunes are dynamic systems that vary spatially and temporally under the action of both natural and anthropogenic factors [9–12].

From the temporal point of view, there are long-term and short-term variations. Long-term variations are correlated to alterations in coastal, wind, and river sedimentary balance [13–16]. Often these alterations are caused by anthropogenic factors such as the construction of buildings, infrastructures, ports and coastal defense works in coastal areas [17], and the construction of hydraulic structures interfere with fluvial dynamics such as levees, dams, and inert drains from riverbeds [18,19]. On the other hand, the short-term variations are mainly related to natural factors such as single sea storm [20], cluster of sea storms [21,22], extreme flood events [23], or concurrent events [24–26] which cause coastal flooding [27]. Instead, a full dune recovery generally can take several decades [28].

From the spatial point of view, the erosive processes of the beach–dune system can vary greatly along the coast for various reasons, the main ones being linked to the geomorphological characteristics of the area and to the variability of the wave climate and of the longshore transport [29,30].

The issue of morphological changes of coastal dunes and of the beach–dune system and the analysis of the related causes is widely analyzed in the scientific literature by analyzing real cases [31,32], with laboratory experiments [33–37] and applying analytical and numerical models [38–40]. Indeed, these variations can be analyzed both in probabilistic terms [41,42] to identify critical thresholds of sea storms that cause dune erosion [43] and in terms of the beach response to the action of the most intense sea storms [44]. From this point of view, monitoring the dune–beach system at different time scales is of particular importance, also using remote sensing [45–49].

The analysis of the causes of erosive phenomena is of fundamental importance to quantify and predict the vulnerability of beach–dune systems on different timescales and to correctly plan any interventions on dunes and beaches [50–60].

Often, the analysis of these causes focuses on extreme events [61] or on clusters of extreme events [62] or on single factors such as coastal erosion. Instead, a complete analysis is one that considers the largest possible number of the main natural and anthropogenic factors that influence the dune systems. Examples of these analyses are by Pye and Neal [31], who analyzed both the effects of sea storms and the effects caused by the construction of a wall and by a dredge spoil dumping. Subsequently, Cohn et al. [63] analyzed morphological and environmental parameters, such as beach slope and total water level (TWL), by numerical modeling with XBeach. Dissanayake et al. [64] have developed a two-step framework that classifies sea storms and analyzes their impacts on beach–dune systems. Hird et al. [65] analyzed both hydrodynamic forcing and river channel evolution over a 10-year time interval. Finally, Sanromualdo-Collado et al. [66] analyzed the impacts caused by the construction of beach equipment, services, and uses on beach–dune systems. The paper, through a case study, analyzes the beach and dune erosive processes, their causes, and the possible interventions. The paper does not analyze a single erosive factor, but it is a complete analysis, which considers the effects of anthropogenic pressure, wave climate and sea storms, and river dynamics. Furthermore, the paper analyzes the effects produced on the dune–beach system by a temporary intervention, carried out urgently due to the damage caused by the storms of winter 2013–14, and by a subsequent intervention. The case study concerns the archaeological site of Kaulon, located on the dune system in the Ionian coast of Calabria (Italy). Kaulon's erosive phenomena have already been analyzed by Barbaro et al. [67,68] but it was an incomplete analysis due to the limited data available. This paper, on the other hand, uses recent and current data and expands the time interval of analysis to better understand the causes of erosive phenomena.

2. Materials and Methods

This section is divided into two parts. The first part describes the geographical, geomorphological, sedimentological, and hydrological characteristics of the study area. The second part describes the methodology used to analyze beach and dune erosion processes and their possible causes.

2.1. Site Description

The site is located on the Ionian coast of Calabria (Southern Italy), near the town of Monasterace Marina and between the mouth of the rivers Assi and Stilaro (Figure 1). The coastline has an inclination of 15° from the North. The site is affected by the prevalent winds that blow from the South and South-Easterly and from the North and North-Easterly directions. The most severe sea storms mainly come from the South and South-Easterly direction, where the fetch is up to 700 km, and they are prevalent in the winter season. The archaeological site extends along the coast for 1 km and consists of an ancient town, a Doric Temple, and a Pillbox (Figure 1) which are inside a sand dune along the coast located at an altitude of about 8 m.

Figure 1. Location of the study area.

The study area is characterized by recent tectonic uplift accompanied by compressive phenomena eastward and distensive and sinking phenomena to the west. The intense and recent tectonic activity has provided a morphology subject to intense erosion and gravitational processes as occurs in many rocks outcropping in other zones of Calabria [69]. The geological formations of the area are sedimentary rocks of Holocene-Recent age overlapped to the oldest sedimentary rocks (Mio-Pliocene-Pleistocene) and to crystalline-metamorphic units of Paleozoic age. Overall, the litho-stratigraphic sequence, starting from the most ancient rocks, can be summarized as follows: clays and polychrome silts (Miocene Inf.-Med.), clays and whitish marl (Lower-Middle Pliocene), conglomerates and sands (Pliocene Sup.-Calabriano), sandstones and clay silts (Middle-Superior Pliocene), conglomerates and mica sands (Sup. Pliocene-Calabriano "Ghiaie di Messina"), conglomerates and sands (Pleistocene), debris fan (Holocene), solifluction products (Holocene), stabilized alluvial deposits (Holocene); coastline and mobile alluvial deposits (Holocene), stabilized sand dunes, mobile sand dunes.

From the geomorphological point of view, the area is characterized by a narrow coastal plain and by hills that connect to the reliefs of the Serre. The coast is characterized by a low beach, a well-developed dune system and a sea bottom with moderate slope (2–3%). From the sedimentological and grain size distribution point of view, the beach sediments are composed of sand and light grey gravels, with D_{50} equal to 5 mm next to the isobaths of +1.0 m, D_{50} = 0.7 mm next to the isobaths −3.0 m, and D_{50} equal to 0.82 mm at the isobaths

−7.0 m [70]. From the hydrological point of view, the hydrography is characterized by a series of rivers (called *fiumare* and typical of Calabria) with dendritic and fin patterns, with high drainage density, perpendicular to the coastline with a typical braided course. Generally, most of the Calabrian rivers are characterized by high slopes, modest time of concentration and torrential regime so floods occur suddenly [71,72]. For these reasons, river sediment transport is mainly related to Soil Erosion by Water (WSE) [73]. The study area and, in general, most of the Calabrian rivers are particularly prone to WSE [74] which also influence shoreline evolution [75,76].

2.2. Methodology

The methodology proposed in this paper aims to evaluate the causes of the erosive processes observed near the archaeological site of Kaulon. To identify these causes, erosive factors and their effects on the study area were examined, according to the flow chart of Figure 2. In detail, three main erosive factors on the beach–dune system were analyzed. These factors are anthropogenic pressure, wave climate and sea storms, and river transport. The effects were assessed in terms of shoreline changes and of damage to the beach–dune system.

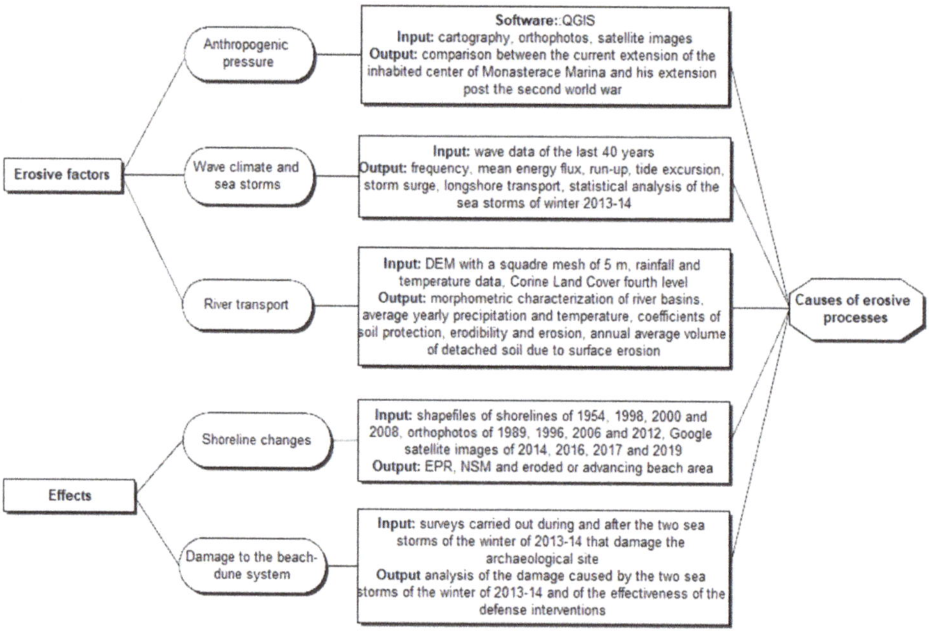

Figure 2. Flow chart of the proposed methodology.

2.2.1. Anthropogenic Pressure

The analysis of the anthropogenic pressure was carried out on QGIS through the overlap and the comparison of various cartography data, which consists of CASMEZ, "*Cassa del Mezzogiorno*", cartography of 1954, available in the Open Data section of the Calabrian Geoportal (http://geoportale.regione.calabria.it/opendata, accessed on 1 December 2021); orthophotos of 1989 and 1996 available as a WMS service in the Open Data section of the Italian Geoportal (http://www.pcn.minambiente.it/mattm/servizio-wms/, accessed on 1 December 2021), and the most recent Google satellite image of 2019. The main aim of this analysis concerns the comparison between the current extension of the inhabited center of Monasterace Marina and his extension post the second world war, before the considerable anthropic pressures observed in most of the Calabrian and Italian territories. Furthermore, by comparing the current extension with that of the 1990s, it was possible to better date the anthropization process.

2.2.2. Wave Climate and Sea Storms

This phase concerns the evaluation of the wave climate and the statistical analysis of the sea storms of winter 2013–2014, which damaged the dune and the archaeological site. Regarding wave climate, the following was calculated: frequency of occurrence of sea state (hereinafter referred to as frequency), mean energy flux, run-up, tide excursion, storm surge, and longshore transport. Frequency and mean energy flux identify the directions of frequent and intense sea storms, respectively. Furthermore, the comparison between the estimated run-up, tide excursion and baric and storm surge values, and the altimetry of the study area was used to evaluate whether these parameters can directly affect the archaeological site. Regarding longshore transport, the estimated value was compared with the river contribution to understand whether the two forcing are on average equivalent or not. Finally, the analysis of the wave climate is the basis for the statistical analysis of the sea storms of winter 2013–2014.

The input data are the wave data from the last 40 years, available in the database developed by the MeteOcean group of the University of Genoa (http://www3.dicca.unige.it/meteocean/hindcast.html, accessed on 15 September 2021). This data was reconstructed from the Climate Forecast System Reanalysis (CFSR) database.

The time series is relative to point 7624, with coordinates 16.638 E and 38.46 N and located at a depth of over 110 m. The time series consists of 332,884 sea states, for each of which significant wave height, mean and peak periods, and wave direction are available. These data were grouped in sectors of 10° each and the time series was analyzed both entirely and divided into intervals of 10 years each.

Starting from time series data, the frequency and the mean energy flux Φ were calculated. The frequency of the generic sector is the ratio between the number of sea states from that sector and the total number of registered sea states. The mean energy flux of the generic sector is the sum of the energy flux of each sea state from that sector, that depends on the specific gravity of the water, on the peak period, on the square of the significant height and on the frequency. Run-up was estimated by the Stockdon et al. model [77]. This model is based on the following equations and depends on the foreshore beach slope β_f, on the significant wave height at deep water H_0, on the wavelength at deep water L_0, and on the Iribarren number ξ_0.

$$R_{u2\%} = 1.1 \{0.35\ \beta_f\ (H_0\ L_0)^{1/2} + \frac{1}{2}\ [H_0\ L_0\ (0.563\ \beta_f^{\,2} + 0.004]^{1/2}\} \text{ for } \xi_0 < 0.3, \quad (1)$$

$$R_{u2\%} = 0.043\ (H_0\ L_0)^{1/2} \text{ for } \xi_0 \geq 0.3, \quad (2)$$

The slope was estimated by analyzing the open access bathymetry available on the EMODNET portal (https://www.emodnet-bathymetry.eu/, accessed on 1 December 2021) while significant wave height and wavelength at deep water were calculated starting from the time series for different return times (1, 10, 20, 50, 100, 200, and 500 years). To estimate the tide excursions, the recordings of the tide gauges of Crotone and Reggio Calabria were analyzed, also the Tide Tables of the Italian Marine Hydrographic Institute [78] and scientific papers were consulted, especially that of Sannino et al. [79]. The storm surge was estimated using the Bretschneider model [80]. This model is based on the following equation and depends on the depth at the shelf edge d_1, on the depth near the coast d_2, on the wind drag coefficient K, and on the length of the wind fetch L. To this value the surge caused by the minimum atmospheric pressure recorded during an atmospheric disturbance was added, considering that a barometric decrease of 1 mbar compared to the normal value of 1013 mbar causes a surge of 1 cm.

$$S/d_1 = ((K\ U^2\ L)/(g\ d_1^{\,2}\ (1-d_1/d_2))\ \ln\ (d_1/d_2) \quad (3)$$

Longshore sediment transport has been evaluated by the Tomasicchio et al. model [81]. This model is of general validity, it is based on a mobility index depending on the granulometry of the beach sediments and was obtained starting from the model of Lamberti and Tomasicchio [82]. The main input data of the Tomasicchio et al. model are significant wave height H_s, peak wave period T_p, off-shore wave angle θ_0, and nominal diameter of the unit D_{n50}. Using these data, characteristic wave height H_k, mean wave period T_m, wave length L, wave celerity c, wave group celerity c_g, characteristic wave height at breaking $H_{k,b}$, characteristic wave angle at breaking $\theta_{k,b}$, modified stability number N_s **, length of displacement l_d, number of displaced particles at the end of 1000 wave attack N_{od}, longshore transport measured as number of units per wave S_N and, finally, longshore transport rate in volume per unit time Q_{LT} were calculated, as shown by the flow chart of Figure 3. Therefore, the granulometry described in Section 2.1 and the wave time series were used as input data.

The statistical analysis of the sea storms of winter 2013–2014 consists of the evaluation of the maximum significant wave height, of the wave direction at the peak of significant wave height, and of the duration of each storm. From these data, the return periods of each sea storm were calculated.

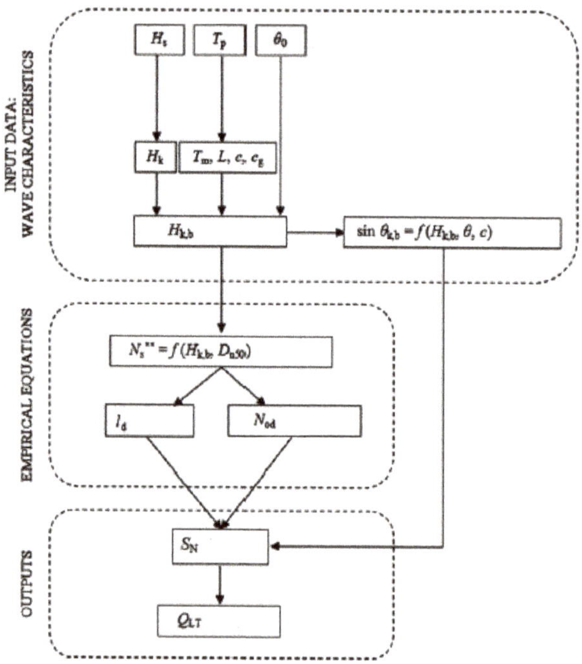

Figure 3. Flowchart of the Tomasicchio et al. model [81].

2.2.3. River Transport

The aim of this phase concerns the analysis of river transport in terms of average volume of detached soil due to surface erosion. This parameter was compared with the longshore transport rate to understand whether the two forcing are on average equivalent or not. River transport was estimated using the Gavrilovic model [83], which is particularly reliable in basins where sediment transport is mainly related to WSEs as in this case study. The Gavrilovic model is based on an analytical equation to determine the annual volume of detached soil due to surface erosion. This equation depends on the morphometric characteristics of the river basin, such as the area, the perimeter, the average slope, the average height, the main stream length, the total stream length, on the average yearly precipitation and temperature, and on some coefficients related to the soil protection (a function of the type of vegetation cover), to the erodibility (a function of type of rock), and to the type of basin erosion.

Preliminarily, it was necessary to identify and morphometrically characterize the basins at the edges of the study area, Assi and Stilaro river, using QGIS. In addition, for each basin land uses were analyzed to estimate the coefficients of soil protection, erodibility and erosion, and rainfall and temperature time series were analyzed to estimate the average yearly precipitation and temperature.

In detail, the identification of the river basins and their morphometric characterization was carried out by starting with the DEM with square mesh of 5 m available in the Open-Data section of the Calabrian Geoportal (http://geoportale.regione.calabria.it/, accessed on 15 September 2021). For each basin, the area, the perimeter, the main stream length, the total stream length, the maximum and the average heights, the average slope, the Horton order, the Gravelius index, and the time of concentration were calculated. The estimation of the last parameter was undertaken using the formulas of Giandotti [84], Kirpich [85], and NRCS [86].

The average yearly precipitation and temperature values of each basin were estimated from the time series of rainfall and temperature records available in the Historical Data section of the Calabrian Multi-Risk Functional Center (http://www.cfd.calabria.it/, accessed on 15 September 2021). Preliminarily, the rainfall and temperature gauges were identified with statistically significant time series in the Assi and Stilaro River basin and in its neighboring areas. For each gauge, the registration period, the number of years available, the elevation, the weight, estimated on QGIS using the Thiessen polygon method [87,88], and the average yearly precipitation and temperature were analyzed. The average yearly precipitation and temperature values of each basin were calculated as a weighted average of the values recorded by each gauge.

The land cover data used was from the Corine Land Cover project fourth level relating to the year 2018 and freely available on the government agency website "Istituto Superiore per la Protezione e la Ricerca Ambientale (ISPRA)" (https://www.isprambiente.gov.it/it/attivita/suolo-e-territorio/copertura-del-suolo/corine-land-cover, accessed on 15 September 2021). A value of each of the coefficients of soil protection, erodibility, and erosion was associated with each category of land use of the Corine Land Cover. The average values of these coefficients of each basin were calculated as a weighted average of the coefficient values of each land use category, considering the area of each land use category as weight.

2.2.4. Shoreline Changes

The analysis of the shoreline changes was carried out through the comparison of various cartography data, which consists of shapefiles of the historical shorelines of 1954, 1998, 2000, and 2008 taken from the Open Data section of the Calabrian Geoportal (http://geoportale.regione.calabria.it/opendata, accessed on 1 December 2021); orthophotos of 1989, 1996, 2006, and 2012 taken from the Open Data section of the Italian Geoportal (http://www.pcn.minambiente.it/mattm/servizio-wms/, accessed on 1 December 2021); and satellite imagery of 2014, 2016, 2017, and 2019 provided by Google Earth.

The analysis was divided into three phases as follows. The first phase concerned the manual digitization of each missing shoreline, using QGIS for orthophotos of 1989, 1996, 2006, and 2012, thus obtaining shapefiles, and using the spatial analysis tools of Google Earth Pro for satellite imagery of 2014, 2016, 2017, and 2019, thus obtaining kml files then saved on QGIS as shapefiles. Both the orthophotos and the satellite images relate to the summer months, between May and September, so all the traced shorelines are relative to the summer profile. The second phase concerned the evaluation of the beach width for each transept. Finally, the last phase concerned the determination of shoreline rates of change using End Point Rate (EPR) and Net Shoreline Movement (NSM) statistics and the estimate of the eroded or advancing beach area.

Regarding the first phase, the digitalization of the missing shorelines was carried out on a scale of 1:1000 on QGIS and at an eye altitude of 200 m, corresponding to a higher scale, on Google Earth Pro. Generally, the uncertainties in the digitization phase concern georeferencing, the orthorectification process, the resolution of the different imagery sets, the digitizing uncertainty, the uncertainty in the identification of the wet/dry line, and any error caused by a variation in some factors affecting the shoreline change such as the seasonal cycle of erosion and deposition, and the impact of storms [89,90]. In this case, the reference line chosen was the wet/dry line. Additionally, the cartography data is all related to the summer period and no storm conditions were observed in any of the data, so the effects of seasonal variation and individual storms on shoreline change are of limited importance. Therefore, the uncertainties in the shoreline position are less than one meter and the shoreline changes were approximated to the meter. This accuracy agrees with the aims of the paper, which concern the evaluation of the erosion and advancement trends, and not their precise quantification. To estimate the tide excursions, the recordings of the tide gauges of Crotone and Reggio Calabria were analyzed, and scientific papers were consulted, especially that of Sannino et al. [79]. Therefore, in the study area the tidal excursion is of the order of tens of centimeters so the effects on the variation of the shoreline position are negligible.

Regarding the second phase, to analyze the shoreline changes of the entire physiographic sub-unit between the mouths of the Stilaro and Assi rivers, 15 transepts were traced, with an average spacing of the order of 200 m. Additionally, the related baselines were identified for each transect. These lines identify the upper limit of the beach and correspond to promenades, roads, and structures. Regarding the third phase, the NSM and EPR and the eroded or advancing beach area between any two successive shorelines were calculated.

2.2.5. Dune System Damage

The archaeological site was damaged by two sea storms, both of which occurred in the winter of 2013–2014. The damage analysis was based on some surveys carried out during and after the sea storms. The aim of this phase concerns the assessment of the damage caused by the two sea storms to the archaeological site and the assessment of the effectiveness of two defense interventions carried out, one between the first and the second sea storm and the other a few years after the second sea storm.

3. Results

This section is divided into five subsections: anthropogenic pressure, wave climate and sea storms, river transport, shoreline changes, and dune system damage, similar to what is described in the methodology section.

3.1. Anthropogenic Pressure

To evaluate and date the anthropogenic pressure in the study area, cartography, orthophotos, and satellite images of different time periods were compared (Figure 4). The older source, from 1954 (CASMEZ cartography), and the most recent source, from 2019 (Google satellite image), were initially compared. The inhabited center of Monasterace Marina is located south of the archaeological site and was almost entirely built in the second half of the last century. Indeed, in 1954 there was only a strip of buildings around the main road. This strip was about 500 m long, was over 100 m from the shoreline behind the beach and there was no waterfront. Instead, the current extension of the built-up part is about 1500 m long, and buildings and promenade have been built instead of a large portion of the beach which has a current width of a few tens of meters.

Then, the 1989 and 1996 orthophotos were compared with each other and with the most recent satellite image. This comparison shows that most of the anthropization process took place between 1954 and 1989. Indeed, in 1989 there are the promenade and numerous buildings, both built where in 1954 there was a beach. Furthermore, the extension of the inhabited center in 1996 is almost the same as the current one, while between 1989 and 1996 only some buildings were built in the southern part of the inhabited center. Therefore, since 1996 the anthropization process has almost stopped. Finally, throughout the time interval analyzed, the anthropization process did not affect the archaeological site.

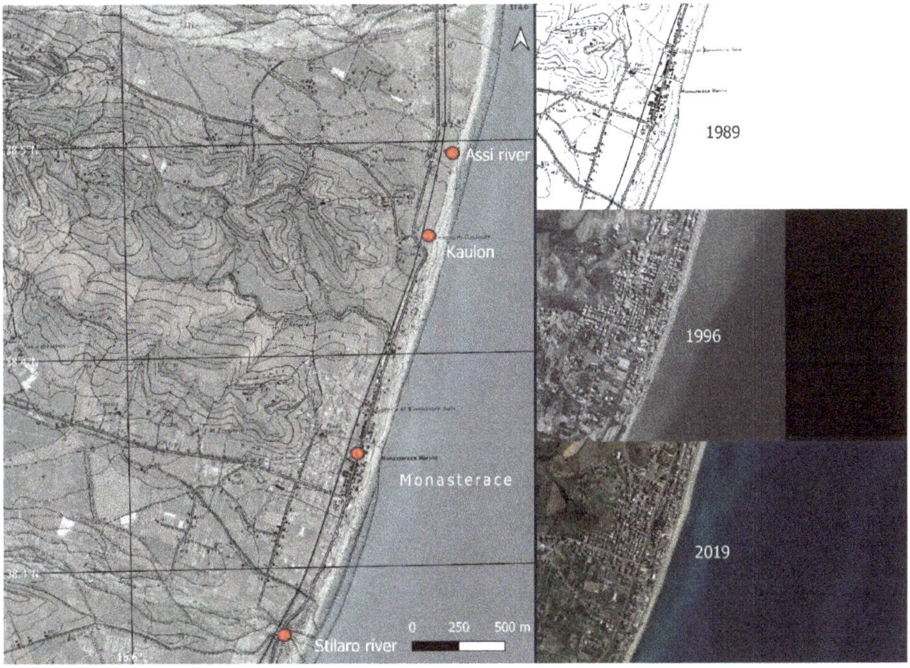

Figure 4. Large panel: overlap between CASMEZ cartography of 1954 and Google satellite image of 2019. Small panels: CASMEZ cartography of 1954 (**up**), orthophotos of 1996 (**center**), Google satellite image of 2019 (**down**).

3.2. Wave Climate and Sea Storms

The analysis of the wave climate of the last 40 years shows that the study area has been affected by frequent sea states from the northeast and south, with a maximum value of about 8% in the sectors centered on 180 and 190° N. Furthermore, the mean energy flux graph has two peaks, one of about 450 N/s in the sector centered on 50° N, also with a relevant value in the adjacent sector centered on 40° N, and with the main peak in the sector centered on 130° N, with value higher than 650 N/s and with relevant values in all sectors between 120° and 160° N. The sea states coming from sectors between 120 and 180° are generated by the Scirocco winds, acting on fetches with lengths of several hundreds of km. Instead, the sea states coming from the sectors around 50° are generated by the winds of Grecale, acting on fetches with lengths of a few hundred km. Therefore, the study area is subject to sea storms both significantly inclined and almost orthogonal with respect to the coast. The wave climate was also analyzed in shorter time intervals, of 10 years each, and the main results are summarized in Table 1. In detail, in each time interval the maximum values of significant wave height $H_{s,max}$, the mean energy flux and the main sector were evaluated. The analysis highlighted that the higher values of $H_{s,max}$, almost 6.7 m, were recorded in the decade of the 1980s, while in the following decades there are decreasing values, up to 5.9 m in the last decade. Instead, the mean energy flux initially grows between the 1980s and the 1990s and then decreases in the following decades. Instead, the main sector is always the one centered on 130°.

Regarding the run-up values, slight variations are observed as the return period varies. Indeed, the values obtained are between 1 m, with a return period of 1 year, and 2 m, with a return period of 500 years. These slight variations depend mainly on the modest foreshore beach slope value, about 3%. The tide excursion and the storm surge were added to this value. The maximum value of the tide excursion recorded by the tide gauges of Crotone and Reggio Calabria and by the Tide Tables of the Italian Marine Hydrographic Institute [78] is just over 20 cm, according to the studies by Sannino et al. [79]. The storm surge calculated with the Bretschneider model [80] is just under 10 cm. To this value the surge caused by the minimum atmospheric pressure recorded during an atmospheric disturbance was added, which in the Mediterranean assumes values no lower than 970 mbar. Therefore, the total storm surge value is of the order of 50 cm. By adding the values of run-up, tide excursion, and storm surge, a value of less than 3 m is obtained. Due to the morphology of the territory, the archaeological site was not directly affected by these parameters as it is located at an altitude of about 8 m while the total value of run-up, tide excursion, and storm surge does not exceed 3 m. Indeed, a barometric decrease of 1 mbar compared to the normal value of 1013 mbar causes a baric surge of 1 cm.

Additionally, the longshore transport evaluated by the Tomasicchio et al. model [81] is about 90,000 m^3/year and is on average directed by South from North direction. However, the study area is subject to sea storms coming from both South-East and North-East directions so there can be a significant longshore transport also from North to South direction.

The statistical analysis of the sea storms that occurred in the winter of 2013–2014 (Table 2), which damaged the dune and the archaeological site, highlighted that these are statistically frequent sea storms. Indeed, the one that occurred between November and December 2013 reached a maximum significant wave height of 5.21 m, lasted 74 h, the direction at the peak of significant wave height was 131° N and had a return period of 4 years. Instead, the storm that occurred between January and February 2014 reached a maximum significant wave height of 5.61 m, lasted 140 h, the direction at the peak of significant wave height was 114° N and had a return period of 7.5 years. The maximum significant wave height reached in this last storm is lower than the maximum significant wave height value observed in the last decade, which is 5.9 m (Table 1). This value was observed at the beginning of that decade and the sea storms after that of February 2014 reached maximum significant wave heights always lower than 5.5 m. Furthermore, in all previous decades the observed maximum significant wave height values are higher than 5.9 m. These considerations explain the low return period values obtained for each of the two sea storms analyzed.

Table 1. Wave climate characteristics in the various time intervals.

Time Interval	H_s Max (m)	Φ (N/s)	Main Sector
Entire	6.67	6167	
1980s	6.67	1598	
1990s	6.26	1679	130
2000s	5.91	1482	
2010s	5.9	1407	

Table 2. Characteristics of the sea storms that damaged the dune and the archaeological site.

Sea Storm	Start	End	H_s Max (m)	Duration (Hour)	Peak Direction	Return Period (Year)
2013	29 November 5:00 PM	2 December 7:00 PM	5.21	74	131	4
2014	29 January 11:00 PM	4 February 7:00 PM	5.61	140	114	7.5

3.3. River Transport

The analysis of river transport, in terms of average volume of detached soil due to surface erosion, was preceded by the perimeter and morphometric characterization of the Assi and Stilaro river basins. This phase has highlighted that the basins are bordering, and both have an elongated shape. Stilaro river, compared to the Assi river, has a greater area, about 100 km^2 instead of over 65 km^2, due to its greater width since the perimeters are almost the same, about 60 km. The largest area is also correlated with a greater length of the tributaries, over 160 km instead of about 100 km, while the main stream lengths are almost the same, about 30 km. The average slope and the Horton order are also the same, 35% and 6, respectively, while the average height and the Gravelius index are greater in the Assi river, about 700 m instead of about 600 m and 2 instead of 1.73, respectively. Finally, the time of concentration is greater in the Stilaro river, 4.3 h instead of 3.5 h.

Regarding the rainfall time series, there are 11 rainfall and temperature gauges in the Assi and Stilaro River basin and in its neighboring areas, six of which are currently active. Among them, the Stilo-Ferdinandea thermo-pluviometric gauge has greater weight in both basins. Furthermore, for the Stilaro basin, also the Stignano gauge has significant weights while for the Assi basin the Santa Caterina dello Ionio gauge, for the rainfall, and the Monasterace-Punta Stilo gauge, for the temperature, have significant weights. The average weighted values of rainfall and temperature are, respectively, about 1260 mm and 14.4 °C in the Stilaro basin and about 1300 mm and 13.4 °C in the Assi basin.

Regarding land uses, there are 21 different uses. In both basins, the wooded and shrub areas are about 60%, with a prevalence of holm oaks and beeches, while the agricultural areas are about 30%, with a prevalence of olive groves. The weighted coefficients of soil protection, erodibility, and erosion are, respectively, 0.38, 1.05, and 0.34 in the Stilaro River basin and 0.36, 1.04, and 0.31 in the Assi River basin.

Finally, the average annual volume of eroded sediments is equal to over 60,000 m^3/year for the Stilaro basin and over 40,000 m^3/year for the Assi basin. Comparing these values to the areas of the two basins, it is observed that the specific erosion is almost the same for both basins, about 650 m^3/year*km^2. This result is congruent with the evident hydrological, climatic, and land use similarities between the two basins.

3.4. Shoreline Changes

The analysis of the shoreline changes in the entire physiographic sub-unit between the Stilaro and Assi river mouths, shown in Table 3 and in Figure 5, highlighted that most of the erosive processes took place in the period between 1954 and 1989. Indeed, in 13 transepts out of 15 the maximum beach width dates back to 1954 and transepts 11 and 12 are exceptions whose maximum beach width dates back to 1989.

In 1954, the beach widths have a maximum value of over 130 m, a minimum value of over 40 m, and an average value of about 85 m. In 1989, on the other hand, the beach widths ranged from 30 to 70 m, with an average value of about 50 m. In this time interval, erosion between 10 and 95 m are observed, with an average value of 35 m and with relatively low speeds due to the considerable temporal extension of the interval. Both in 1954 and 1989 the maximum erosions are observed in the external transepts, located near the Stilaro and Assi river mouths. In terms of beach area, over 100,000 m^2 were lost in this time interval.

Another significant erosive phase is observed between 1989 and 1996, with erosion in all transepts with values of the order of tens of meters up to a maximum of 45 m and with an average value of 17 m. In 1996, the beach width was between 20 and 55 m approximately, with an average value of 32 m which is also the lowest value observed in the various years examined. In terms of beach area, about 60,000 m^2 were lost in this time interval.

Regarding the subsequent time intervals, in the interval 1996–1998 the evolutionary trend is advancement, with a maximum value of about 15 m and an average value of 5 m. Only the transepts near the mouth of the Stilaro river show slight erosion. In terms of beach area, there has been an increase of 20,000 m^2 in this time interval. In the 1998–2000 interval, on the other hand, all the transepts are in slight advancement, with a maximum value of less than 10 m and an average value of 4 m. In terms of beach area, there has been an increase of about 20,000 m^2 in this time interval.

In these first four time intervals, the evolutionary trends are substantially homogeneous in the whole study area. Instead, in the new century, in each time interval analyzed, a very variable trend is observed along the various transepts.

Between 2000 and 2006 and between 2006 and 2008, erosive processes prevail. In the first time interval, the average value of the beach width increased by 2 m compared to the previous interval, but, in terms of beach area, about 5000 m^2 were lost in this time interval. In the second time interval, the average value of beach width decreased by 8 m and, in terms of beach area, about 20,000 m^2 were lost in this time interval.

Between 2008 and 2012 equilibrium conditions prevails. Indeed, eight transepts are advancing, and seven transepts are in erosion and the eroded areas are of the same order of magnitude as the increased ones.

Table 3. Beach area variations of each time interval. Legend: red = erosion; green = accretion; negative = erosion; positive = accretion.

Time interval	54–89	89–96	96–98	98–00	00–06	06–08	08–12	12–14	14–16	16–17	17–19
Area (m^3 × 10^3)	−102	−60	20	18	−5	−18	0	2	−2	10	6

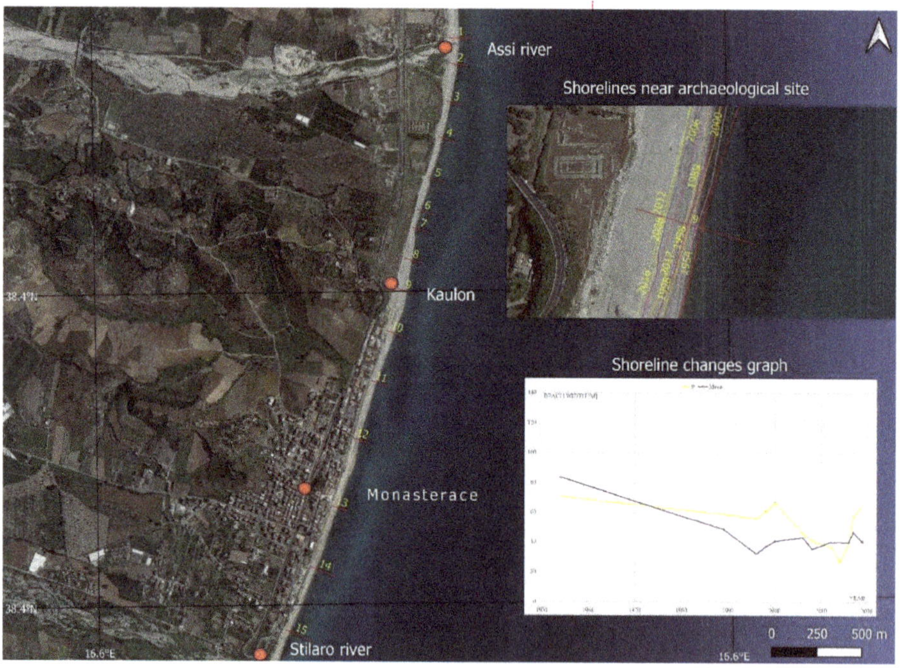

Figure 5. Large panel: analyzed transepts. Small panels: shorelines near the archaeological site (**up**) and graph of the shoreline changes in the transept at the archaeological site, yellow line, and of the average changes in all analyzed transepts, purple line (**down**).

Between 2012 and 2014 and between 2014 and 2016, the global evolutionary trend would seem to be in equilibrium, as the variation of the beach width is equal to 0 in both intervals. Additionally, in terms of beach area, there has been an increase of only 1000 m^2 in the first time interval and an erosion of only 1000 m^2 in the second time interval. However, by analyzing the individual transepts, it is observed that between 2012 and 2014 seven transepts are in erosion, with a maximum value exceeding 10 m, seven transepts are advancing, with a maximum value of about 30 m, and only one transept is stable. Between 2014 and 2016 seven transepts are in erosion, with a maximum value of about 20 m, seven transepts are advancing, with a maximum value exceeding 15 m, and only one transept is stable. Therefore, in these time intervals the solid material has spatially redistributed between the various transepts and the transepts where the greatest variations are observed in the external ones, located at the Stilaro and Assi rivers mouths.

Between 2016 and 2017, on the other hand, advancement processes prevail, with an average value of 7 m and, in terms of beach area, there has been an increase of 10,000 m^2 in this time interval.

Instead, between 2017 and 2019 erosive processes prevail again with an average erosion of 7 m and, in terms of beach area, there has been an erosion of over 5000 m^2. Additionally, it should be noted that the most recent average beach width is the same as in 2000 and less than half than that of 1954.

Finally, by analyzing the evolutionary trends of transept 9, located in front of the archaeological site, erosion is observed up to 1996, with a value equal to 15 m, then advancement is observed up to 2000 of a value equal to 10 m. Subsequently, an erosive process begins, the maximum of which occurs in 2014, with a value of 40 m and with an erosion than 30 m. From 2014 to 2019, the beach advances and the most recent width is just over 70 m, the same value as in 1954.

3.5. Dune System Damage

The archaeological site was damaged by two sea storms, both of which occurred in the winter of 2013–2014. The first storm that occurred in early December 2013, eroded much of the beach below the archaeological site and caused the partial collapse of the dune, with the sliding of some archaeological finds into the beach below (Figure 6a). After this sea storm, in early February 2014 another sea storm caused further damage to the dune and to the archaeological site (Figure 6b). These damages are like those observed after the first storm but did not cause significant retreat of the dune system. Following the sea storm of December 2013, the Provincial Administration of Reggio Calabria financed in urgency the construction of a barrier. The barrier has a trapezoidal shape, is 3 m in height, 6 m in width, and 30 m long, and was built on the beach a few meters from the sea, at about 8 m from the dune. The height between the ground level and the top of the barrier is 1.5 m (Figure 7a). This barrier, however, represents a temporary intervention, with limited effectiveness as it is much shorter than the entire dune to be protected and as the sea storms crossed it (Figure 7b) as occurred during the storm of February 2014. To solve these problems, an intervention extended to the entire archaeological site was carried out between 2017 and 2019 (Figure 8b). In detail, a barrier was built at the foot of the dune that protects the dune from the direct action of sea storms and reduces the erosion on the foot. Up to now, this intervention has effectively protected the beach–dune system.

(a) (b)

Figure 6. Doric Temple. (**a**) After the December 2013 storm. (**b**) After the February 2014 storm.

(a) (b)

Figure 7. (**a**) Barrier made in January 2014. (**b**) Storm of February 2014 (blue arrows) that crosses the barrier (red arrows).

(a) (b)

Figure 8. (**a**) Google satellite image of 2014 (scale 1:1000, North oriented). (**b**) Google satellite image of 2019 (scale 1:1000, North oriented).

4. Discussion

To evaluate the effects produced by the three examined erosive factors, which are anthropogenic pressure, wave climate, and sea storms and river transport, the shoreline changes and the damage to the dune system were analyzed, also evaluating the effectiveness of the defense interventions. The main causes of the erosive processes were identified through the cross analysis of erosive factors and their effects.

The analysis of the shoreline changes has shown that most of the erosive processes in the study area are concentrated between 1954 and 1996. In this period, the inhabited center of Monasterace has considerably expanded and buildings and a promenade have been built instead of large beach areas. Similar situations have also been observed in many other Calabrian territories. Anthropogenic pressure has not affected only the coastal areas but also many rivers. Indeed, from the 60s of the last century up to the 90s, numerous weirs were built in most of the Calabrian rivers following disastrous floods. Among the effects produced by them, it should be mentioned the immobilization of significant quantities of sediments, which have reduced the river transport. In the study area, the greatest shoreline retreats between 1954 and 1996 have been observed at the Assi and Stilaro river mouths and at the inhabited center. Therefore, the erosive processes observed between 1954 and 1996 are mainly related to anthropogenic factors.

In the last 25 years, there has been an alternation between erosions and accretions, and the current average beach width is the same as in 2000 and less than half than that of 1954. Instead, by moving the analysis scale from the entire sub-unit to the single transepts, a homogeneous trend is observed in line with the global one up to 2000. After, the trend becomes very variable even in the time intervals where the average variation of the beach area is nothing or small. The situation that in the same time interval there are transepts in erosion and in advancement, but with zero or small variations in the total area of the beach, highlights that the sediments are redistributed within the sub-unit. Additionally, in this time interval, no significant anthropogenic interventions were observed along coasts and rivers of the study area. Therefore, anthropogenic pressure does not appear to be the cause of these evolutionary trends, so to identify the possible causes of this phenomenon, natural factors were analyzed, especially the balance between longshore and river sediment transport. In detail, the mean longshore transport estimated with the Tomasicchio et al. [78] model is about 90,000 m^3/year and is directed from South to North so the Stilaro River would be the main source of sediment for the study area. However, this area is also subject to sea storms from the North-East directions, so the Assi River is also a source of sediments. The mean river transport, estimated with the Gavrilovic model [80], is over 60,000 m^3/year for the Stilaro River and over 40,000 m^3/year for the Assi River. Globally, the river contribution afferent to the examined physiographic sub-unit is approximately 100,000 m^3/year. Therefore, longshore and river transport are on average of the same magnitude. This consideration allows us to hypothesize that the evolutionary trend of the last 25 years is mainly related to natural factors that alter the sedimentary balance, such as the action of single sea storms or particularly rainy or dry periods. Regarding the latter parameter, this hypothesis is also reinforced by the following considerations: the river sediment transport is mainly related to WSE, significant anthropic interventions are absent, and the sediments are redistributed along the study area, and the greatest variations are observed in the external transepts, near the Stilaro and Assi river mouths.

Regarding the damage to the dune system, the sea storms that occurred in the winter of 2013–2014 were analyzed. These storms occurred when the beach width was very small. From a statistical point of view, the two sea storms had return periods of 4 and 7.5 years, therefore corresponding to statistically frequent events. Between the two sea storms, the one that most damaged the coast was the second as it was characterized by a long duration, 140 h, and coming from almost orthogonal directions with respect to the coast itself. Therefore, these sea storms are not extreme events, so the damage is probably related to the action of each sea storm and is accentuated from the fact that these events occurred when the beach width was a few tens of meters, much lower than that observed in the other time intervals analyzed. The storms of winter 2013–2014 did not only affect the coast of Monasterace but also affected several locations in Western Europe causing beach and dune erosion [46,54]. In summary, the main causes of the erosive processes of the dune–beach system are anthropic in the last century and natural in the new century.

Finally, after analyzing the main causes of erosive processes, the effectiveness of two different interventions was analyzed. The first intervention is a barrier that was built in urgency between the two sea storms and was located on the beach between the shoreline and the dune. This barrier, however, represents a temporary intervention, with limited effectiveness as the sea storms crossed it (Figure 7b) as occurred during the storm of February 2014. Furthermore, the barrier has a length of 30 m while the archaeological site has a length of over 250 m so that a significant portion of the dune has remained unprotected (Figure 8a). Instead, the barrier built after 2017 was located close to the dune and extends the entire dune length. Thus, the problem of the wave motion crossing was solved, and the entire dune was protected. The validity of this intervention is highlighted by the fact that after its realization no further damage was reported. Instead, temporary and urgent interventions such as the one at the beginning of 2014 often are not very effective.

5. Conclusions

The paper, through a case study concerns the archaeological site of Kaulon, located on a dune in the Ionian coast of Calabria (Italy), analyzed the beach and dune erosive processes, their causes, and possible interventions.

The methodology proposed in this paper does not focus on a single erosive factor but evaluates the effects caused by three main factors on the beach–dune system. These factors are anthropogenic pressure, wave climate, and sea storms and river transport. On the other hand, the effects were assessed in terms of shoreline changes and damage to the dune system.

The analysis of the shoreline evolutionary trends carried out at a physiographic sub-unit scale highlighted a possible correlation between erosive processes and anthropogenic pressures in the second half of the last century. On the other hand, in the new century the evolutionary trends are mainly correlated to natural factors. Among the natural factors, the most relevant were the sea storms of winter 2013–2014 which damaged the archaeological site. River transport is also relevant both due to the peculiarities of the Calabrian rivers and due to the shoreline changes observed near the Assi and Stilaro river mouths.

In addition to this analysis, the effects produced by the construction of two interventions to protect the dune and the archaeological site were also evaluated. In detail, a barrier of modest width was initially built, located in the stretch of beach between the dune and the shoreline. Subsequently, a barrier was built at the foot of the dune, which extends along the entire length of the archaeological site.

However, the initial barrier did not serve to protect the dune itself which, on the contrary, was more damaged by the second storm and the wave motion bypassed the barrier itself. Instead, the barrier at the foot of the dune that extends along the entire length of the archaeological site seems to be a suitable intervention to protect the archaeological site from the action of sea storms as evidenced by the absence of new damage from its construction to date.

Finally, the analysis described in the paper highlighted how temporary interventions, often punctual and carried out in urgency, are not resolutive of the erosive problem but, instead, can accentuate it. Therefore, extended interventions, that consider the causes of erosive phenomena, are more suitable. This analysis is of interest for the protection and management of coastal areas and can be easily replicated in any location as it requires free access data and uses free access software and models present in scientific literature and characterized by ease of application.

Author Contributions: Conceptualization, G.B.; methodology, G.B., G.F., G.C.B. and F.F.; software, G.F. and G.C.B.; validation, G.B., G.F., G.C.B. and F.F.; formal analysis, G.F. and G.C.B.; investigation, G.F. and G.C.B.; resources, G.F. and G.C.B.; data curation, G.F. and G.C.B.; writing—original draft preparation, G.F.; writing—review and editing, G.B., G.F., G.C.B. and F.F.; visualization, G.F.; supervision, G.B.; project administration, G.B. All authors have read and agreed to the published version of the manuscript.

Funding: This research received no external funding.

Institutional Review Board Statement: Not applicable.

Informed Consent Statement: Not applicable.

Acknowledgments: Part of this work was carried out within the SIMONA Project (Systems and technologies for the monitoring of cultural areas in underwater and terrestrial environment, POR Calabria 2007/2013). Furthermore, the authors thank Giovanni Besio for providing the wave data of the MeteOcean database.

Conflicts of Interest: The authors declare no conflict of interest.

References

1. Fryberger, S.G.; Dean, G. Dune forms and wind regime. In *A Study of Global Sand Seas*; McKee, E.D., Ed.; United States Geological Survey Professional Paper; US Government Printing Office: Washington, DC, USA, 1979; Volume 1052, pp. 137–169.
2. Jay, H. Beach–Dune Sediment Exchange and Morphodynamic Responses: Implications for Shoreline Management, the Sefton Coast, NW England. Ph.D. Thesis, University of Reading, Reading, UK, 1998.
3. Sabatier, F.; Anthony, E.J.; Héquette, A.; Suanez, S.; Musereau, J.; Ruz, M.-H.; Regnauld, H. Morphodynamics of beach/dune systems: Examples from the coast of France. *Géomorphol. Relief Processus Environ.* **2009**, *15*, 3–22. [CrossRef]
4. Harley, M.; Ciavola, P. Managing local coastal inundation risk using real-time forecasts and artificial dune placements. *Coast. Eng.* **2013**, *77*, 77–90. [CrossRef]
5. Luijendijk, A.; Hagenaars, G.; Ranasinghe, R.; Baart, F.; Donchyts, G.; Aarninkhof, S. The State of the World's Beaches. *Sci. Rep.* **2018**, *8*, 6641. [CrossRef] [PubMed]
6. Mentaschi, L.; Vousdoukas, M.I.; Pekel, J.-F.; Voukouvalas, E.; Feyen, L. Global long-term observations of coastal erosion and accretion. *Sci. Rep.* **2018**, *8*, 12876. [CrossRef] [PubMed]
7. Vousdoukas, M.I.; Ranasinghe, R.; Mentaschi, L.; Plomaritis, T.A.; Athanasiou, P.; Luijendijk, A.; Feyen, L. Sandy coastlines under threat of erosion. *Nat. Clim. Chang.* **2020**, *10*, 260–263. [CrossRef]
8. European Commission. Article 17 Technical Report 2001–2006; European Topic Centre on Biological Diversity. 2008. Available online: https://ec.europa.eu/environment/nature/knowledge/rep_habitats/index_en.htm#heading2001/06 (accessed on 20 October 2021).
9. Pye, K. Physical and human influences on coastal dune development between the Ribble and Mersey estuaries, Northwest England. In *Coastal Dunes: Processes and Morphology*; Nordstrom, K.F., Psuty, N.P., Carter, R.W.G., Eds.; Wiley: Chichester, UK, 1990; pp. 339–359.
10. Sancho, F.; Abreu, T.; D'Alessandro, F.; Tomasicchio, G.R.; Silva, P.A. Surf hydrodynamics under collapsing coastal dunes. *J. Coast. Res.* **2011**, *64*, 144–148.
11. Anthony, E.J. Storms, shoreface morphodynamics, sand supply, and the accretion and erosion of coastal dune barriers in the southern North Sea. *Geomorphology* **2013**, *199*, 8–21. [CrossRef]
12. Feagin, R.; Furman, M.; Salgado, K.; Martinez, M.; Innocenti, R.; Eubanks, K.; Figlus, J.; Huff, T.; Sigren, J.; Silva, R. The role of beach and sand dune vegetation in mediating wave run up erosion. *Estuar. Coast. Shelf Sci.* **2019**, *219*, 97–106. [CrossRef]
13. Pye, K.; Blott, S. Decadal-scale variation in dune erosion and accretion rates: An investigation of the significance of changing storm tide frequency and magnitude on the Sefton coast, UK. *Geomorphology* **2008**, *102*, 652–666. [CrossRef]
14. Swann, C.; Bride, K.; Spore, N. *Coast Foredunes: Identifying Coastal, Aeolian and Management Interactions Driving Morphological State Change*; ERDC/CHL TR-17873; U.S. Army Corps of Engineers: Washington, DC, USA, 2014.
15. Walker, I.J.; Davidson-Arnott, R.G.D.; Bauer, B.O.; Hesp, P.A.; Delgado-Fernandez, I.; Ollerhead, J.; Smyth, T.A.G. Scale-dependent perspectives on the geomorphology and evolution of beach-dune systems. *Earth-Sci. Rev.* **2017**, *171*, 220–253. [CrossRef]
16. Tomasicchio, G.R.; Francone, A.; Simmonds, D.J.; D'Alessandro, F.; Frega, F. Prediction of Shoreline Evolution. Reliability of a General Model for the Mixed Beach Case. *J. Mar. Sci. Eng.* **2020**, *8*, 361. [CrossRef]
17. Miduri, M.; Foti, G.; Puntorieri, P. Impact generated by Marina of Badolato on adjacent coasts. In Proceedings of the 13th International Congress on Coastal and Marine Sciences, Engineering, Management & Conservation (MEDCOAST), Mellieha, Malta, 31 October–4 November 2017; Volume 2, pp. 935–945.
18. Zema, D.A.; Bombino, G.; Boix-Fayos, C.; Tamburino, V.; Zimbone, S.M.; Fortugno, D. Evaluation and modeling of scouring and sedimentation around check dams in a Mediterranean torrent in Calabria, Italy. *J. Soil Water Conserv.* **2014**, *69*, 316–329. [CrossRef]
19. Foti, G.; Barbaro, G.; Manti, A.; Foti, P.; La Torre, A.; Geria, P.F.; Puntorieri, P.; Tramontana, N. A methodology to evaluate the effects of river sediment withdrawal: The case study of the Amendolea River in southern Italy. *Aquat. Ecosyst. Health Manag.* **2020**, *23*, 465–473. [CrossRef]
20. Dissanayake, P.; Brown, J.; Karunarathna, H. Impacts of storm chronology on the morphological changes of the Formby beach and dune system, UK. *Nat. Hazards Earth Syst. Sci.* **2015**, *15*, 1533–1543. [CrossRef]
21. Splinter, K.D.; Carley, J.T.; Golshani, A.; Tomlinson, R. A relationship to describe the cumulative impact of storm clusters on beach erosion. *Coast. Eng.* **2014**, *83*, 49–55. [CrossRef]
22. Dissanayake, P.; Brown, J.; Wisse, P.; Karunarathna, H. Comparison of storm cluster vs isolated event impacts on beach/dune morphodynamics. *Estuar. Coast. Shelf Sci.* **2015**, *164*, 301–312. [CrossRef]
23. Wernette, P.; Houser, C.; Lehner, J.; Evans, A.; Weymer, B. Investigating the Impact of Hurricane Harvey and Driving on Beach-Dune Morphology. *Geomorphology* **2020**, *358*, 107119. [CrossRef]
24. Barbaro, G.; Petrucci, O.; Canale, C.; Foti, G.; Mancuso, P.; Puntorieri, P. Contemporaneity of Floods and Storms. A Case Study of Metropolitan Area of Reggio Calabria in Southern Italy. In *New Metropolitan Perspectives. ISHT 2018. Smart Innovation, Systems and Technologies*; Springer: Cham, Switzerland, 2019; Volume 101, pp. 614–620. [CrossRef]
25. Canale, C.; Barbaro, G.; Foti, G.; Petrucci, O.; Besio, G.; Barillà, G.C. Bruzzano river mouth damage due to meteorological events. *Int. J. River Basin Manag.* **2021**, 1–17. [CrossRef]
26. Canale, C.; Barbaro, G.; Petrucci, O.; Fiamma, V.; Foti, G.; Barillà, G.C.; Puntorieri, P.; Minniti, F.; Bruzzaniti, L. Analysis of floods and storms: Concurrent conditions. *Ital. J. Eng. Geol. Environ.* **2020**, *1*, 23–29. [CrossRef]

27. Barbaro, G.; Foti, G.; Nucera, A.; Barillà, G.C.; Canale, C.; Puntorieri, P.; Minniti, F. Risk mapping of coastal flooding areas. Case studies: Scilla and Monasterace (Italy). *Int. J. Saf. Secur. Eng.* **2020**, *10*, 59–67. [CrossRef]
28. Castelle, B.; Marieu, V.; Bujan, S.; Splinter, K.D.; Robinet, A.; Sénéchal, N.; Ferreira, S. Impact of the winter 2013–2014 series of severe Western Europe storms on a double-barred sandy coast: Beach and dune erosion and megacusp embayments. *Geomorphology* **2015**, *238*, 135–148. [CrossRef]
29. Houser, C. Alongshore variation in the morphology of coastal dunes: Implications for storm response. *Geomorphology* **2013**, *199*, 48–61. [CrossRef]
30. de Winter, R.; Gongriep, F.; Ruessink, B. Observations and modeling of alongshore variability in dune erosion at Egmond aan Zee, The Netherlands. *Coast. Eng.* **2015**, *99*, 167–175. [CrossRef]
31. Pye, K.; Neal, A. Coastal dune erosion at Formby Point, north Merseyside, England: Causes and Mechanisms. *Mar. Geol.* **1994**, *119*, 39–56. [CrossRef]
32. Molina, R.; Manno, G.; Re, C.L.; Anfuso, G. Dune Systems' Characterization and Evolution in the Andalusia Mediterranean Coast (Spain). *Water* **2020**, *12*, 2094. [CrossRef]
33. van Thiel de Vries, J.S.M.; Van Gent, M.R.A.; Walstra, D.J.R.; Reniers, A.J.H.M. Analysis of dune erosion processes in large-scale flume experiments. *Coast. Eng.* **2008**, *55*, 1028–1040. [CrossRef]
34. Tomasicchio, G.R.; D'Alessandro, F.; Barbaro, G. Composite modelling for large-scale experiments on wave–dune interaction. *J. Hydraul. Res.* **2011**, *49*, 15–19. [CrossRef]
35. Tomasicchio, G.R.; Sánchez-Arcilla, A.; D'Alessandro, F.; Ilic, S.; James, M.R.; Sancho, F.; Fortes, C.J.; Schüttrumpf, H. Large-scale experiments on dune erosion processes. *J. Hydraul. Res.* **2011**, *49*, 20–30. [CrossRef]
36. D'Alessandro, F.; Tomasicchio, G.R. Wave–dune interaction and beach resilience in large-scale physical model tests. *Coast. Eng.* **2016**, *116*, 15–25. [CrossRef]
37. Palmsten, M.; Splinter, K.D. Observations and simulations of wave runup during a laboratory dune erosion experiment. *Coast. Eng.* **2016**, *115*, 58–66. [CrossRef]
38. Larson, M.; Erikson, L.; Hanson, H. An analytical model to predict dune erosion due to wave impact. *Coast. Eng.* **2004**, *51*, 675–696. [CrossRef]
39. Roelvink, D.; Reniers, A.; Van Dongeren, A.; De Vries, J.V.T.; McCall, R.; Lescinski, J. Modelling storm impacts on beaches, dunes and barrier islands. *Coast. Eng.* **2009**, *56*, 1133–1152. [CrossRef]
40. D'Alessandro, F.; Tomasicchio, G.R.; Musci, F.; Ricca, A. Dune erosion physical, analytical and numerical modelling. In Proceedings of the 33rd International Conference on Coastal Engineering, Santander, Spain, 1–6 July 2012. [CrossRef]
41. Esteves, L.S.; Brown, J.M.; Williams, J.J.; Lymbery, G. Quantifying thresholds for significant dune erosion along the Sefton Coast, Northwest England. *Geomorphology* **2012**, *143–144*, 52–61. [CrossRef]
42. Li, F.; van Gelder, P.; Vrijling, J.; Callaghan, D.; Jongejan, R.; Ranasinghe, R. Probabilistic estimation of coastal dune erosion and recession by statistical simulation of storm events. *Appl. Ocean Res.* **2014**, *47*, 53–62. [CrossRef]
43. Furmańczyk, K.; Dudzińska-Nowak, J.; Paplińska-Swerpel, B.; Brzezowska, N.; Furmańczyk, K. Critical storm thresholds for the generation of significant dune erosion at Dziwnow Spit, Poland. *Geomorphology* **2012**, *143–144*, 62–68. [CrossRef]
44. Tătui, F.; Vespremeanu-Stroe, A.; Preoteasa, L. Alongshore variations in beach-dune system response to major storm events on the Danube Delta coast. *J. Coast. Res.* **2014**, *70*, 693–699. [CrossRef]
45. Saye, S.E.; van der Wal, D.; Pye, K.; Blott, S.J. Beach–dune morphological relationships and erosion/accretion: An investigation at five sites in England and Wales using LIDAR data. *Geomorphology* **2005**, *72*, 128–155. [CrossRef]
46. Pye, K.; Blott, S.J. Assessment of beach and dune erosion and accretion using LiDAR: Impact of the stormy 2013–14 winter and longer term trends on the Sefton Coast, UK. *Geomorphology* **2016**, *266*, 146–167. [CrossRef]
47. Lerma, A.N.; Ayache, B.; Ulvoas, B.; Paris, F.; Bernon, N.; Bulteau, T.; Mallet, C. Pluriannual beach-dune evolutions at regional scale: Erosion and recovery sequences analysis along the aquitaine coast based on airborne LiDAR data. *Cont. Shelf Res.* **2019**, *189*, 103974. [CrossRef]
48. Pagán, J.I.; Bañón, L.; López, I.; Bañón, C.; Aragonés, L. Monitoring the dune-beach system of Guardamar del Segura (Spain) using UAV, SfM and GIS techniques. *Sci. Total Environ.* **2019**, *687*, 1034–1045. [CrossRef]
49. Andriolo, U.; Gonçalves, G.; Sobral, P.; Fontán-Bouzas, Á.; Bessa, F. Beach-dune morphodynamics and marine macro-litter abundance: An integrated approach with Unmanned Aerial System. *Sci. Total. Environ.* **2020**, *749*, 141474. [CrossRef] [PubMed]
50. D'Alessandro, F.; Tomasicchio, G.R.; Frega, F.; Carbone, M. Design and management aspects of a coastal protection system. A case history in the South of Italy. *J. Coast. Res.* **2011**, *64*, 492–495.
51. van Rijn, L. Coastal erosion and control. *Ocean Coast. Manag.* **2011**, *54*, 867–887. [CrossRef]
52. Suanez, S.; Cariolet, J.-M.; Cancouët, R.; Ardhuin, F.; Delacourt, C. Dune recovery after storm erosion on a high-energy beach: Vougot Beach, Brittany (France). *Geomorphology* **2012**, *139–140*, 16–33. [CrossRef]
53. Scott, T.; Masselink, G.; O'Hare, T.; Saulter, A.; Poate, T.; Russell, P.; Davidson, M.; Conley, D. The extreme 2013/2014 winter storms: Beach recovery along the southwest coast of England. *Mar. Geol.* **2016**, *382*, 224–241. [CrossRef]
54. Castelle, B.; Bujan, S.; Ferreira, S.; Dodet, G. Foredune morphological changes and beach recovery from the extreme 2013/2014 winter at a high-energy sandy coast. *Mar. Geol.* **2017**, *385*, 41–55. [CrossRef]
55. Smith, E.R.; D'Alessandro, F.; Tomasicchio, G.R.; Gailani, J.Z. Nearshore placement of a sand dredged mound. *Coast. Eng.* **2017**, *126*, 1–10. [CrossRef]

56. Williams, A.; Rangel-Buitrago, N.; Pranzini, E.; Anfuso, G. The management of coastal erosion. *Ocean Coast. Manag.* **2018**, *156*, 4–20. [CrossRef]
57. Bazzichetto, M.; Sperandii, M.; Malavasi, M.; Carranza, M.; Acosta, A. Disentangling the effect of coastal erosion and accretion on plant communities of Mediterranean dune ecosystems. *Estuar. Coast. Shelf Sci.* **2020**, *241*, 106758. [CrossRef]
58. D'Alessandro, F.; Tomasicchio, G.R.; Francone, A.; Leone, E.; Frega, F.; Chiaia, G.; Saponieri, A.; Damiani, L. Coastal sand dune restoration with an eco-friendly technique. *Aquat. Ecosyst. Health Manag.* **2020**, 417–426. [CrossRef]
59. Fernández-Montblanc, T.; Duo, E.; Ciavola, P. Dune reconstruction and revegetation as a potential measure to decrease coastal erosion and flooding under extreme storm conditions. *Ocean Coast. Manag.* **2020**, *188*, 105075. [CrossRef]
60. Leone, E.; Kobayashi, N.; Francone, A.; Bartolo, S.; Strafella, D.; D'Alessandro, F.; Tomasicchio, G. Use of Nanosilica for Increasing Dune Erosion Resistance during a Sea Storm. *J. Mar. Sci. Eng.* **2021**, *9*, 620. [CrossRef]
61. Cohn, N.; Brodie, K.L.; Johnson, B.; Palmsten, M.L. Hotspot dune erosion on an intermediate beach. *Coast. Eng.* **2021**, *170*, 103998. [CrossRef]
62. Baldock, T.; Gravois, U.; Callaghan, D.; Davies, G.; Nichol, S. Methodology for estimating return intervals for storm demand and dune recession by clustered and non-clustered morphological events. *Coast. Eng.* **2021**, *168*, 103924. [CrossRef]
63. Cohn, N.; Ruggiero, P.; García-Medina, G.; Anderson, D.; Serafin, K.A.; Biel, R. Environmental and morphologic controls on wave-induced dune response. *Geomorphology* **2019**, *329*, 108–128. [CrossRef]
64. Dissanayake, P.; Brown, J.; Sibbertsen, P.; Winter, C. Using a two-step framework for the investigation of storm impacted beach/dune erosion. *Coast. Eng.* **2021**, *168*, 103939. [CrossRef]
65. Hird, S.; Stokes, C.; Masselink, G. Emergent coastal behaviour results in extreme dune erosion decoupled from hydrodynamic forcing. *Mar. Geol.* **2021**, *442*, 106667. [CrossRef]
66. Sanromualdo-Collado, A.; García-Romero, L.; Peña-Alonso, C.; Hernández-Cordero, A.I.; Ferrer-Valero, N.; Hernández-Calvento, L. Spatiotemporal analysis of the impact of artificial beach structures on biogeomorphological processes in an arid beach-dune system. *J. Environ. Manag.* **2021**, *282*, 111953. [CrossRef]
67. Barbaro, G.; Foti, G.; Sicilia, C.L. Erosive Phenomena in the Proximity of Kaulon Archaeological Park: Origins and Remedies. *Procedia Soc. Behav. Sci.* **2016**, *223*, 714–719. [CrossRef]
68. Barbaro, G.; Foti, G.; Sicilia, C.L. Fenomeni erosivi in prossimità del parco archeologico di kaulon (RC): Cause e rimedi. *LaborEst* **2016**, *12*, 75–79. [CrossRef]
69. Mandaglio, M.C.; Gioffrè, D.; Pitasi, A.; Moraci, N. Qualitative Landslide Susceptibility Assessment in Small Areas. *Procedia Eng.* **2016**, *158*, 440–445. [CrossRef]
70. Calabria Region. *Indagine Conoscitiva Dello Stato Delle COSTE Calabresi, predisposizione di una Banca Dati Dell'evoluzione del Litorale e Individuazione Delle Aree a Rischio e Delle Tipologie di Intervento. Studi su Aree Campione e Previsione Delle RELATIVE opere di Difesa*; Final Report; Calabria Region: Catanzaro, Italy, 2003. (In Italian)
71. Sabato, L.; Tropeano, M. Fiumara: A kind of high hazard river. *Phys. Chem. Earth Parts A/B/C* **2004**, *29*, 707–715. [CrossRef]
72. Sorriso-Valvo, M.; Terranova, O. The Calabrian fiumara streams. *Z. Für Geomorphol.* **2006**, *143*, 109–125.
73. Barbaro, G.; Foti, G.; Mandaglio, G.; Mandaglio, M.; Sicilia, C.L. Estimate of sediment transport capacity in the basin of the Fiumara Annunziata (RC). In Proceedings of the Atti del 86° Congresso Nazionale della Società Geologica Italiana, Arcavacata di Rende (CS), Italy, 18–20 September 2012; Volume 21, pp. 696–697.
74. Terranova, O.; Antronico, L.; Coscarelli, R.; Iaquinta, P. Soil erosion risk scenarios in the Mediterranean environment using RUSLE and GIS: An application model for Calabria (southern Italy). *Geomorphology* **2009**, *112*, 228–245. [CrossRef]
75. Barbaro, G.; Bombino, G.; Foti, G.; Borrello, M.M.; Puntorieri, P. Shoreline evolution near river mouth: Case study of Petrace River (Calabria, Italy). *Reg. Stud. Mar. Sci.* **2019**, *29*, 100619. [CrossRef]
76. Foti, G.; Barbaro, G.; Bombino, G.; Fiamma, V.; Puntorieri, P.; Minniti, F.; Pezzimenti, C. Shoreline changes near river mouth: Case study of Sant'Agata River (Reggio Calabria, Italy). *Eur. J. Remote Sens.* **2019**, *52*, 102–112. [CrossRef]
77. Stockdon, H.F.; Holman, R.A.; Howd, P.A.; Sallenger, A.H. Empirical parameterization of setup, swash, and runup. *Coast. Eng.* **2006**, *53*, 573–588. [CrossRef]
78. Istituto Idrografico della Marina. *Tavole di Marea e Delle Correnti di Marea*; Istituto Idrografico della Marina Italiana: Genova, Italy, 2020; p. 144. ISBN 97888II3133. (In Italian)
79. Sannino, G.; Carillo, A.; Pisacane, G.; Naranjo, C. On the relevance of tidal forcing in modelling the Mediterranean thermohaline circulation. *Prog. Oceanogr.* **2015**, *134*, 304–329. [CrossRef]
80. Bretschneider, C.L. Engineering aspects of hurricane surge. In *Estuary and Coastline Hydrodynamics*; Ippen, A.T., Ed.; McGraw-Hill: New York, NY, USA, 1966.
81. Tomasicchio, G.R.; D'Alessandro, F.; Barbaro, G.; Malara, G. General longshore transport model. *Coast. Eng.* **2013**, *71*, 28–36. [CrossRef]
82. Lamberti, A.; Tomasicchio, G.R. Stone mobility and longshore transport at reshaping breakwaters. *Coast. Eng.* **1997**, *29*, 263–289. [CrossRef]
83. Gavrilovic, S. *Méthode de la Classification des Bassins Torrentiels et Équations Nouvelles Pour le Calcul des Hautes Eaux et du Débit Solide*; Vadoprivreda: Belgrado, Serbia, 1959. (In French)
84. Giandotti, M. *Previsione Delle Piene e delle Magre dei corsi d'acqua*; Servizio Idrografico Italiano, Memorie e Studi Idrografici: Roma, Italy, 1934; Volume 8, pp. 8–13. (In Italian)

85. Kirpich, Z.P. Time of concentration of small agricultural watersheds. *Civ. Eng.* **1940**, *10*, 362.
86. Natural Resources Conservation Service (NRCS). *Pondsplanning, Design Construction. Agriculture Handbook*; United States Department of Agriculture (USDA): Washington, DC, USA, 1997.
87. ASCE. *Hydrology Handbook*; ASCE Manuals and Reports on Engineering Practice n. 28; ASCE: Reston, VA, USA, 1996. [CrossRef]
88. Fiedler, F.R. Simple, Practical Method for Determining Station Weights Using Thiessen Polygons and Isohyetal Maps. *J. Hydrol. Eng.* **2003**, *8*, 219–221. [CrossRef]
89. Boak, E.H.; Turner, I. Shoreline Definition and Detection: A Review. *J. Coast. Res.* **2005**, *214*, 688–703. [CrossRef]
90. Hapke, C.J.; Himmelstoss, E.A.; Kratzmann, M.G.; List, J.H.; Thieler, E.R. *National Assessment of Shoreline Change: Historical Shoreline Change along the New England and Mid-Atlantic Coasts*; Open-File Report 1118; US Geological Survey: Reston, Virginia, 2010. [CrossRef]

Journal of
Marine Science and Engineering

MDPI

Article

The Dune Engineering Demand Parameter and Applications to Forecasting Dune Impacts

Matthew S. Janssen * and Jon K. Miller

Department of Civil, Environmental, and Ocean Engineering, Stevens Institute of Technology, Hoboken, NJ 07030, USA; jmiller@stevens.edu
* Correspondence: mjanssen@stevens.edu

Abstract: Breaching or overtopping of coastal dunes is associated with greater upland damages. Reliable tools are needed to efficiently assess the likelihood of dune erosion during storm events. Existing methods rely on numerical modeling (extensive investment) or insufficiently parameterize the system. To fill this gap, a fragility curve model using a newly developed dune Engineering Demand Parameter (EDP) is introduced. Conceptually, the EDP is similar to the Shield's parameter in that it represents the ratio of mobilizing terms to stabilizing terms. Physically, the EDP is a measure of storm intensity over the dune's resilience. To highlight potential applications, the proposed EDP fragility curve models are fit to a spatially and temporally robust dataset and used to predict dune response subjected to varying storm intensities including both extratropical and tropical storm. This approach allows for the probabilistic prediction of dune impacts through an innovative, computationally efficient model. Several different forms of the EDP are tested to determine the best schematization of the dune resilience. The final recommended EDP is the Peak Erosion Intensity (PEI) raised to the fourth power over the product of the dune volume and berm-width squared. Including both storm intensity and resilience terms in the EDP enables comparison of different beach configurations in different storm events fulfilling a need existing vulnerability assessors cannot currently account for directly.

Keywords: erosion potential; dune erosion; New Jersey; fragility curves; probabilistic dune response; coastal hazards

Citation: Janssen, M.S.; Miller, J.K. The Dune Engineering Demand Parameter and Applications to Forecasting Dune Impacts. *J. Mar. Sci. Eng.* **2022**, *10*, 234. https://doi.org/10.3390/jmse10020234

Academic Editors: Felice D'Alessandro, Giuseppe Roberto Tomasicchio and Ferdinando Frega

Received: 1 December 2021
Accepted: 1 February 2022
Published: 9 February 2022

1. Introduction

1.1. Motivation

The performance of the combined beach and dune system subjected to coastal storms is of importance to engineers, coastal managers and underwriters alike. Breaches, overwash or significant overtopping of a dune are associated with greater impacts to upland infrastructures. This has in part led to the development and widespread use of categorical scales for predicting/assessing storm impact such as the Sallenger [1] Storm Impact Scale. Alternatively, dune and beach performance can be predicted numerically using models of varying complexity and fidelity (e.g., DUROS, SBEACH, XBeach, CShore, etc.). However numerical methods require significant effort through model initialization (pre-event bathymetry and topography), calibration and, for the 2D models over large domains, significant computational run times. These traits require significant investment and preclude the widespread implementation in real-time forecasting applications. Furthermore, performance of cross-shore models is highly dependent on the calibration and even with a calibrated model, the skill of a model may not be uniform if applied over wide range of storm severity (e.g., moderate to severe events) [2]. Fragility curves, which have been used extensively in the field of seismic design, have more recently been applied to coastal engineering problems [3,4]. These methods allow for the consideration of probabilistic estimates and can be extended beyond the binary 'survive' or 'fail' models.

Despite extensive advancements in numerical simulations, there is still a demonstrated need to predict dune impacts in storm events using analytic or index-based equations. However, the approaches available in the existing literature have one or more deficiencies that can broadly be categorized as follows:

1. The parameters account for either the dune resilience parameters or the storm intensity, but not both;
2. Provide a categorical response only (e.g., the scale is not a continuous number which limits the usefulness in statistical fitting);
3. Incomplete parameterization of the resilience terms (e.g., consider dune volume but not position of the dune);
4. Incomplete consideration for storm intensity parameters (e.g., account for peak storm intensity but not duration effects).

A summary of current approaches is presented in the following subsections, where they are grouped according to the interdependence of beach width and volume in the design of coastal defense projects (Section 1.2), widely used parameters to describe dune resilience (Section 1.3) and lastly parameterization of storm intensity (Sections 1.4 and 1.5).

Here, the dune EDP is proposed to fulfill the need for a single variable that simultaneously parameterizes both the storm intensity and the resilience of a beach and dune in a physically meaningful way. The EDP is ideally suited to applications within a fragility curve framework by preserving the underlying physics and not exclusively relying on a statistical fit. The parameter considers both the mobilizing terms (intensity) and the stabilizing terms (resilience) to capture the demand placed on the dune. Characterizing the demand enables better estimation of the vulnerability over a range of storm intensities and beach configurations. The intended application of the EDP is in the evaluation of dunes with incomplete or partially defined data (e.g., cases where the dune volume may be estimated but offshore bathymetry is unknown). While the methods do not, and should not, replace numerical modeling, they have uses in forecasting, rapid or regional scale assessments, and as a design tool in conceptual design.

1.2. Quantifying the Combined Effects of the Beach Width and Dune Volume

The protective benefit afforded by the dunes and berm in the broader beach system are deeply intertwined. However, despite the consensus that dunes are a primary defense feature to coastal storms, the design and quantification of the protection afforded by dunes and, more specifically, the combined effect of beach width and dune volume is limited in the existing literature. Consider that in Dean's textbook [5], appropriately considered the seminal work in the beach nourishment design, the benefit of dunes is not included in either the chapter or appendix dedicated to quantification of the benefit of nourishment. Instead, the quantification of benefit is focused exclusively on beach width. Additional gold standards for consulting engineering including [6–8] provide an excellent description of the physical processes and role of dunes as a reserve volume in the beach–dune system. However, none provide explicit guidance on the design of dunes in relationship to the beach width. Bruun [9] highlighted the necessity for the beach and dune to be considered comprehensively and described considerations for geometric shape and placement. The often cited FEMA 540-rule [10] is used in assessment of existing dunes for the National Flood Insurance program, though has been adapted to provide guidance for construction as well [11]. Some of the most relevant quantification on dune design stems from The Netherlands starting with Vellinga [12] and resulting in technical guidance, e.g., [13]. A more expansive summary can be found in [14].

1.3. Resilence or Vulnerbility Assessors

One of the first comprehensive methods to identify areas with increased susceptibly to storm-driven morphology changes (e.g., breaches, overtopping, etc.) at a regional scale was performed by Morgan and Stone [15]. Morgan and Stone developed the Storm Wave Susceptibility Quotient (SWS), an index used to quantify the vulnerability of sandy coasts located along Florida's barrier islands. The schematization of the wave climate was simple but precludes assessment to a specific storm or event.

First-order estimates, such as the 540-rule rule [10] are delightfully simple to apply. However, they poorly resolve the risk of dune failure for return periods above or below 100-years, fail to account for the position of the dune (setback) and do not quantify the uncertainty and spread in possible outcomes. Consequently, guidance for best practice has been to double the expected median eroded volumes to account for the significant uncertainty [11].

Sallenger [1] introduced the Storm Impact Scale for barrier islands that categorically classified impacts for both tropical and extra-tropical storms based on wave runup. Four regimes are classified as 'swash', 'impact', 'overwash' and 'inundation'. Each regime characterizes when different physical processes dominate, and the magnitude of impacts change dramatically. The category is determined between the ratio of the maximum run-up elevation and the dune toe and crest elevations. The scale is useful in qualitatively describing the dominant physics, but it does not consider duration, the volume of the dune, nor does it a provide the magnitude of the expected consequences (e.g., dune volume loss). The scale has been tested in hindcasting historical storms by Stockdon, et al. [16]. Leaman, et al. [17] expanded this scale to create a storm hazard matrix including the effects of storm duration and thus erosion-dominated events.

The Sallenger [1] Storm Impact Sale has been used by the U.S. Geological Survey (USGS) to predict shoreline changes through the Coastal Change Hazard Portal [18]. The portal is unique in that it is one of the few available sources for forecasting storm impacts during storm events. When applied through forecasting, the model uses LiDAR measurements for pre-storm dune elevations and real-time forecasts of wave and water level conditions. However, consistent with the original Storm Impact Scale, the USGS model is based on peak run-up and does not consider duration effects. This shortcoming has been acknowledged by USGS and has been attributed as a potential cause of the underprediction of overwash regimes [19]. At the time of this writing, updates to the Coastal Change Hazard Portal are expected to be released in the near future to improve the performance. Recent advances include applications of machine learning to estimate the likely beach impacts [20] by combining both the storm intensity and vulnerability of the beach–dune system.

Judge, et al. [21] assessed the vulnerability parameters to characterize the performance of coastal dunes subjected to a Category 3 (SSHWS) hurricane at 110 profiles along the coast of North Carolina. Dune performance was assessed categorically as 'failed' (greater than 50% eroded volume) or 'survived' (less than 50% eroded volume). Vulnerability parameters included median erosion [10], erosion potential [22], Strom Impact Scales [1], and traditional dune geometries including crest height, dune centroid and a new Erosion Resistance (*ER*) term (Equation (1)). Use of the *ER* term improved the prediction of survival and failure classification. The *ER* parameter is a surrogate for mass moment of inertia and accounts for the dune's cross-sectional area (A) and the centroid (Z_c) in reference to mean high water.

It is worth noting alongshore transects were spaced at approximately 300 m. Along-shore variation in storm intensity was not quantified. While the paper includes estimates of Hurricane Fran's storm surge, wave heights and duration, these parameters are not directly accounted for in the *ER* parameter.

$$ER_{DUNE} = (Z_c)^2(A) \tag{1}$$

1.4. Storm Intensity Measures

Numerous indices or scales exist to quantify and communicate storm intensity. However, most of these indices cannot be used in fragility functions. Often, the scales are not continuous and, in some cases, are not representative of the physics of the hazard. For example, application of the ubiquitous Saffir–Simpson Hurricane Wind Scale (SSHWS) applied to dune erosion is neither continuous (although wind-speed is), nor does it represent the principal hazard (elevated water levels, wave heights) as it relates to dune erosion. It is therefore unsurprising that the SSHWS storm classification is poorly corelated with dune erosion, e.g., [23] and only loosely correlated with monetary damages [24].

A summary of available storm intensity parameters is presented in Table 1. Note they are separated by storm type (i.e., tropical, extratropical or both) and intensity measure (peak, cumulative or both). The only parameters suitable for use with fragility functions that have a continuous numeric scale, are physically based and consider both duration and peak effects are the COSI [25] and SEI parameters. While either could be utilized within a fragility function framework, SEI was chosen due to its demonstrated use in literature including spatial and temporal considerations.

Table 1. Summary of available measures of Storm Intensity and key characteristics. Note variables are identified in 'Parameters Considered' column.

Parameter	Storm Type	Intensity Measure	Equation	Output	Units	Parameters Considered	Source
Saffir–Simpson Hurricane Wind Scale	Tropical	Peak	N/A	Category (1–5)	-	Windspeed	Schott et al. (2012) [26]
Hurricane Intensity Index (*HII*) & Hurricane Hazard Index (*HHI*)	Tropical	Peak	$HII = \left(\frac{V_{max}}{V_{max0}}\right)^2$ $HHI = \left(\frac{R}{R_0}\right)^2\left(\frac{V_{max}}{V_{max0}}\right)^3\left(\frac{S_0}{S}\right)$	-	-	Windspeed (V), Radius (R) and translation speed (S), 0 designates reference values	Kantha (2006) [27]
Integrated Kinetic Energy (*IKE*)	Tropical	Peak	$IKE = \int_v \frac{1}{2}\rho U^2 dV$	-	-	Wind speed (U), Storm Volume (V)	Powell and Reinhold (2007) [28]
Hurricane Severity Index	Tropical	Peak	Size + Intensity	50-point scale	-	Wind speed, Radius	Hebert et al. (2010) [29]
Surge Scale (*SS*)	Tropical	Peak	$SS = (2.43E-4)\Delta p L_{30m}\psi x\left(\frac{R_{33}}{L_{30m}}\right)$	surge	m	Pressure (Δp), Wind Radius (R_{33}), Bathymetry ($L30_m$), Storm size (ψ)	Irish and Resio (2010) [30]
Cyclone Damage Potential Index (*CDP*)	Tropical	Peak	$CDP = 4\frac{\left[\left(\frac{v_m}{65}\right)^3 + 5\left(\frac{R_h}{50}\right)\right]}{v_t}$	-	-	Wind speed (v_m), Radius (R_h), forward speed (v_t)	Done et al. (2015) [31]
Dolan and Davis Scale	Extra-tropical	Peak	$P = (H_s)^2 t_d$	-	m^2-s	Wave Height (H_s) duration (t_d)	Dolan and Davis (1992) [32] Mendoza et al. (2011) [33]
Shoreline Risk Index	Extra-tropical	Peak	$I = SH(t_D)^{0.3}$	-	m^2-$s^{0.3}$	Wave height (H), Surge (S), number of tidal cycles (t_D)	Kriebel et al. (1996) [22]
Storm Erosion Potential Index (*SEIP*)	Extra-tropical	Cumulative	$SEPI = \sum_{t=0}^{t_D} S_{2SD}(t)\, H_{MHHW}(t)\Delta t$	-	m^3	Storm surge (S and H), tide duration (t)	Zhang et al. (2001) [34]
Balsillie Regression Analysis	Both	Peak	$Qavg = \frac{1}{1622}\left(g^{1/2} t_r S^2\right)^{4/5}$	Volume (Q)	m^3	Surge (S), Tide Rise (t_r)	Balsillie (1986) [35]
Maximum Wave Run-up (R_{max})	Both	Peak	$R_{max} = H_0 2.32 \xi_o$ where $\xi_o = \tan\frac{\beta}{\sqrt{\frac{H_o}{L_o}}}$	Wave Run-up (R)	m	Wave height (H_0), period (T_p), water level, shoreline slope (β)	Kraus and Wise (1993) [36]
Coastal Storm Impulse Parameter (*COSI*)	Both	Cumulative	$I_s = \int\left[fp_{(t)} + M_{(t)}\right]d_t$	Impulse (I_s)	kg-s per m	Total Horizontal momentum, surge ($fp_{(t)}$) and Wave ($M_{(t)}$) (wave height, period, surge, duration)	Basco and Mhmoudpour (2012) [25]
Storm Erosion Potential Index	Both	Both	$\text{SEI} = \int_t \text{IEI}(t_i) = W_*(t_i)\left[\frac{0.068H_b(t_i)+S(t_i)}{B+1.28H_b(t_i)}\right]$	cross-shore distance	m	Water level (S), wave height (H_b) and duration (t), Berm elevation (B)	Miller and Livermont (2008) [37] Lemke and Miller (2020) [38]

1.5. Storm Erosion Index (SEI)

SEI developed by Miller and Livermont [37] and refined by Lemke and Miller [38] combines the three primary drivers of coastal erosion (wave height, total water level, and storm duration) into a physically meaningful index. The index is both continuous and physically based, enabling evaluation of storms based on their erosion potential. The storm's erosion potential is defined by the sum of an instantaneous erosion intensity (IEI) over the duration of the storm (t_d):

$$\text{SEI} = \sum_{t_d} \text{IEI}(t_i) = \sum_{t_d} W_*(t_i)\left[\frac{0.068H_b(t_i) + S(t_i)}{B + 1.28H_b(t_i)}\right] \quad (2)$$

where H_b is the depth-limited breaking wave height (H_b = 0.8 h_b), W^* is the width of the active surfzone (approximated as the distance to the breakpoint), B is the berm height, S is the water level height above the mean sea level, and t_i is a time index.

The foundation of the Storm Erosion Index (SEI) is the physical response of a beach profile due to increased water levels. SEI is based on a form of the well-known Bruun Rule [39] modified by Dean and Dalrymple [6] to predict the equilibrium shoreline recession due to an increase in water level, and cross-shore varying wave set-up due to breaking waves. Miller and Dean [40] utilized a time-varying form of this modified Bruun Rule to predict equilibrium shoreline change based on wave height, water level, and storm duration. Miller and Livermont [37] suggested that the time varying form could be interpreted as the Instantaneous Erosion Intensity (IEI) and that the maximum value of IEI represented a measure of storm intensity they termed the Peak Erosion Intensity (PEI).

Notably, IEI has units of length and represents a theoretical landward progression of the equilibrium shoreline position if storm conditions were to persist indefinitely. While it is not realistic to expect SEI or IEI to directly represent shoreline change, it provides a quantifiable metric for erosive potential; including both instantaneous and duration effects. Though subtle, this is important for parameterization of the physics involved. Numerous studies have found the primary factor for predicting structural damages is the cross-shore distance from the mean sea level elevation or shoreline (e.g., Walling, et al. [41], FEMA [11], Dean [5]). Thus, using a quantifiable metric that represents a theoretical landward progression of the shoreline position in terms of a length scale provides a physically meaningful parameter.

Unlike many other indices, IEI/PEI/SEI makes no distinction between tropical and extratropical storms. It has been successfully applied to a number of locations, including but not limited to, the Gulf of Mexico [42]; Atlantic coast of Florida [43,44]; Gulf coast of Florida [45]; North and South Carolina [46]; New Jersey [37,38,47,48]. SEI has been shown to be more closely related to observed erosion than traditional indices.

2. Study Area and Storm Climate

The New Jersey coast provides a rich blend of high-quality datasets over a spatially diverse domain exposed to both tropical and extra-tropical storms. The oceanfront coast of New Jersey extends roughly 210 km over a predominately sandy coast including natural and nourished beaches, and both undeveloped and developed barrier islands. Lemke and Miller [38] characterized the climate of the region's storms through the lens of the erosion potential over a 34-year period. The historical record of storms [38] is provided at 13 Segments located along the New Jersey Coast. An extensive collection of beach surveys is available through the New Jersey Beach Profile Network (NJBPN) collected bi-annually by the Coastal Research Center at Stockton University. The extents of these segments and the locations of the individual profiles from the NJBPN are shown in Figure 1.

Typical attributes of the beach including dune crest elevations, berm widths, slopes and dune volumes associated with these segments are presented in Figure 2. Median grain sizes range from 0.2 mm to 0.6 mm. Typically, the region experiences three to four storms per year [38]. A list of notable storms is provided in Table 2 with corresponding averaged return periods based on SEI. The aggregated observed impacts of each storm (limited to dune volume loss) is presented in Figure 3.

Storms within the region have been classified into Categories representing intensity [38]. The scale ranges from 1 (low intensity) to 5 (high intensity). The numerical break points for each category and the associated annual probability of exceedance are presented in Table 3. These Categories are used throughout this paper.

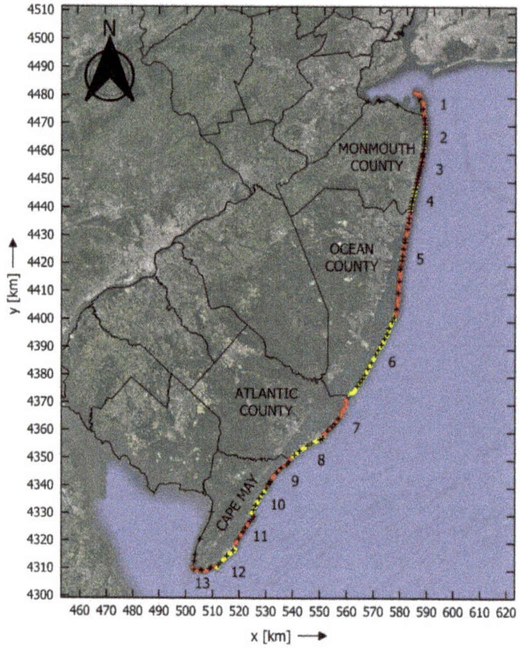

Figure 1. Study area location. Alternating red and yellow lines depict NJ Shoreline Segments utilized by Lemke and Miller [38] to produce the storm erosion potential climatology. Segments are numbered from north to south with #1 at Sandy Hook and #13 at Cape May. NJ coastal are counties labeled accordingly. Black + markers depict NJBPN survey locations. Map coordinate system is NAD83/UTM Zone 18N in kilometers. Courtesy of [20].

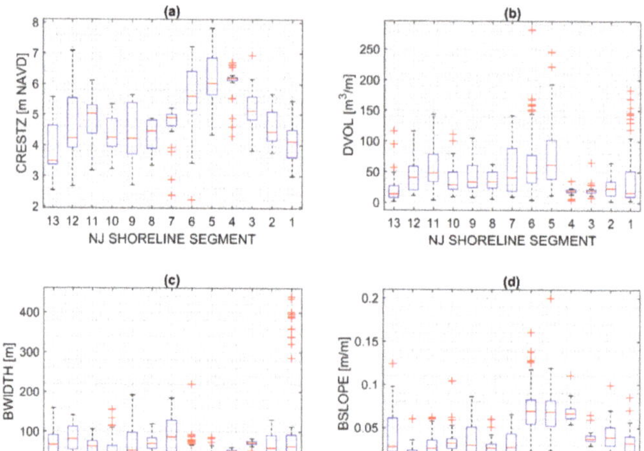

Figure 2. Spatial variance for select Resilience parameters by segment. (**a**) Dune crest elevation; (**b**) Dune volume; (**c**) Berm width; (**d**) Beach slope. Presented by segment, south (13) to north (1). Red horizontal bar—median value, blue box—25th and 75th percentiles. Outliers defined by 1.5 × IQR away from 25th/75th percentiles. Courtesy of [48].

Table 2. Average SEI and PEI values, and associated return periods (t_r), for eighteen historical storms in New Jersey. Storm data from [20]. Return periods are based on frequency of occurrence curves by [49].

Event	SEI	SEI t_r [yr]	PEI	PEI t_r [yr]
October 1991	1534	5.2	67.3	5.2
January 1992	1017	2.2	64.4	4.1
December 1992	3326	24	90.8	20
December 1994	869	1.7	57.5	2.3
January 1996	719	1.3	51.1	1.4
February 1998	1434	4.5	67.0	5.0
September 2003	1040	2.3	53.8	1.8
12 October 2005	1779	6.8	54.3	1.8
25 October 2005	866	1.7	60.0	1.4
September 2006	964	2.0	50.3	1.4
November 2007	670	1.2	60.4	2.9
May 2008	998	2.1	53.4	1.7
September 2008	1714	6.4	54.3	1.8
November 2009 (Vets)	2986	18	74.8	9.1
March 2010	1220	3.2	58.7	2.5
September 2010	581	1.0	62.9	3.6
August 2011 (Irene)	788	1.4	73.3	8.2
October 2012 (Sandy)	3056	19	119	49

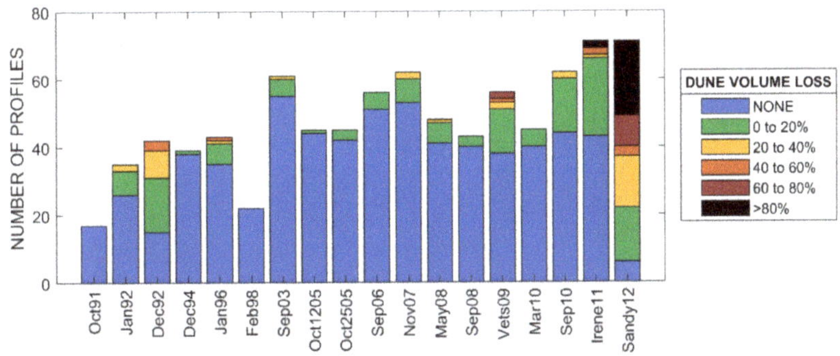

Figure 3. Aggregated summary of quantity of profiles available and corresponding dune volume loss percentage by storm for the New Jersey coast. Courtesy of [20].

Table 3. Storm Categories based on SEI and PEI including parameter value, return period (Tr) and annual probability of exceedance. Storm categories determined by [38], return period calculated from [49].

	SEI			PEI		
Category	**Value**	**Tr**	**Annual Prob.**	**Value**	**Tr**	**Annual Prob.**
2	730	1	77.1%	23	<1.0	>99.9%
3	1510	5	19.8%	58	2	41.5%
4	2300	10	9.5%	93	22	4.7%
5	3080	19	5.2%	128	62	1.6%

3. Methodology

This paper can be broken down into two components:

1. Development of the EDP;
2. Application of the EDP to develop probabilistic fragility curves of dune impacts in storm events.

This workflow is visually summarized in Figure 4. This paper makes extensive use of two existing datasets characterizing the storm intensity measures (IM) [38] and beach morphology parameters [20]. Morphology parameters include both the Resilience (R_f) parameters (e.g., beach characteristics including width, volume, etc.) and the Impact (i.e., observed dune volume loss for a given storm). The data are recast into the EDP by combining the IM and R_f from previously published datasets. The EDP is then used as a physically based parameter to fit fragility curves to predict the observed Impact classes. The EDP is the argument, and the probability of class is the return.

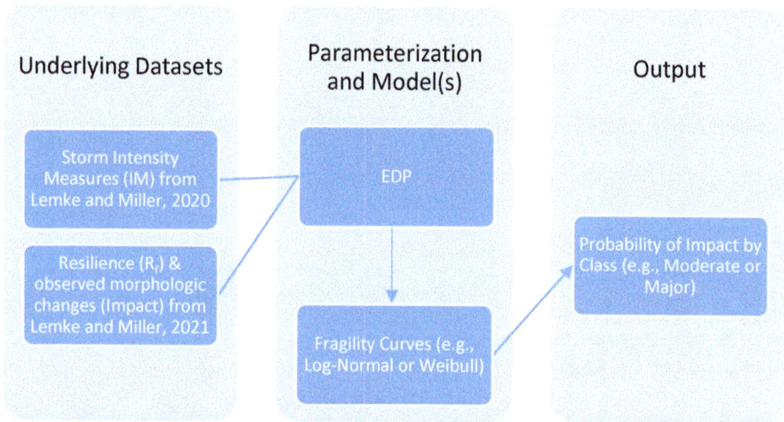

Figure 4. Conceptual flowchart to highlight sources of previously published data used to generate the EDP and subsequent fragility curves.

3.1. *Development of the Engineering Demand Parameter*

In the most general from, the EDP is a non-dimensional parameter representing the ratio of Storm Intensity ('IM') to the Resilience factors ('R_f'). The general form of the EDP is presented in Equation (3) with arbitrary variables. In theory, any combination of any R_f or IM parameters may be included in an EDP with corresponding power variables (e.g., 'i', 'j', 'k') selected to ensure non-dimensionality.

In all, 21 different IM, R_f and Impact parameters were considered. Reduction of variables was selected based on two guiding principles. The first was to ensure the most physically meaningful parameters were included both from a qualitative and quantitative means. The second was to the extent possible, include only parameters that are easily measured or estimated without specialized tools or equipment. This is a necessary compromise to make the curves more useful in real-time forecasting with limited time and resources to collect data.

Quantitative comparison was completed using Spearman correlation between the parameters and impact (i.e., dune loss percentage). In the cases of highly correlated parameters (e.g., berm volume and berm width) only the most easily measured parameter was used. A summary of parameters used is presented in Table 4.

Several forms of the EDP were considered. The EDPs and underlying parameters are presented in the third column of Table 5. Care has been taken to ensure the selection of parameters used in the various cases of the EDP are physically meaningful and not based exclusively on best fit.

$$\begin{gathered} \mathrm{EDP} = \frac{\mathrm{IM}}{R_f} = \frac{\mathrm{IM}^i}{R_{fa}{}^j,\ R_{fb}{}^k \ldots R_{fn}{}^m} \\ \textit{with}\ \mathrm{IM} = \textit{Storm Intensity Measure}\ (\text{e.g., PEI}) \\ R_f = \text{Resilience Factor (e.g., dune volume, berm width, etc.)} \end{gathered} \tag{3}$$

Table 4. Reduced parameters of interest. Note FB denotes freeboard or vertical distance between parameter and mean sea level, WL—water level, vol—volume, Hb—breaking wave height, cumE—cumulative energy (H^2), Z denotes elevation, Max denotes the maximum hourly value over a single storm duration and Bslope/Islope denote beach and intertidal slope.

Type	Parameter Name	Symbol/Abbreviation	Units	Highly Correlated
Intensity Measure	Peak Erosion Intensity	PEI	Length (m)	MaxWL, MaxHb, cumE, DL, DLvol
Resilience	Berm Width	Bwidth	Length (m)	Berm Vol
Resilience	Crest Width (cross-shore distance from dune toe to crest)	Fwidth	Length (m)	Fvol, Dvol, $Fslope^{-1}$
Resilience	Dune volume (volume from toe to heel)	Dvol	Volume per length (m^3/m)	CrestZ, ToeZ, CrestFB
Resilience	Foredune volume (volume from toe to crest)	Fvol	Volume per length (m^3/m)	Fwidth, CrestZ, Dvol, CrestFB, $Fslope^{-1}$
Resilience	Dune Crest Elevation	CrestZ	Length (m)	Dvol, Fvol, ToeZ, CrestFB, ToeFB, Bslope,
Resilience	Dune Toe Elevation	ToeZ	Length (m)	CrestFB, ToeFB, d50, ISlope, BSlope, CrestZ
Impact	Dune Loss Percent	DL	% (-)	DLvol

Table 5. Considered EDPs, underlying parameters (IM and R_f) and physical meaning.

Case	Parameters	EDP	Physical Proxy
1	PEI, Berm Width	$\frac{\text{PEI}}{(Bwidth)}$	Setback
2	PEI, Berm Width, Dune Crest Width	$\frac{\text{PEI}}{(Bwidth+Fwidth)}$	Setback
3	PEI, Dune Volume	$\frac{\text{PEI}^2}{(Dvol)}$	Volume
4	PEI, Foredune Volume	$\frac{\text{PEI}^2}{(Fvol)}$	Volume | '540-rule'
5	PEI, Berm Width and Dune Volume	$\frac{\text{PEI}^2}{(Bwidth^2+Dvol)}$	Shear
6	PEI, Berm Width and Dune Volume	$\frac{\text{PEI}^3}{(Bwidth \times Dvol)}$	Moment
7	PEI, Berm Width and Dune Volume	$\frac{\text{PEI}^4}{(Bwidth^2 \times Dvol)}$	Simplified Mass-moment of Inertia
8	PEI, Berm Width and Dune Volume	$\frac{\text{PEI}^4}{\left[\left(\sqrt{(Bwidth+Fwidth)^2+\left(\frac{1}{3}(CrestZ-ToeZ)\right)^2}\right)^2 \times Dvol\right]}$	Mass-moment of Inertia

3.2. Physical Meaning of Various EDP Cases

EDP cases 1 through 4 isolate the Resilience as function of a single parameter (i.e., setback or volume). Two common measures of dune volume are tested. Cases 5 and 6 model the resilience of the dune in a conceptually similar method to shear (L^2) and moment (L^3). Cases 7 and 8 model the dune as a mass-moment of inertia (L^4) with the distinction being Case 7 that uses a simplified measure of the moment arm (neglects vertical component of the moment arm), whereas Case 8 estimates the centroid of the dune by modeling the dune as a triangular shape. The vertical location of the centroid is then estimated as one third the difference between the crest and toe elevations. The resultant centroid is estimated using the Pythagoras theorem; adding the berm width and the crest width together. The equation shown in Case 8 in Table 5 shows the redundant square of the square-root for clarity. This is a slightly simplified approach of the Erosive Resistance (*ER*) parameter introduced by Judge, Overton and Fisher [21].

The significance of the physical proxy (shear, moment, etc.) has implications to potential optimization of the dune configuration. Consider an analogy between a dune and a beam. If shear controls, then the cross-sectional area of the beam must be increased. Now, consider if the moment controls and two separate beams, each with identical cross-sectional areas but with different W-shapes. The beam with a deeper section will be able to span a greater distance even if the total material is the same by taking advantage of the geometric properties. Now, consider the berm (depth) and dune volume (area) as the web and the flange of a beam, respectively. If the volume of the dune is analogous to the cross-sectional area of steel and both an indicator of material cost, then a more efficient geometric shape can potentially imply more cost-effective dunes.

3.3. The Fragility Model

The fragility models have been developed using the EDP as the argument. The fragility curves are fit to a long-term and spatially diverse record. The observational record was developed and first presented by Lemke [20,48]. This record is the combination of two datasets parameterizing both storm intensity and dune morphology, respectively. Historical storms are extracted from the dataset published by Lemke and Miller [38]. Dune morphology, including pre- and post-event indicators, are based on bi-annual beach surveys. The combination of each of these respective datasets allows for the characterization of the beach parameters pre and post storm, quantifiable measures of storm impacts and parameterization of the responsible storm intensity.

This dataset provides a unique long-term (>20-year) and spatial diverse (200-km) dataset to fit the model to. However, the dataset has some limitations identified by Lemke [48]. The median time between the 'storm' and pre/post- surveys was approximately 5 months. This introduces some uncertainty around the quantification of the loss (e.g., percent volume loss) and the resilience parameters (particularly berm width which fluctuates seasonally) that were present at the time of the storm.

To reduce the influence of this uncertainty, the fragility models are fit to predict categorical response of the dune impacts for a given storm intensity in lieu of a regression analysis (e.g., predicting percent dune loss or volume loss). Categories for response are based on the percentage of dune volume loss as presented in Table 6 and are identical to those established by Lemke and Miller [20]. The Major threshold was based on empirical evidence in the aftermath of Hurricane Sandy dunes with volume losses greater than 40% generally exhibited evidence of overwash, as characterized by a lowering of the post-storm dune crest elevation. The onset of overwash is attributed with greater impacts (damages) to upland assets and, thus, a useful delineation in the prediction of storm damages.

Table 6. Classification of storm-induced dune impacts based on quantitative changes.

Damage Class	Definition
Major	Dune volume loss > 40%
Moderate	Dune volume loss 5–40%
Minor	Dune volume loss < 5%

3.4. Fragility Function and Fitting Techniques

Two forms of fragility functions were considered. The first, is a fragility curve based on a log-normal cumulative distribution function. This form is commonly used to represent fragility functions for structural dynamic analysis, particularly in applications predicting the probability of collapse during seismic events, e.g., [50]. Herein, the return of the argument is the probability of the respective damage classification. Each categorical classification (e.g., Moderate, Major) has its own curve which is a function of the EDP.

The log-normal distribution is shown in Equation (4). The left-hand side of the equation states that the probability of category can be estimated by the EDP. The probability is estimated as the log-normal cdf evaluated at a given EDP and two fit parameters, θ and β. The parameters θ and β define the median of the distribution (i.e., the EDP measure corresponding to 50% probability classification) and the standard distribution (commonly referred to as the dispersion), respectively.

These parameters are estimated by fitting observed classifications using a multiple stripe analysis (MSA). In MSA, the data are binned into groups for similar intensity measures. The percentage of observed moderate and major categories within each bin is plotted against the EDP and are then fit using the Maximum Likelihood Estimation (MLE) method.

Using MSA, the curve is fit to the percent observations corresponding to each classification in each bin. The advantage is that all data from the record are retained. In traditional curve fit failure analysis, only the cases in which an observed failure (or classification in this context) are used. For the dataset used herein, approximately 15% of the available data are classified as Moderate or Major.

$$P\left(C\left|\frac{\text{IM}}{R_f} = \text{EDP}\right.\right) = \Phi\left(\frac{\ln\left(\frac{\text{EDP}}{\theta}\right)}{\beta}\right) \tag{4}$$

As an alternative to the log-normal distribution, a Weibull distribution was fit to the Moderate and Major classifications using only the observations corresponding to each classification. The CDF of the two-parameter Weibull function, as modified to use the EDP, is shown in Equation (5):

$$P\left(C\left|\frac{\text{IM}}{R_f} = \text{EDP}\right.\right) = 1 - exp\left(\frac{-\text{EDP}}{\lambda}\right)^{\kappa} \tag{5}$$

The two fit variables of the distribution are the scale (λ) and shape (κ). The scale factor shifts the distribution, with the shape parameter influencing the shape. Parameters are estimated using the MLE fitting techniques.

The Weibull distribution is one of the most commonly applied distributions to model predicted failure. The distribution has been shown to be remarkably flexible, with the ability to characterize the failure based on multiple inputs (e.g., time to failure, length of elongation, stress, cycles, number of units, etc.). There are only two limitations in fitting a Weibull distribution to the data that are of particular concern to the data presented.

- The confidence of the fit distribution is proportional to the number of samples at failure. With the consensus that more than 20 samples are needed to have reasonable performance. Each sample must be independent.
- The samples must be tested to failure, with the demand (i.e., value of 'x') at failure known.

Two separate distributions and fitting techniques were used to assess the influence of asymmetry in the distribution. Advantages and limitations of each method is discussed further in Section 5.2.

3.5. Data Screening, Fitting Techniques and Sensitivity Analysis

The original data [20] contained 865 datapoints and were screened to eliminate smaller local storms and beaches that were clearly manipulated between surveys.

The first condition was the identification of morphologic changes but for a storm intensity so low (i.e., low water level and/or wave heights) that the storm fails to meet the criteria of a storm based on the definitions provided by Lemke and Miller [38]. This occurs when the spatial variation in the storm is such that it may qualify as a storm in either the northern sections or southern limits of the state, but not both. This phenomenon occurred only for two storms in the dataset and reduced the dataset to 852 points.

The second screened condition were profiles with volume changes that indicated post-storm manipulation. Volume changes indicating greater than or equal to 10% net positive volume change (i.e., highly accretional events) were flagged as likely attributed to beach scrapping or manipulation. Beach management within New Jersey typically falls under local jurisdictions, with many local municipalities opting, or having previously experimented, with beach scraping and winter dunes.

To reduce the influence of potential Type II errors attributed to beach manipulation, all cases in which the dune loss returned a post-storm accretion of 10% or greater were removed from the dataset. This amounted to 172 datapoints. The value of 10% was chosen based on engineering judgement. The idea being that collision regime and/or berm rollover events that could lead to accretional volumes at the toe of the dune would be allowed to remain in the dataset while egregious volume accretions that would likely be attributed to a measurement error or anthropogenic beach management activities (e.g., beach scraping) would be removed from the dataset. This is a slight departure from the screening process presented in Lemke [48].

3.6. *Comparision of the EDPs and Evaluation of Curves*

3.6.1. Log-Normal and MSA Curves

For the MSA, the dataset was split into a training and test dataset using a 70–30 split. Evaluation metrics included: Mean Absolute Error (MAE); Bias; and Root Mean Squared Error (RMSE). Sensitivity analysis on bin width was performed including variable and constant bin widths. The selected model was determined by minimizing the RMSE for training and testing datasets for each classification. MAE and Bias for the test and training set are reported for context.

The error is calculated based on difference between the expected (predicted) probability of damage based on class and the observed fraction of damage class corresponding to the bin. The error is calculated at each bin. Due to the variable width of the bin, the median value of the EDP within the bin is used; not the mean value EDP of the bin. For small values of EDP, in the well-resolved portion of the data, the mean EDP of the bin and median EDP of the data contained within the bin are nearly identical. That is to say, the EDPs of the binned samples are relatively evenly distributed over the bin. However, for large values of EDP, the number of observations within each bin decreases while the bin width increases. In these cases, the median EDP of the samples within the bin are used to better characterize the data with the training and testing datasets.

3.6.2. Weibull Distribution

For the Weibull distribution, the curves were fit using only the observations corresponding to classification (i.e., Moderate or Major). A 70–30 split in the data would not be possible while maintaining sufficient datapoints for fitting each class.

4. Results

4.1. *Preliminary Analysis*

The observed dune loss percent vs. storm intensity for the entire dataset is presented in Figure 5. Colored dots denote observed Major (red) and Moderate (blue) damage classifications. Minimal impacts are denoted by x's. Negative dune loss denotes accretion of the dune. (Potential causes and the treatment of these datapoints are discussed in Section 3.5). While the storm intensity measure alone (PEI) appears to separate the data, use of the EDP (Figure 6) improves the clustering by damage classifications. In Figure 6, the storm Categories (measure of PEI alone) are converted into an EDP using the median dune volume and berm width. Other EDPs (Cases 1–6, 8) had similar effect. However, generally speaking, clustering was improved for EDP cases that considered both berm width and dune volumes.

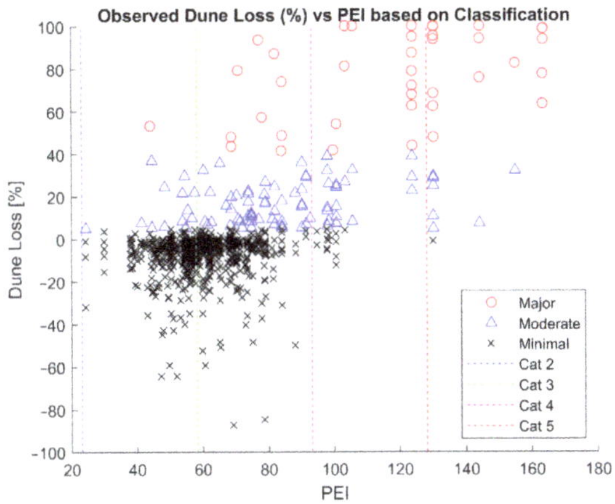

Figure 5. Observed dune loss percentage as a function of storm intensity (PEI) alone; each data point represents observations from a single profile. X-axis is storm intensity (PEI). Storm categories are denoted by colored vertical dashed lines. The Y-axis is dune loss percentage. Red circles indicate Major loss (>40%), blue circles indicate Moderate (>5% but <40%). Negative dune loss indicates accretion.

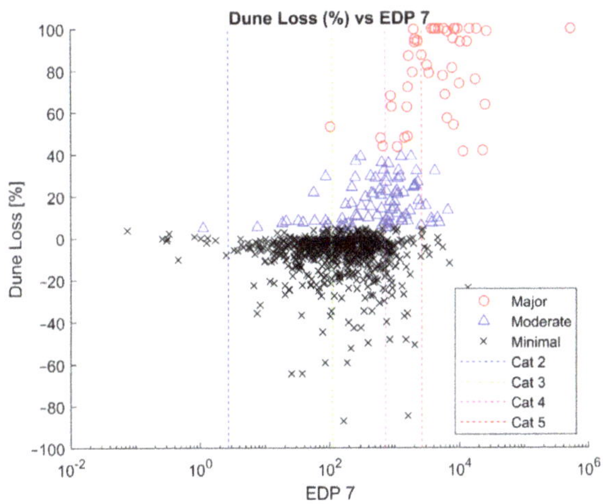

Figure 6. Observed dune loss percentage as a function of EDP Case 7; each data point represents observations from a single profile. X-axis is the EDP. Storm categories are denoted by colored vertical dashed lines. Storm category PEI values converted to an EDP using median resilience parameters. The Y-axis is dune loss percentage. Red circles indicate Major loss (>40%), blue circles indicate Moderate (>5% but <40%). Negative dune loss indicates accretion.

Low values of EDP indicate a low demand placed on a beach through either extremely resilient beaches (i.e., high volume and wide berms) or low intensity storms (i.e., low PEI and, therefore, low water levels and wave heights). Conversely, high EDP values denote high intensity storms or beach–dune systems that are very vulnerable.

Data corresponding to each classification were extracted and plotted in a Q-Q plot. This was done for all EDP cases. Figures 7 and 8 illustrate these results using the Weibull distribution for Moderate and Major classification, respectively. The performance of the Major classification (Figure 8) is distorted by a single outlier. Removal of this datapoint improves the fit drastically (Figure 9). Justification for this is discussed in Section 5.1.

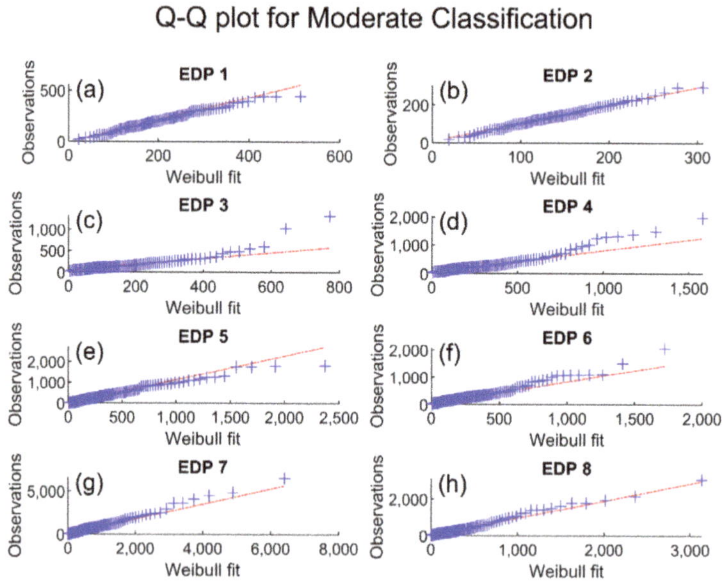

Figure 7. Weibull distribution quantile plots for all observations corresponding to Moderate classification. (**a**) EDP 1—Berm Width; (**b**) EDP 2—Berm and Dune width; (**c**) EDP 3—Dune Volume; (**d**) EDP 4—Foredune Volume; (**e**) EDP 5—Shear; (**f**) EDP 6—Moment; (**g**) EDP 7—Simplified Mass-moment of Inertia; (**h**) EDP 8—Mass-moment of Inertia.

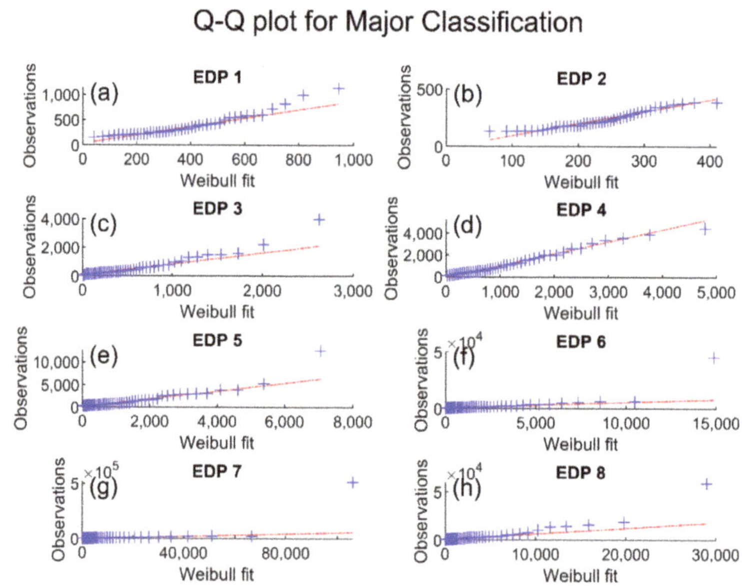

Figure 8. Weibull distribution quantile plots for all observations corresponding to Major classification. (**a**) EDP 1—Berm Width; (**b**) EDP 2—Berm and Dune width; (**c**) EDP 3—Dune Volume; (**d**) EDP 4—Foredune Volume; (**e**) EDP 5—Shear; (**f**) EDP 6—Moment; (**g**) EDP 7—Simplified Mass-moment of Inertia; (**h**) EDP 8—Mass-moment of Inertia.

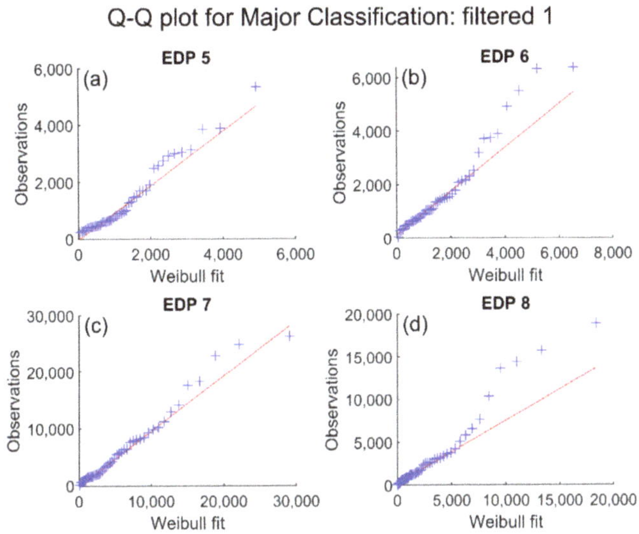

Figure 9. Weibull distribution quantile plots removing maximum observation corresponding to Major classification with a single datapoint removed. Only EDP cases 5 to 8 are shown for clarity. (**a**) EDP 5—Shear; (**b**) EDP 6—Moment; (**c**) EDP 7—Simplified Mass-moment of Inertia; (**d**) EDP 8—Mass-moment of Inertia.

4.2. EDP Curves

The selected curves for both the log-normal and Weibull distributions are shown in Figures 10 and 11, respectively. For brevity, only EDP Case 7 is shown. Note in Figure 10, the fit curve describes the onset of damage. That is to say, the predicted probability and fraction of observed classifications are in better agreement for low values of EDP than at the upper EDP values.

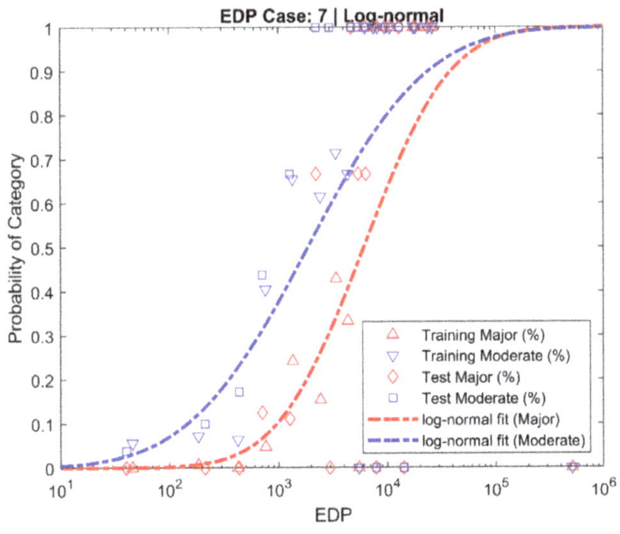

Figure 10. Example log-normal distribution for EDP 7 for calibration/training (triangular shapes) and test datasets (square/diamond). Dashed curves represent MLE log-normal fit curves with red denoting Major and blue denoting Moderate classification.

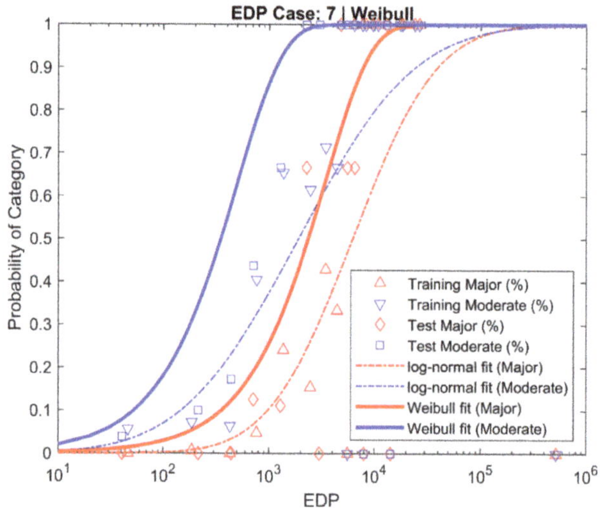

Figure 11. Comparation of Weibull (solid line) vs. log-normal distribution (dashed) for EDP 7. Observed fractions of classification are denoted by shapes and retained from previous figure to illustrate performance. Red denotes Major, Blue denotes Moderate classification.

4.3. Comparison of EDPs and Fragility Functions

To enable comparison of the EDPs and the respective fitting techniques, the entire dataset (combined training and test) is then re-grouped by storm intensity corresponding to the PEI Categories defined by Lemke and Miller [38]. The mean hindcasted probabilities and the observed fractions of each classification are presented. Figure 12 illustrates all EDP cases using the log-normal distribution. In Figure 13, the most physically relevant EDPs (5–8) are shown for both the log-normal MSA and the Weibull fitting technique.

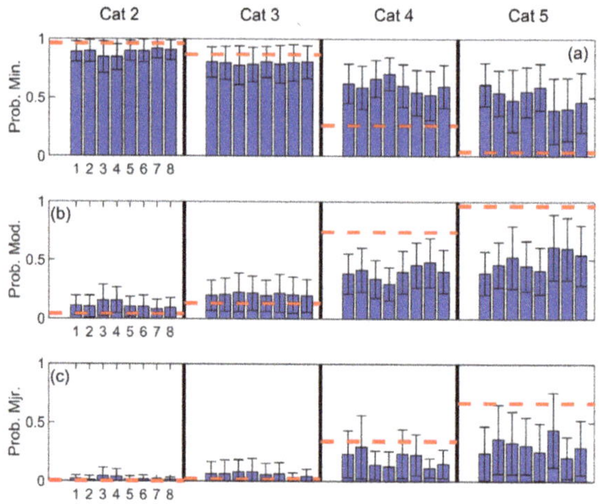

Figure 12. Log-normal predicted and observed fractions of damage classification by storm category (vertical columns). Each subplot corresponds to a damage classification: (**a**) Minimal; (**b**) Moderate; (**c**) Major. Vertical bars correspond to EDP case (1 to 8 in each box), whiskers are standard deviation in predicted probabilities for storms within the Category. Dashed red horizontal bar is the fraction of observed classifications corresponding to the appropriate storm category.

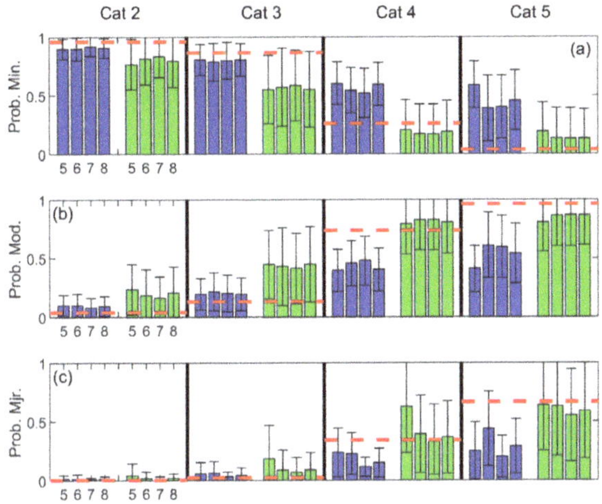

Figure 13. Comparison of Weibull (green/right) and log-normal (blue/left) models by storm category (vertical columns). Each subplot corresponds to a damage classification: (**a**) Minimal; (**b**) Moderate; (**c**) Major. Each vertical bar corresponds to an EDP case and model. Note only EDPs 5–8 are shown. Whiskers are standard deviation in predicted probabilities for the corresponding category of storms. Dashed red horizontal bar is the actual observed fraction corresponding to the appropriate storm category.

Both figures display a 3 × 4 matrix, each row corresponds to the classification (i.e., Minimal, Moderate and Major). Each column corresponds to the storm category (i.e., Category 2, 3, 4 and 5). Each EDP case is represented by a vertical bar. The vertical bar represents the estimated mean probability of the damage classification for the storms contained within the respective storm category groups. The whiskers represent the standard deviation in the predicted probabilities. The dashed red horizontal bar represents the fraction of observed classification within the storm category.

The estimated best fit parameters are shown in Tables 7 and 8 for the log-normal and Weibull distributions, respectively. Estimates of errors for the combined training and test datasets are shown in Table 9.

Table 7. Best Estimate Normal Distribution parameters.

EDP	Moderate		Major	
	θ (Median)	β (Dispersion)	θ (Median)	β (dispersion)
5	1220	1.80	2058	1.10
6	702	1.51	1694	1.15
7	1839	1.99	45629	2.50
8	1260	2.34	7915	2.05

Table 8. Best Estimate Weibull Distribution parameters.

EDP	Moderate		Major	
	λ (Scale)	κ (Shape)	λ (Scale)	κ (Shape)
5	366	1.30	635	3.04
6	282	1.78	1012	2.16
7	570	1.14	4428	1.32
8	243	1.13	1697	1.35

Table 9. Error metrics for log-normal MSA curves.

	MAE		Bias		RMSE	
	Moderate	**Major**	**Moderate**	**Major**	**Moderate**	**Major**
1	26%	16%	−21%	−16%	35%	26%
2	23%	13%	−19%	−13%	30%	20%
3	26%	17%	−19%	−16%	32%	24%
4	26%	15%	−19%	−14%	33%	21%
5	25%	16%	−20%	−16%	33%	25%
6	15%	9%	−11%	−8%	19%	12%
7	15%	18%	−12%	−18%	19%	28%
8	19%	15%	−16%	−15%	25%	22%

5. Discussion

5.1. General Assessment of the EDP

The relatively strong performance of a wide range of EDP cases in the Q-Q plots shown in Figures 7–9 indicates the strength of the EDP in parameterizing the physical demands placed on the beach and dune systems subjected to coastal storms. The dataset is robust by combing both tropical and extra-tropical storms which extends the applicability of the presented EDP.

However, the observational dataset has one import shortcoming—the temporal resolution. It is impossible to discern the timestep within the storm that leads to onset of the classification. Consideration for these cases can be depicted in the removal of a single point from Figures 8 and 9. Note that removal of the datapoint prevents over-estimation of the resilience of the beach–dune system. If the removed datapoint is used as an evaluation point, the performance of the curves is substantially better than if the point is used to define the onset of failure. The distinction is due to the use of the cdf used in the validation of the curves vs. the assignment of failure at a defined EDP in the pdf. A subtle distinction, however important.

The EDP represents the ratio of the peak intensity of the storm to the resiliency of the beach and dune system. Consider a theoretical beach with low resiliency terms subjected to a moderate event (PEI ~ 80) that results in an observed Major classification. Now, consider the identical beach subjected to a more intense storm (say, PEI ~ 100) which results in Major classification. However, using EDP 7, the change in EDP is ~2.5 larger for the second event. The classification is assigned to this PEI. In reality, the beach–dune system is subjected to a ramp up in the IEI over the course the storm. The limitation of the dataset is that it is impossible to discern at which timestep the classification occurs. This is a limitation of the dataset and not the EDP.

The implicit assumption is that the damage classification occurs at or nearly at the peak of the storm. Mathematically, that is to say, the IEI resulting in classification is approximately equal to the PEI. This is likely true for most events. However, for high value EDPs, the likelihood increases that classification occurs at some IEI value less than the PEI. This is the justification for removing the highest EDP values in the Weibull curve fitting (i.e., Figures 8 and 9).

5.2. Fragility Model Performance

When presented using the Q-Q plots, outliers discussed above are more readily seen. The concern becomes more apparent using the MSA analysis when individual events become obscured by the binning procedure and data aggregation. Uncertainties within the dataset can lead to miss-binning. The influence of this was attempted to be counteracted through extensive sensitivity testing of the bin widths, including variable bin width sizing which improved the fit slightly. However, the underprediction of Major classification seen in Figures 10 and 12 is a significant impediment to the applicability of the curves if not accounted for.

Herein, the underprediction has been addressed by fitting a Weibull distribution and removing the MSA. Use of the Weibull distribution allows for the more flexibility in the shape of the distribution over a log-normal distribution. The performance for Major impacts in severe storms (e.g., Cat 4 or 5) is markedly improved in Figure 12.

Consider the two assumptions stated for the Weibull distribution. The presented dataset has 865 independent observations, but only contains 88 instances of moderate and 45 instances of major impacts. The number of observations (sample size) for each category therefore exceeds the minimum number required even if ~20 observations would be removed. However, the combined samples only account for roughly 15% of the total data. Undoubtedly, meaningful information contained with those datapoints is ignored by this fitting process.

5.3. Limitations of the Curves and Potential Improvements

Reduction of data is less than ideal, but necessary due to limitations of the temporal resolution of the underlying observational data. The presented curves, particularly the Weibull fits, are still useful given these limitations. This is particularly true because they fill a niche exhibiting ease of application while providing probabilistic outputs. Consider the proposed curves in comparison to the widely used methods such as the 540-rule [10].

It is postulated that the fragility curves using EDPs, particularly Cases 6, 7 and 8 could be improved through the use numerical or physical tests that can resolve the temporal response of the beach–dune system within the storm duration. Resolution on a storm time scales (i.e., hourly) through numerical or physical modeling could better quantify duration effects. The observational dataset used herein could then be used exclusively for evaluation of the experimentally determined curves.

5.4. Comparing EDP Cases and the Physical Meaning

Despite the limitations of the dataset, the strength of the EDP is the physically meaningful proxy. Similar to the methods presented by Judge, Overton and Fisher [21], the EDPs are presented based on a physical surrogate. The results herein further substantiate the findings by Judge, Overton and Fisher [21] that the resilience of the dune can be modeled using the volume and berm widths as a surrogate for the mass-moment of inertia (EDP Cases 7 and 8). With an eye towards extending the usefulness of the curves without the need for specialized measurements, a simplified version (EDP 7) was tested with comparable performance. In the simplified estimation of the centroid of the dune, the vertical component is omitted. Thus, for cases where the berm width >> dune toe elevation, the performance should be comparable. It should be noted that in the derivation of SEI, the berm elevation is included, thus somewhat accounting for the erosive potential of the storm as related to the position of the dune toe and berm height.

The performance gap between EDP 6 and 7 narrows and eventually even reverses for high intensity storms (Figure 13, Table 9). Under Swash or Collision regimes, the berm width would need to be significantly eroded to allow higher wave heights to impact the dunes. However, during significantly elevated high-water levels associated with Overwash and/or Inundation regimes, the berm will be fully submerged. It is postulated that protective benefit of the berm is therefore reduced, while the importance of the dune volume is magnified. This is conceptually similar to the reduced protection afforded by a drowned offshore bar (or coastal structure) during elevated water levels. There appears to be some evidence to support this in the performance of the single-parameter EDPs. EDPs using setback alone (EDP 1 and 2) appeared better than volume alone (EDP 3 and 4) for lower magnitude events (Category 2 and 3). However, the performance was narrowed or even reversed for Category 5 storms (Figure 12). Admittedly, this is not exhaustive, but conceptually shows the flexibility of the EDP in parameterizing the physics governing the demands placed on a beach and dune system during erosive events.

6. Conclusions

6.1. Thematic

The Engineering Demand Parameter has been introduced to address the need for a physically based, continuous parameter that simultaneously accounts for the most important Resilience terms (e.g., dune volume and setback) and the storm intensity (e.g., erosive potential). The advantage of the presented form of the EDP is that it can, if needed, be easily modified where additional parameters may be needed. The only explicit assumptions are:

1. The EDP must represent the ratio of storm intensity over the resilience terms;
2. Terms in numerator and denominator must be combined in a way that produces a non-dimensional parameter.

To demonstrate the potential usefulness of the parameter, the EDP has been applied to a categorical fragility model. The model is fit to a robust (but imperfect) observational dataset that includes a wide range of storm intensities (moderate to severe); tropical and extratropical storms; and a wide variation in beach/dune morphology. The relative skill of the EDP is indicative of the robustness of the approach in parameterizing the physics in a simple, yet meaningful way.

Several forms of the EDP with different fitting techniques and distributions were tested. Curves were generated using MSA with log-normal distribution and a Weibull distribution. Both curves were fit using MLE techniques.

6.2. Recommended EDP and Fragility Function

Based on the observed data, the recommended form the EDP is Case 7, which is the PEI [37] over a modified version of the Erosive Resistance parameter [21]. In the event Major impacts are the primary concern, the Weibull distribution (i.e., $\lambda = 4428$, $\kappa = 1.32$) is the recommended form.

In general, the hindcasted probabilities of Major (i.e., dune volume loss > 40%) impacts are well resolved by the Weibull distribution for all storm intensities (e.g., Cat 2 through Cat 5). However, if the prediction of the onset of moderate impacts (i.e., dune volume loss > 5% but <40%) during lower intensity storms (e.g., Cat 2 or 3) is of paramount concern, the curves fit using the log-normal and MSA analysis are superior. The reason is not the distribution but rather from the fitting technique. When using MSA the curve is fit to the fraction of observations corresponding to the classification in the record. Consequently, the fit dataset retains both 'failed' and 'resilient' observations whereas the Weibull distributions only consider a limited subset of the full dataset and only those resulting in the appropriate classification.

6.3. Limitations, Compromises and Potential Improvements

The development of the EDP has been specifically tailored for the application of fragility function in a forecasting application. This need has led to necessary compromises in the selection of underlying parameters. Care has been taken to include Resilience parameters (e.g., volume, berm width) that are easily estimated with simple tools or remote sensing (e.g., satellite, aerial photographs, UAVs, etc.). This has led to a conscious decision to remove factors that are important (e.g., crest, toe, centroid elevations, slope, etc.) but require expensive or specialized equipment (e.g., RTK GPS). Attempts to quantify the impact of these omission (e.g., EDP Case 7 and 8) suggest the increased complexity may not be warranted for the curves intended application (e.g., forecasting)—following the adage of "everything must be made as simple as possible, but not simpler."

At extreme events with high storm intensity, the resilience may decrease from EDP Case 7 to 6 (i.e., the product of the dune volume and berm width vs. berm width squared). It is hypothesized that at high water levels (therefore high PEI values and high EDPs), the berm may be fully submerged, allowing higher wave heights to impact the dunes. This is conceptually similar to attenuating protection afforded by a drowned offshore bar during elevated water levels. Admittedly, this has not been conclusively tested, but warrants further investigation.

Author Contributions: Conceptualization, M.S.J.; methodology, M.S.J. and J.K.M.; formal analysis, M.S.J.; data curation, M.S.J.; writing—original draft preparation, M.S.J.; writing—review and editing, J.K.M.; visualization, M.S.J.; supervision, J.K.M.; project administration, J.K.M. All authors have read and agreed to the published version of the manuscript.

Funding: This research was funded by the New Jersey Department of Environmental Protection (NJDEP) through the New Jersey Coastal Protection Technical Assistance Service (N.J.S.A. 18A:64L-1).

Institutional Review Board Statement: Not applicable.

Informed Consent Statement: Not applicable.

Data Availability Statement: Publicly available datasets were analyzed in this study. Details regarding these sources can be found at The Stockton University Coastal Research Center (Welcome-Coastal Research Center I Stockton University) and in Lemke and Miller [20] and Lemke and Miller [38].

Acknowledgments: The authors would like to acknowledge and express sincere gratitude to Laura Lemke who graciously provided unfettered access and insights into her datasets. The authors would also like to thank the reviewers whose meaningful feedback improved the presentation of the paper.

Conflicts of Interest: The authors declare no conflict of interest. The funders had no role in the design of the study; in the collection, analyses, or interpretation of data; in the writing of the manuscript, or in the decision to publish the results.

References

1. Sallenger, A.H. Storm impact scale for barrier islands. *J. Coast. Res.* **2000**, *16*, 890–895.
2. Kalligeris, N.; Smit, P.B.; Ludka, B.C.; Guza, R.T.; Gallien, T.W. Calibration and assessment of process-based numerical models for beach profile evolution in southern California. *Coast. Eng.* **2020**, *158*, 103650. [CrossRef]
3. Tomiczek, T.; Kennedy, A.; Zhang, Y.; Owensby, M.; Hope, M.E.; Lin, N.; Flory, A. Hurricane Damage Classification Methodology and Fragility Functions Derived from Hurricane Sandy's Effects in Coastal New Jersey. *J. Waterw. Port Coast. Ocean Eng.* **2017**, *143*, 17. [CrossRef]
4. Tomiczek, T.; Kennedy, A.; Rogers, S. Collapse Limit State Fragilities of Wood-Framed Residences from Storm Surge and Waves during Hurricane Ike. *J. Waterw. Port Coast. Ocean Eng.* **2014**, *140*, 43–55. [CrossRef]
5. Dean, R.G. *Beach Nourishment*; World Scientific: Singapore, 2003; Volume 18, p. 420.
6. Dean, R.G.; Dalrymple, R.A. *Coastal Processess with Enginerring Applications*; Cambridge University Press: New York, NY, USA, 2004; p. 475.
7. Kamphuis, J.W. *Introduction to Coastal Engineering and Management*; World Scientific: Singapore, 1999.
8. Reeve, D.; Chadwick, A.; Fleming, C. *Coastal Engineering: Processes, Theory and Design Practice*, 3rd ed.; Spon Press: New York, NY, USA, 2012.
9. Bruun, P. Dunes—Their Function and Design. *J. Coast. Res.* **1998**, *26*, 26–31.
10. Hallermeier, R.J.; Rhodes, P.J. *Description and Assessment of Coastal Dune Erosion*; Dewberry & Davis, Inc.: Fairfax, VA, USA, 1986.
11. FEMA. *Coastal Construction Manual Principles and Practices of Planning, Siting, Designing, Construction and Maintaining Residential Buildings in Coastal Areas*, 4th ed.; FEMA: Washington, DC, USA, 2011.
12. Vellinga, P. *Beach and Dune Erosion during Storm Surges*; Hydraulics Laboratory: Delft, The Netherlands, 1982.
13. ENW. *Technical Report on Dune Erosion (in Dutch)*; Delft Hydraulics: Delft, The Netherlands, 2007.
14. Van Koningsveld, M.; Otten, C.J.; Mulder, J.P.M. Dunes: The Netherland's Soft but Secure Sea Defences. In Proceedings of the 18th World dreding congress (WODCON XVIII), Lake Buena Vista, Florida, USA, 27 May–1 June 2007; WODA: Lake Buena Vista, Florida, USA, 2007; pp. 167–184.
15. Morgan, J.P.; Stone, G.W. A technique for quantifying the coastal geomorphology of Florida's barrier island and sandy beaches. *Shore Beach* **1985**, *53*, 19–26.
16. Stockdon, H.F.; Sallenger, A.H.; Holman, R.A.; Howd, P.A. A simple model for the spatially-variable coastal response to hurricanes. *Mar. Geol.* **2007**, *238*, 1–20. [CrossRef]
17. Leaman, C.K.; Harley, M.D.; Splinter, K.D.; Thran, M.C.; Kinsela, M.A.; Turner, I.L. A storm hazard matrix combining coastal flooding and beach erosion. *Coast. Eng.* **2021**, *170*, 104001. [CrossRef]
18. United States Geological Survey (USGS). *Coastal Change Hazards Portal*; USGS: Reston, VA, USA, 2019.
19. United States Geological Survey (USGS). Before and After: Coastal Change Caused by Hurricane Michael. Available online: https://www.usgs.gov/news/featured-story/and-after-coastal-change-caused-hurricane-michael (accessed on 1 December 2019).
20. Lemke, L.; Miller, J.K. Role of Storm Erosion Potential and Beach Morphology in Controlling Dune Erosion. *J. Mar. Sci. Eng.* **2021**, *9*, 1428. [CrossRef]
21. Judge, E.K.; Overton, M.F.; Fisher, J.S. Vulnerability indicators for coastal dunes. *J. Waterw. Port Coast. Ocean Eng. Asce* **2003**, *129*, 270–278. [CrossRef]
22. Kriebel, D.L.; Dalrymple, R.A.; Pratt, A.P.; Sakovich, V. Shoreline risk index for noreasters. In Proceedings of 1996 Conference on Natural Disaster Reduction, Washington, DC, USA, 3–5 December 1996; pp. 251–252.

23. Ellis, J.T.; Harris, M.E.; Román-Rivera, M.A.; Ferguson, J.B.; Tereszkiewicz, P.A.; McGill, S.P. Application of the Saffir-Simpson Hurricane Wind Scale to Assess Sand Dune Response to Tropical Storms. *J. Mar. Sci. Eng.* **2020**, *8*, 670. [CrossRef]
24. The National Oceanic and Atmospheric Administration, N. Billion-Dollar Weather and Climate Disasters: Events. Available online: https://www.ncdc.noaa.gov/billions/events/US/1980-2019 (accessed on 14 October 2021).
25. Basco, D.R.; Mahmoudpour, N. The modified Coastal Storm Impulse (COSI) parameter and quantification of fragility curves for coastal design. *Coast. Eng. Proc.* **2012**, *1*. [CrossRef]
26. Schott, T.; Landsea, C.; Hafele, G.; Lorens, J.; Taylor, A.; Thurm, H.; Ward, B.; Willis, M.; Zaleski, W. *The Saffir-Simpson Hurricane Wind Scale*; NOAA/National Weather Service: Silver Spring, MD, USA, 2012.
27. Kantha, L. Time to replace the Saffir-Simpson hurricane scale? *Eos Trans. Am. Geophys. Union* **2006**, *87*, 3–6. [CrossRef]
28. Powell, M.D.; Reinhold, T.A. Tropical cyclone destructive potential by intergrated kinetic energy. *Bull. Am. Meteorol. Soc.* **2007**, *88*, 513–526. [CrossRef]
29. Hebert, C.; Weinzapfel, B.; Chambers, M. Hurricane Severity Index: A new way of estimating a tropical cyclones destructive potential. In Proceedings of the 29th Conference on Hurricanes and Tropical Meteorology, Tucson, AZ, USA, 10–14 May 2010.
30. Irish, J.L.; Resio, D.T. A hydrodynamics-based surge scale for hurricanes. *Ocean Eng.* **2010**, *37*, 69–81. [CrossRef]
31. Done, J.M.; PaiMazumder, D.; Towler, E.; Kishtawal, C.M. Estimating impacts of North Atlantic tropical cyclones using an index of damage potential. *Clim. Change* **2018**, *146*, 561–573. [CrossRef]
32. Dolan, R.; Davis, R.E. An intensity scale for atlantic coast northeast storms. *J. Coast. Res.* **1992**, *8*, 840–853.
33. Mendoza, E.T.; Jimenez, J.A.; Mateo, J. A coastal storms intensity scale for the Catalan sea (NW Mediterranean). *Nat. Hazards Earth Syst. Sci.* **2011**, *11*, 2453–2462. [CrossRef]
34. Zhang, K.; Douglas, B.C.; Leatherman, S.P. Beach Erosion Potential for Severe Nor'easters. *J. Coast. Res.* **2001**, *17*, 309–321.
35. Balsillie, J.H. Beach and coast erosion due to extreme event impact. *Shore Beach* **1986**, *54*, 15.
36. Kraus, N.C.; Wise, R.A. Simulation of 4 January 1992 storm erosion at Ocean City, Maryland. *Shore Beach* **1993**, *61*, 34–41.
37. Miller, J.K.; Livermont, E. An index form Predicting Storm Erosion Due to Increased Waves and Water Levels. In Proceedings of the Solutions to Coastal Disasters, Turtle Bay, Oahu, HI, USA, 13–16 April 2008; p. 16.
38. Lemke, L.; Miller, J.K. Evaluation of storms through the lens of erosion potential along the New Jersey, USA coast. *Coast. Eng.* **2020**, *158*, 103699. [CrossRef]
39. Bruun, P. Sea-level rise as a cause of shore erosion. *J. Waterw. Port Coast* **1962**, *88*, 117–132. [CrossRef]
40. Miller, J.K.; Dean, R.G. A simple new shoreline change model. *Coast Eng.* **2004**, *51*, 531–556. [CrossRef]
41. Walling, K.; Herrington, T.O.; Miller, J.K. Hurricane Sandy damage comparison: Oceanfront houses protected by a beach and dune system with vs. without a rock seawall. *Shore Beach* **2016**, *84*, 35–41.
42. Janssen, M.S.; Lemke, L.; Miller, J.K. Application of Storm Erosion Index (SEI) to parameterize spatial storm intensity and impacts from Hurricane Michael. *Shore Beach* **2019**, *87*, 41–50. [CrossRef]
43. Wehof, J.; Miller, J.K.; Engle, J. Application of the Storm Erosion Index (SEI) to three unique storms. In Proceedings of the 34th International Conference on Coastal Engineering, Seoul, Korea, 15–20 June 2014.
44. Miller, J.K.; Wehof, J. *Evaluation of Storm Severity Based on the Storm Erosion Index along the Southeast Atlantic Coast of Florida in the wake of Hurricane Sandy. Technical Report Prepared for the Jacksonville District of the USACE*; Stevens Insitute of Technology, Davidson Laboratory: Hoboken, NJ, USA, 2013; p. 14.
45. Cheng, J.; Cossu, F.C.; Wang, P. Factors controlling longshore variations of beach changes induced by Tropical Storm Eta (2020) along Pinellas County beaches, west-central Florida. *Shore Beach* **2021**, *89*, 10. [CrossRef]
46. Miller, J.K. *Evaluation of Storm Severity Based on the Storm Erosion Index along the Atlantic Coasts of North Carolina and South Carolina. Technical Report Prepared for the Charleston District of the USACE*; Stevens Insitute of Technology, Davidson Laboratory: Hoboken, NJ, USA, 2015; p. 14.
47. Lemke, L.; Miller, J.K. Development of a storm erosion climatology for the New Jersey Coast, US. In Proceedings of the 36th International Conference on Coastal Engineering, Baltimore, MD, USA, 30 December 2018; p. 14.
48. Lemke, L. *Hybrid Approach for Evaluating the Erosion Potential of Coastal Storms*; Stevens Institute of Technology: Hoboken, NJ, USA, 2021.
49. Janssen, M.S. *Risk-Based Assessment of Coastal Defense Projects: Quantifying Hazard, Vulnerability, and Parametric Design Applications*; Stevens Institute of Technology: Hoboken, NJ, USA, 2022.
50. Baker, J.W. Efficient analytical fragility function fitting using dynamic structural analysis. *Earthq. Spectra* **2015**, *31*, 579–599. [CrossRef]

Journal of
Marine Science and Engineering

MDPI

Review

Beach–Dune System Morphodynamics

Felice D'Alessandro [1,*], Giuseppe Roberto Tomasicchio [2], Ferdinando Frega [3], Elisa Leone [2], Antonio Francone [2], Daniela Pantusa [4], Giuseppe Barbaro [5] and Giandomenico Foti [5]

1 Department of Environmental Science and Policy, University of Milan, Via Celoria 2, 20133 Milano, Italy
2 Department of Engineering for Innovation, University of Salento, EUMER Campus Ecotekne, Via Monteroni, 73100 Lecce, Italy; roberto.tomasicchio@unisalento.it (G.R.T.); elisa.leone@unisalento.it (E.L.); antonio.francone@unisalento.it (A.F.)
3 Department of Civil Engineering, University of Calabria, Ponte P. Bucci, 87036 Arcavacata di Rende, Italy; ferdinando.frega@unical.it
4 Faculty of Engineering, eCampus University, 22060 Novedrate, Italy; daniela.pantusa@uniecampus.it
5 DICEAM Department, Mediterranea University of Reggio Calabria, Via Graziella Loc. Feo di Vito, 89122 Reggio Calabria, Italy; giuseppe.barbaro@unirc.it (G.B.); giandomenico.foti@unirc.it (G.F.)
* Correspondence: felice.dalessandro@unimi.it

Abstract: Beach–dune system morphodynamics is probably one of the most classical coastal engineering problems. While the topic has been studied extensively and literature is plentiful of considerable research contributions, from the authors' knowledge the subject is still challenging for coastal and environmental sciences. As a part of the Special Issue entitled "Beach–dune system morphodynamics" of this Journal, the present paper reviews traditional issues and design advances building bridges between potential risks and adaptation measures. The benefits of nature-based and hybrid solutions and the need for multidisciplinary studies and approaches to promote sustainable and resilient conservation of the coastal environment are emphasized. Considering the importance and complexity of the subject, this work cannot be fully complete. It is limited to providing a general overview and outlining some important directions intending to serve as a springboard for further research in the field of beach–dune system morphodynamics.

Keywords: wave–beach/dune interaction; resistance; resilience; adaptation; nature-based solutions; hybrid approaches

Citation: D'Alessandro, F.; Tomasicchio, G.R.; Frega, F.; Leone, E.; Francone, A.; Pantusa, D.; Barbaro, G.; Foti, G. Beach–Dune System Morphodynamics. *J. Mar. Sci. Eng.* **2022**, *10*, 627. https://doi.org/10.3390/jmse10050627

Academic Editor: Marcello Di Risio

Received: 25 March 2022
Accepted: 30 April 2022
Published: 5 May 2022

1. Introduction

Coastal dunes are known both for their aesthetic qualities and also their functions in ecological systems, providing a unique habitat that has a high value due to their biodiversity of flora and fauna (Figure 1) [1]. They also represent the land's boundary with the sea and act as a protective natural barrier against flooding due to storm surges and wave attacks for millions of people living near the coasts around the world.

Beach–dune systems are highly dynamic features, exposed and vulnerable to extreme events. Their evolution is mainly governed by the mutual and complex exchange of sediment through hydrodynamic and aeolian processes. In particular, during storm-induced surges, the action of waves and wind can cause severe erosion of beach dunes with widespread morphological changes and possible damage to nearby infrastructures [2]. Therefore, to reduce this vulnerability, it is imperative for coastal engineers to study the exposure of beach–dune systems, their sensitivity to hazards-driven changes, and their adaptive capacity to determine safety adaptation measures. Furthermore, despite the progress made in the literature, it is essential to improve the accuracy of empirical, numerical, and physical models for predicting wave-dune interaction phenomena.

Figure 1. A photo of a vegetated dune on the beach of "Alimini" in Salento, Adriatic coast, southern Italy. Photo courtesy of Giuseppe Roberto Tomasicchio.

From an adaptation perspective, both nature-based and hybrid solutions are increasingly seen as effective strategies [1,3–5]. Recent advances in coastal research and ecological sciences are being adopted to find innovative environmentally friendly approaches to increase the resistance and resilience of beach–dune systems. In this context, there is an urgent need to support viable policy responses and identify appropriate actions to mitigate risks and improve the environmental and social resilience of coastal communities that also impact ocean-based economic activities, including tourism and recreation [6].

This review paper is organized as follows: First, the main characteristics (morphology, resistance, resilience) of the beach–dune system are introduced. The interaction processes between wave, beach, and dune are described, followed by a brief discussion of the modeling approaches to predict beach–dune profile evolution. Adaptation perspectives are then presented, including both nature-based and hybrid solutions. Finally, the work ends with conclusions and some recommendations.

2. The Beach-Dune System

The beach–dune system is a natural, dynamic environment whose morphology is determined by a variety of factors including climate variability, relative sea level, wind and wave energy, sediment supply, and vegetation [7].

Basically, dunes can be divided into primary and secondary dunes. Primary dunes consist of sand blown directly from the active beach, while secondary dunes are formed by the subsequent alteration of primary dunes. Primary dunes are those dunes closest to the shoreline, dynamically associated with beach processes, and significantly affected by wave action: they are the subject of this paper (Figure 2).

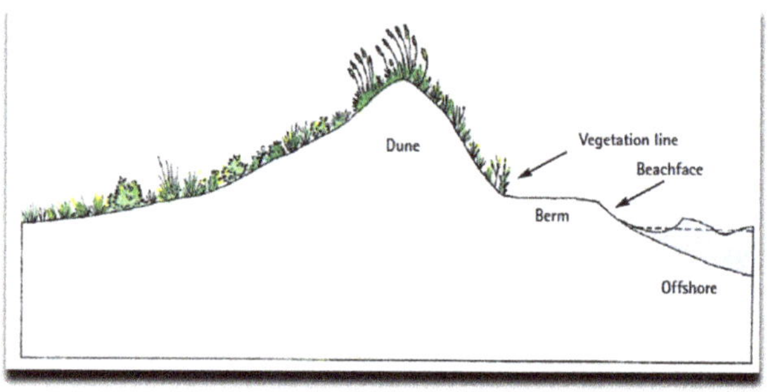

Figure 2. Schematic cross-section of a typical beach–dune profile.

Dunes are a natural habitat for a variety of species, increase the recreational value of an area, and attract tourism [8]. In general, the beach–dune profile is part of a "sand-sharing" system influenced by variations in wave energy, sea-level rise, currents, and mobilization of sand from one zone to another [9,10]. The beach–dune profile at a given location is usually directly related to the sediment that is added or removed from a coastal system over time (the sediment budget [11]).

Dunes are dynamic in that they can increase in length due to wind action, increase in height due to the accumulation of sand grains, or change shape due to wind speed and direction and the type of vegetation [12]. The near-surface wind vector (speed, direction) is thus the most influential factor in the shape and size of dunes and the main driver of erosion and accretion patterns which is included in standard formulations of aeolian sediment transport models [13,14].

Wind speed is usually reduced by vegetation, which favors the deposition and retention of wind-blown sand with leaves, branches, and roots. However, wind can also accelerate locally in gaps between plants, especially tussocky plants. Vegetation is thus a crucial element in the evolution and dynamics of the beach–dune system. Studies of the effects of vegetation on the transport of wind-blown sand have shown that when the wind blows over a smooth, unobstructed surface, the shear stress acts uniformly over the entire surface, but when non-erodible roughness elements are present, some of the shear stress is absorbed by the roughness elements on the underlying erodible surface. The degree of protection is a function of their size, geometry, and spacing [15,16]. Field and wind tunnel experiments indicate that the most important influence of vegetation depends on the threshold wind shear velocity for transport [17–19]. Ecologists have studied the response of plant communities (e.g., spatial sorting, zonation, and diversity) to physical and chemical gradients (including sand burial, wind action, salt spray, soil moisture, groundwater salinity, soil pH, and nutrients), many of which are influenced by topography [20–23]. Although the importance of dune-forming grasses [24,25] has been recognized, most geomorphological research has focused on the role of sand supply (a combination of sediment budget and wind-transport potential) as a function of beach morphology and wind regime in the process of dune formation [26,27].

Beach–dune systems are subject to recurrent natural disturbances (wind speed and direction, sediment supply characteristics, vegetation availability, duration, and intensity of storms) and human impacts that exert strong pressures that alter their dynamics and reduce resistance and resilience. The former refers to the "ability of a system to withstand a force without any change", while the latter is a "measure of its ability to respond to the consequences of a disturbance and return to its original state" [28]. Resistance is the capacity exercised before the system is perturbed, while resilience is measured once the perturbation has occurred: it indicates the extent to which the system is capable of restoring its dynamic equilibrium.

3. Wave–Dune Interaction Processes

Storms can have varying degrees of impact on the beach–dune system. A scale for the impact of storms based on the morphological response of beach–dune systems to hydrodynamic conditions was proposed by [29]. The scale consists of four regimes of wave-dune interaction processes: (1) swash regime, where wave run-up is confined to the foreshore; (2) collision regime, where wave run-up erodes the base of the dune, and dune sediment is transported offshore; (3) overwash regime, in which the wave overtops the dune crest, and the overwash sediment is deposited on the backside of the dune; (4) breaching and inundation regime, in which the dune crest is generally overtopped and the beach topography subsequently flattens. A schematic of storm impact regimes is given in Figure 3.

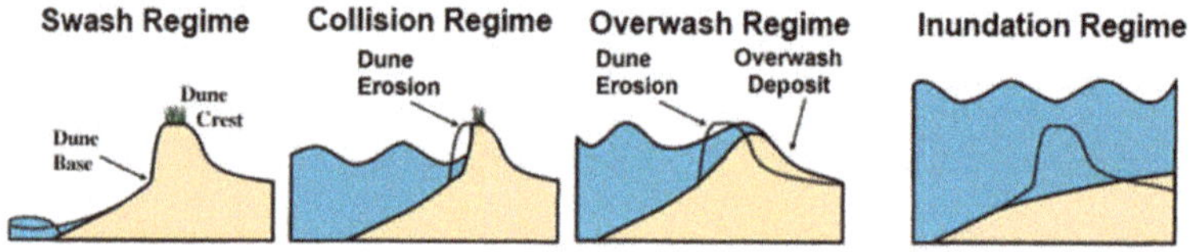

Figure 3. Schematic of storm impact regimes.

As an illustrative case, Figures 4 and 5 show the wave-dune interaction processes observed during large-scale laboratory experiments performed by the authors. The physical model tests have been conducted at the CIEM wave flume of the Universitat Politecnica de Catalunya in Barcelona, within the framework of the Hydralab-III European "BOB" project [2,30–33]. Specifically, as shown in Figure 4, waves broke directly on the sloping dune front (Figure 4a). As a result, the dune front steepened until it was nearly vertical (Figure 4b). The geometry of the dune front changed due to scarping and slumping. Eroded sand from the dune was transported offshore by a strong undertow and sediment suspension in the proximity of the dune was high, resulting in a large offshore transport capacity [34] (Figure 4c). The incoming breaking waves that encountered the outgoing reflected broken waves (wave collision) generated a large amount of turbulence (Figure 4d). However, the eroded sand remained in the region of the active profile and formed a new profile with a bar where the waves broke. The new profile was more efficient in dissipating the energy associated with the incoming waves. The underwater bar prevented the offshore sand from migrating. Consequently, dune erosion rates decreased as the storm was in progress (Figure 4e). The dune face retreated under the impacting waves: part of the dune face collapsed, and large chunks of sediment fell or slid down (Figure 4f). In agreement with [35], the observed low run-up did not allow overtopping of the waves.

When severe waves attack the beach, they often have enough energy to overtop the dune crest and cause landward transport of sediment (Figure 5a,b). A rapid increase in wave overtopping and sediment overwash, combined with the erosion of the dune crest, determined the dune breaching (Figure 5c,d). Similar behavior was also observed during small-scale flume tests conducted by [36], where dune breaching was caused by some overtopping waves at the location along the coast where the dune crest had previously been lowered by minor wave overwash.

Erosion of beach–dune systems can occur over several hours and days during a storm. However, resilience, as the ability of a dune to recover after a storm, varies [5]. It can take years, even decades, and is influenced by many factors depending on the response of the beach–dune system to sea-level rise, changes in the frequency and/or magnitude of storm surges [37], sediment supply [38–40], beach width, wetting and drying cycles, and the growth and condition of existing vegetation, and its ability to retain wind-blown sediment. In the case of a collision regime, the recovery process can fully restore the pre-storm beach–dune profile because no sediment has been lost from the beach–dune system. On the other hand, overwashed and breached dunes can only recover through aeolian processes or engineered (nature-based and/or hybrid) restoration efforts.

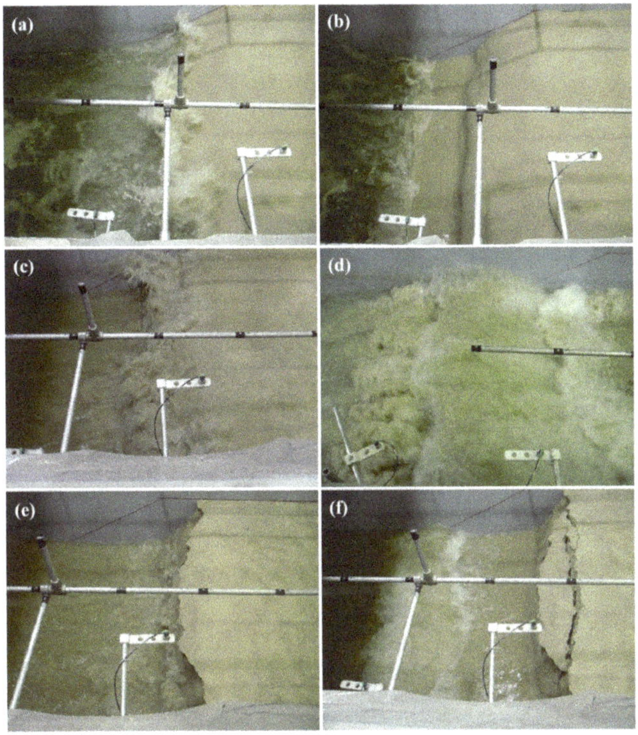

Figure 4. Observed collision regime. (**a**) Wave breaking on the sloping dune front, (**b**) formation of the dune scarp, (**c**) wave breaking and high sediment suspension in the proximity of the dune, (**d**) wave collision, (**e**) dune erosion, (**f**) collapsing of the dune face. Hydralab-III BOB project at CIEM wave flume. Photo courtesy of Felice D'Alessandro.

Figure 5. Observed overwash (**a**,**b**) and breaching (**c**,**d**) regimes. Hydralab-III BOB project at CIEM wave flume. Photo courtesy of Felice D'Alessandro.

4. Modeling Beach–Dune Profile Evolution

Modeling the evolution of beach–dune profiles has reached a verifiable level of maturity in recent decades thanks to well-defined equations, established numerical solutions, and high-quality laboratory and field data. Furthermore, remote sensing methods are useful tools for ecosystem monitoring, as they allow to capture a wide range of properties of vegetation [41–45].

Empirical models are primarily used to estimate the volume of erosion associated with a storm. An empirical model evaluates the dune profile before the storm and the erosion volume based on the profile after the storm. They offer advantages in that their simplicity makes them easy to apply, which is valuable in the early stages of a project when approximate estimates are needed [2]. In the early 1980s, based on extensive physical model experiments, Vellinga [46] presented an empirical model for predicting the dune erosion profile after a storm, taking into account storm surge levels, wave conditions, and sediment grain size characteristics. Larson and Kraus [47] proposed the semi-empirical model SBEACH based on equilibrium equations. Although the model of [47] does not fully resolve the hydrodynamics of the surf zone and the sediment transport processes, it overcomes some limitations of previous models and is able to reproduce bar formation. In [48], an analytical model for predicting the erosion and retreat of coastal dunes was developed based on a wave impact approach, where the dune profile evolution can be estimated by waves directly impacting the dune [49].

Process-based models describe the temporal and spatial evolution of the beach–dune profile based on numerical formulations of the prevailing physical processes. XBeach [50], CSHORE [51], and CROSMOR [52] are well-proven process-based models adopted to estimate the short-term evolution of beach–dune systems under storm conditions. XBeach solves the 2DH equations for wave propagation, current velocity, sediment transport, and bathymetry evolution, for time-varying hydrodynamic conditions. The model was developed to simulate the various effects of a storm as defined in [29]. In CSHORE, numerical integration of the depth-averaged energy, momentum, and continuity equations leads to predictions of wave height, water level, and wave-induced steady currents. The model accounts for wave-current interaction, wave-induced sediment transport, cross-shore bedload, and suspended sediment transport rates. CROSMOR simulates wave propagation, transformation, and refraction along the beach. Statistical parameters are calculated from the results of individual waves. The model accounts for wave-induced set-up, set-down, and breaking of associated long- and cross-shore currents. The depth-averaged return flow is derived from linear mass transport and water depth below the wave trough. The sediment transport rate (bedload and suspended sediment) is determined based on the calculated wave height, depth-averaged long- and cross-shore velocities, orbital velocities, friction factors, and sediment parameters. A detailed explanation of the structure and capabilities of these models is beyond the scope of this article. The interested reader is referred to [50–52].

Since collecting detailed information in the field during storm-induced surges and strong waves is a difficult and risky endeavor, physical model experiments in wave flumes remain fundamental to investigating wave-beach–dune interaction phenomena. Large-scale experiments on dune erosion under collision regimes have been conducted by [35,46,53–60]. Similar experiments have been performed in Germany [61] and in the USA [62]. Physical model tests carried out in the past have generally focused on the study of seaward profile evolution and morphodynamic processes under the given wave and water level conditions. Refs. [35,63] conducted large-scale experiments to investigate the key physical processes driving dune erosion during severe storm events. These experiments updated the results of [29] to include the effects of the wave period. Although accurate modeling of wave overwash processes is critical for predicting dune profile evolution and assessing inundation risk [64–67], observations of overtopping and overwash were not made in the earlier experiments. Despite the importance of beach–dune profile changes during overwash regimes, few small-scale [68–71] and large-scale [2,72–74] laboratory experiments have been conducted. On the other hand, laboratory studies of the same regimes proposed by [29] that also account for vegetation are extremely limited [75–77].

5. Adaptation of Beach-Dune Systems to Climate Change Impacts

Beach–dune systems are particularly vulnerable to the effects of climate change, especially erosion and flooding, because they are subject to multiple drivers that can occur on different time scales [78]. In addition, interdependence among climate drivers (e.g., storm surges and sea-level rise) can lead to extreme compound events [79]. There are several multivariate approaches to account for interdependencies among climate drivers. Copulas are widely used as an efficient tool to study the statistical behavior of dependent variables [64–67].

Sea level rise is generally considered the most important climate driver in coastal areas. This has led to improved projections. Among the key aspects of making accurate forecasts is understanding the long-term variability of sea level. Extensive studies have been conducted in the literature to explore sea-level fluctuations. The IPCC warns that under current trends, projected increases in mean sea level for the year 2100, relative to the time series of tide gauge records from 1986 to 2005, are 400, 470, 480, and 840 mm, for the RCP2.6, RCP4.5, RCP6.0, and RCP8.5 Representative Concentration Pathways scenarios, respectively. As a result, the global mean sea level has been rising over the past century and is projected to rise even faster in the 21st century.

In this context, sea-level rise and changes in storminess patterns could lead to a more frequent occurrence of high-water levels and wave storms which, in turn, could produce an amplification of coastal flood risk, especially in low-lying coastal landscapes. Therefore, the greater impact on beaches and dune systems will increase the important role of flood protection measures.

Restoration of beach–dune systems is increasingly seen as an effective measure for adapting and enhancing resilience to the adverse effects of climate change. The process of restoration consists of preventing degradation and supporting the recovery of a damaged or degraded ecosystem in terms of its health, integrity, and sustainability [80]. In recent decades, measures to mitigate the effects of climate change have been proposed, consisting of hard coastal structures or soft defenses to prevent beach and dune erosion. Although research on hard coastal structures (e.g., dikes and revetments) has continuously improved design approaches [81–85] and risk analysis techniques [64–67,86–97], significant changes in geomorphological processes and habits are inevitable.

Recent advances in coastal engineering and environmental science are being implemented to minimize the risk of flooding and improve coastal resilience [98–102]. Currently, experiments are being conducted with hybrid systems consisting of traditional structures covered with layers of sand in an attempt to mimic dunes. They combine the advantages of both hard and soft coastal protection solutions [4,5,103–108] and are therefore valuable in areas with limited space and allow for innovation in coastal design. Although such hybrid systems have the potential to become an alternative to conventional coastal protection systems, there are still many unknowns regarding the interaction between the soft and hard structural components and their effectiveness in mitigating storm surges and preventing flooding that requires focused research efforts to establish acceptable design guidelines.

In recent years, a growing number of approaches have emerged as important nature-based solutions [2,3,109,110]. Indeed, coastal protection measures based on natural approaches have gained prominence since they were included in several European directives [111–113] as possible disaster risk reduction strategies. For example, dune restoration projects combine coastal habitat recovery, vegetation replanting, and beach nourishments with sustainable and soft solutions inspired and supported by nature. They are particularly suitable where the application of traditional engineering methods might conflict with the intrinsic value of the ecosystem and its biodiversity, such as in coastal natural areas or reserves.

Nature-based solutions emphasize the role that plants play in the geomorphology of coastal dunes, as has been recognized for more than a hundred years [114–116]. According to [117], coastal dunes and their associated vegetation communities act as barriers to storm surges and prevent or reduce flooding and coastal erosion. Vegetated dune systems also play an important role in promoting a positive coastal sediment budget. Various studies describe how plants can contribute to beach recovery after storms, as vegetation retains sediment transported by wind [117–119].

In many cases, these approaches have been implemented as demonstration projects in protected areas. For example, Figure 6 shows the results of an innovative and environmentally friendly technique for the consolidation of coastal dunes studied in recent field experiments conducted by the authors on the Adriatic coast in Salento (southern Italy) [2].

Figure 6. Aerial view of the intervention site (**a**), visual observation of the restored dune six months after the intervention (**b**), and detail of the consolidated dune face (**c**). Photo courtesy of Giuseppe Roberto Tomasicchio.

In addition to reducing wind speed and the deposition of drifting sand, which can be achieved through vegetation and/or conventional techniques (e.g., dune fencing), an alternative solution to reducing the near-surface wind impact and the effects of frequent minor storms is to strengthen the dune face. Specifically, injecting a mineral colloidal silica-based grout into the most exposed side of the dune, which is native material, allows the mechanical strength of the non-cohesive sediment to be increased. The colloidal silica-based grout is a non-toxic, transparent mineral suspension of nanometric particles of silica (SiO_2), which looks like "liquid sand" and resembles the material found in nature. It has a low viscosity that allows it to easily penetrate very fine sand and silty soil. The advantage over traditional dune remediation solutions is that the proposed intervention does not impact the environment and does not increase the footprint of the dune, thus avoiding further obstacles. This aspect can be particularly important when the rehabilitation intervention is carried out on a stretch of beach that is valuable for tourism and recreation [2].

The reliability of this solution has been verified by laboratory experiments conducted in the 2D wave flume of the EUMER laboratory of the University of Salento (Italy) [3]. The observations have shown that the mineral colloidal silica increases the mechanical strength of non-cohesive sediments reducing the volume of dune erosion and improving the resistance and longevity of the beach–dune system.

Although various nature-based solutions can be found on many coasts around the world, in-depth studies of efficiency, vulnerabilities, and natural dynamics are still lacking. To overcome these knowledge gaps, scientific insight from systematic research is urgently needed to identify measures that combine spatial and statistical analysis, remote sensing, field monitoring techniques, and numerical modeling to create a novel understanding of nature-based coastal adaptation to erosion and flooding risks in low-density coastal areas.

6. Conclusions

Although there is much relevant research in the field of beach–dune system morphodynamics, the problem remains challenging for coastal and environmental sciences. Therefore, it is critical to improve the knowledge, tools, and expertise needed to develop new and sustainable ways to mitigate the negative impacts of climate change.

This overview provides the basis for a general summary of recommendations for further research and practice.

- It should be recognized that the most severe climate-related impacts are caused by the interaction of multiple hazards. A systematic research program is overdue, and more comprehensive multidimensional approaches to risk are needed. Such approaches must account for the non-stationarity of risk components and allow for the quantification of uncertainty. In addition, new methods and tools are needed to evaluate the multivariate analysis of extremes.
- A major challenge is the development and implementation of nature-based adaptation measures, whose behavior needs to be better understood. Improved understanding and application of this knowledge, as well as the implementation of innovative monitoring systems, will be a critical component of coastal adaptation planning, likely reducing the need for expensive engineering options at some sites and providing a complementary tool for hybrid engineering design.
- More research on the interaction between the swash and soil mechanics in the dune face seems critical to better understand dune erosion during storm surges.

Furthermore, research efforts should be made to:

- Better use remote sensing technologies to measure indicators of wind erosion potential.
- Integrate wind erosion models with other models, including those for driving forces (e.g., hydrodynamics) and controls (e.g., plant growth).
- Develop databases for numerical models to expand their applications.

With such complex issues, greater collaboration and synchronization across multiple research areas are needed to provide multidisciplinary and comprehensive studies and approaches. At the same time, it is crucial to propose methods for assessing the value of ecosystem services to promote investment in coastal environmental protection and improve the efficiency of natural resource use.

Author Contributions: Conceptualization, F.D. and G.R.T.; writing—original draft preparation, F.D.; writing—review and editing, F.D., G.R.T., F.F., G.B., A.F., E.L., G.F. and D.P. All authors have read and agreed to the published version of the manuscript.

Funding: This research has been funded by Regione Puglia through the grant to the budget of the project titled "Sperimentazione di tecnologie innovative per il consolidamento di dune costiere (INNO-DUNECOST)", POR Puglia FESR FSE 2014–2020—Sub-Azione 1.4.B, Contract n. RM5UKM2.

Institutional Review Board Statement: Not applicable.

Informed Consent Statement: Not applicable.

Data Availability Statement: Not applicable.

Acknowledgments: This paper presents some results of the research project BOB, which was supported by the European Community Sixth Framework Programme (FP6) through a grant to the Integrated Infrastructure Initiative budget HYDRALAB-III within Transnational Access Activities, Contract No 022441. The authors acknowledge the colleagues and friends involved in the BOB project for their important contributions. A special thanks to Bradley Johnson of USACE for his support. Furthermore, the authors are grateful to Professor Nobuhisa Kobayashi of the University of Delaware for the fruitful and valuable discussions on this topic that we have had over these years. Last but not least, the authors thank the anonymous reviewers who provided helpful and constructive comments that have been included in the paper.

Conflicts of Interest: The authors declare no conflict of interest.

References

1. D'Alessandro, F.; Tomasicchio, G.R.; Francone, A.; Leone, E.; Frega, F.; Chiaia, G.; Saponieri, A.; Damiani, L. Coastal sand dune restoration with an eco-friendly technique. *Aquat. Ecosyst. Health Manag.* **2020**, *23*, 417–426. [CrossRef]
2. D'Alessandro, F.; Tomasicchio, G.R. Wave-dune interaction and beach resilience in large-scale physical model tests. *Coast. Eng.* **2016**, *116*, 15–25. [CrossRef]
3. Leone, E.; Kobayashi, N.; Francone, F.; De Bartolo, S.; Strafella, D.; D'Alessandro, F.; Tomasicchio, G.R. Use of nanosilica for increasing dune erosion resistance during a sea storm. *J. Mar. Sci. Eng.* **2021**, *9*, 620. [CrossRef]
4. Almarshed, B.; Figlus, J.; Miller, J.; Verhagen, H.J. Innovative coastal risk reduction through hybrid design: Combining sand cover and structural defenses. *J. Coast. Res.* **2020**, *36*, 174–188. [CrossRef]
5. Figlus, J.; West, N.A.; Almarshed, B.; Jonkman, S.N. Conceptual design and physical model study of core-enhanced dunes as hybrid coastal defense structures. In Proceedings of the Coastal Structures and Solutions to Coastal Disasters Joint Conference, Boston, MA, USA, 9–11 September 2015; pp. 65–73.
6. D'Alessandro, F. Rural revival and coastal areas: Risks and opportunities. In *Interdisciplinary Approaches to Climate Change for Sustainable Growth, Natural Resource Management and Policy 47*; Valaguzza, S., Hughes, M.A., Eds.; Springer Nature: Cham, Switzerland, 2022; pp. 93–105.
7. Sloss, C.R.; Shepherd, M.; Hesp, P. Coastal Dunes: Geomorphology. *Nat. Educ. Knowl.* **2012**, *3*, 2.
8. Bruun, P. Dunes-their function and design. *J. Coast. Res.* **1998**, *26*, 26–31.
9. Hanley, M.E.; Hoggart, S.P.G.; Simmonds, D.J.; Bichot, A.; Colangelo, M.A.; Bozzeda, F.; Heurtefeux, H.; Ondiviela, B.; Os-trowski, R.; Recio, M.; et al. Shifting sands? Coastal protection by sand banks, beaches and dunes. *Coast. Eng.* **2014**, *87*, 136–146. [CrossRef]
10. Psuty, N. The coastal foredune: A morphological basis for regional coastal dune development. In *Coastal Dunes, Ecological Studies 171*; Martínez, M.L., Psuty, N.P., Eds.; Springer: Berlin/Heidelberg, Germany, 2004; pp. 11–27.
11. Psuty, N. Sediment budget and dune/beach interaction. *J. Coast. Res.* **1988**, *3*, 1–4.
12. Henrico, I.; Ledwaba, T.; van Zyl, G. Measuring the effect of wind-driven processes on coastal dunes: A study of the Atlantis and Geelbek dune fields along the West Coast of South Africa. *Spat. Inf. Res.* **2020**, *28*, 569–577. [CrossRef]
13. Bagnold, R.A. *The Physics of Blown Sand and Desert Dunes*; Chapman and Hall: London, UK, 1941.
14. Kawamura, R. *Study of Sand Movement by Wind*; Reports of the Institute of Science and Technology University of Tokyo: Tokyo, Japan, 1951; Volume 5, pp. 95–112. (In Japanese)
15. Lyles, L.; Schrandt, R.L.; Schneidler, N.F. How aerodynamic roughness elements control sand movement. *Trans. Am. Soc. Agric. Eng.* **1974**, *17*, 134–139. [CrossRef]
16. Marshall, J.K. Drag measurements in roughness arrays of varying densities and distribution. *Agric. Meteorol.* **1971**, *8*, 269–292. [CrossRef]
17. Musick, H.B.; Gillette, D.A. Field evaluation of relationships between a vegetation structural parameter and sheltering against wind erosion. *Land Degrad. Rehabil.* **1990**, *2*, 87–94. [CrossRef]
18. Musick, H.B.; Trujillo, S.M. Wind-tunnel modelling of the influence of vegetation structure on saltation threshold. *Earth Surf. Processes Landf.* **1996**, *21*, 589–606. [CrossRef]
19. Stockton, P.H.; Gillette, D.A. Field measurements of the sheltering effect of vegetation on erodible land surfaces'. *Land Degrad. Rehabil.* **1990**, *2*, 77–86. [CrossRef]
20. Oosting, H. Tolerance to salt spray of plants of coastal dunes. *Ecology* **1945**, *26*, 85–89. [CrossRef]
21. Barbour, M.; Jong, T.; Pavlik, B. Marine beach and dune plant communities. In *Physiological Ecology of North American Plant Communities*; Chabot, B., Mooney, H., Eds.; Chapman & Hall and Methuen: New York, NY, USA, 1985; pp. 296–322.
22. Sykes, M.T.; Wilson, J.B. An experimental investigation into the response of some New Zealand sand dune species to salt spray. *Ann. Bot.* **1988**, *62*, 159–166. [CrossRef]
23. Kim, D.; Yu, K. A conceptual model of coastal dune ecology synthesizing spatial gradients of vegetation, soil, and geomor-phology. *Plant. Ecol.* **2009**, *202*, 135–148. [CrossRef]
24. Hesp, P.A. Foredunes and blowouts: Initiation, geomorphology and dynamics. *Geomorphology* **2002**, *48*, 245–268. [CrossRef]
25. Durán, O.; Moore, L.J. Vegetation controls on the maximum size of coastal dunes. *Proc. Natl. Acad. Sci. USA* **2013**, *110*, 17217–17222. [CrossRef]
26. Short, A.D.; Hesp, P.A. Wave, beach and dune interactions in southeastern Australia. *Mar. Geol.* **1982**, *48*, 259–284. [CrossRef]
27. Hesp, P. Surfzone, beach, and foredune interactions on the Australian South-East Coast. *J. Coast. Res.* **1988**, *3*, 15–23.
28. Martínez, M.L.; Hesp, P.A.; Gallego-Fernández, J.B. Coastal dune restoration: Trends and perspectives. In *Restoration of Coastal Dunes*; Springer: Berlin/Heidelberg, Germany, 2013; pp. 323–339.
29. Sallenger, A.H., Jr. Storm impact scale for barrier islands. *J. Coast. Res.* **2000**, *16*, 890–895.
30. Tomasicchio, G.R.; D'Alessandro, F.; Barbaro, G. Composite modelling for large-scale experiments on wave-dune interactions. *J. Hydraul. Res.* **2011**, *49*, 15–19. [CrossRef]
31. Tomasicchio, G.R.; Sanchez Arcilla, A.; D'Alessandro, F.; Ilic, S.; James, M.; Fortes, C.J.E.M.; Sancho, F.; Schüttrumpf, H. Large-scale flume experiments on dune erosion processes. *J. Hydraul. Res.* **2011**, *49*, 20–30. [CrossRef]
32. Sancho, F.; Abreu, T.; D'Alessandro, F.; Tomasicchio, G.R.; Silva, P.A. Surf hydrodynamics under collapsing coastal dunes. *J. Coast. Res.* **2011**, *64*, 144–148.

33. D'Alessandro, F.; Tomasicchio, G.R.; Musci, F.; Ricca, A. Dune erosion physical, analytical and numerical modelling. In Proceedings of the 33rd International Conference on Coastal Engineering, Santander, Spain, 1–6 July 2012.
34. Van Thiel de Vries, J.S.M. Dune Erosion during Storm Surges. Ph.D. Thesis, Delft University of Technology, Delft, The Netherlands, 2009.
35. van Gent, M.R.A.; van Thiel de Vries, J.S.M.; Coeveld, E.M.; de Vroeg, J.H.; van de Graaff, J. Large-scale dune erosion tests to study the influence of wave periods. *Coast. Eng.* **2008**, *55*, 1041–1051. [CrossRef]
36. Kobayashi, N.; Agarwal, A.; Johnson, B.D. Longshore current and sediment transport on beaches. *J. Waterw. Port Coast. Ocean Eng.* **2007**, *133*, 296–306. [CrossRef]
37. Houser, C.; Wernette, P.; Rentschlar, E.; Jones, H.; Hammond, B.; Trimble, S. Post-storm beach and dune recovery: Im-plications for barrier island resilience. *Geomorphology* **2015**, *234*, 54–63. [CrossRef]
38. Bauer, B.O.; Davidson-Arnott, R.G.D. A general framework for modeling sediment supply to coastal dunes including wind angle, beach geometry, and fetch effects. *Geomorphology* **2003**, *49*, 89–108. [CrossRef]
39. Robin, G.D.D.-A. Conceptual model of the effects of sea level rise on sandy coasts. *J. Coast. Res.* **2005**, *21*, 1166–1193.
40. Woodruff, J.D.; Irish, J.L.; Camargo, S.J. Coastal flooding by tropical cyclones and sea-level rise. *Nature* **2013**, *504*, 44–52. [CrossRef]
41. Díaz-Delgado, R.; Lucas, R.; Hurford, C. *The Roles of Remote Sensing in Nature Conservation. A Pratical Guide and Case Studies*, 1st ed.; Springer: Cham, Switzerland, 2017; pp. 1–318.
42. Adam, E.; Mutanga, O.; Rugege, D. Multispectral and hyperspectral remote sensing for identification and mapping of wetland vegetation: A review. *Wetl. Ecol. Manag.* **2010**, *18*, 281–296. [CrossRef]
43. Lawley, V.; Lewis, M.; Clarke, K.; Ostendorf, B. Site-based and remote sensing methods for monitoring indicators of vegetation condition: An Australian review. *Ecol. Indic.* **2016**, *60*, 1273–1283. [CrossRef]
44. Marzialetti, F.; Giulio, S.; Malavasi, M.; Sperandii, M.G.; Acosta, A.T.R.; Carranza, M.L. Capturing Coastal Dune Natural Vegetation Types Using a Phenology-Based Mapping Approach: The Potential of Sentinel-2. *Remote Sens.* **2019**, *11*, 1506. [CrossRef]
45. Valentini, E.; Taramelli, A.; Cappucci, S.; Filipponi, F.; Nguyen Xuan, A. Exploring the Dunes: The Correlations between Vegetation Cover Pattern and Morphology for Sediment Retention Assessment Using Airborne Multisensor Acquisition. *Remote Sens.* **2020**, *12*, 1229. [CrossRef]
46. Vellinga, P. Beach and Dune Erosion during Storm Surges. Ph.D. Thesis, Delft University of Technology, Delft, The Netherlands, (Publication 372, Delft Hydraulics). 1986.
47. Larson, M.; Kraus, N.C. *SBEACH: Numerical Model for Simulating Storm-Induced Beach Change Report 1: Empirical Foundation and Model Development*; Technical Report CERC-89-9; U.S. Army Corps of Engineers, Waterways Experiment Station, Coastal Engineering Research Center (CERC): Vicksburg, MS, USA, 1989; p. 115.
48. Larson, M.; Erikson, L.; Hanson, H. An analytical model to predict dune erosion due to wave impact. *Coast. Eng.* **2004**, *51*, 675–696. [CrossRef]
49. Overton, M.; Fisher, J.; Fenaish, T. Numerical analysis of swash forces on dunes. In Proceedings of the Coastal Sediments '87, New Orleans, LA, USA, 12–13 May 1987; pp. 632–641.
50. Roelvink, D.R.; Van Dongeren, A.; Van Thiel de Vries, J.; Mc Call, R.; Lescinski, J. Modelling storm impacts on beaches, dunes and barrier islands. *Coast. Eng.* **2009**, *56*, 1133–1152. [CrossRef]
51. Johnson, B.D.; Kobayashi, N.; Gravens, M.B. *Cross-Shore Numerical Model CSHORE for Waves, Currents, Sediment Transport and Beach Profile Evolution*; U.S. Army Corps of Engineers Engineer Research and Development Center Coastal Hydraulics Laboratory: Vicksburg, MS, USA, 2012; 147p.
52. van Rijn, L.C.; Walstra, D.J.R.; Grasmeijer, B.; Sutherland, J.; Pan, S.; Sierra, J.P. The predictability of cross-shore bed evolution of sandy beaches at the time scale of storms and seasons using process-based profile models. *Coast. Eng.* **2003**, *47*, 295–327. [CrossRef]
53. Delft Hydraulics. *Dune Erosion During an Extreme Storm Surge, Noorderstrand Schouwen*; Research Report M1797; Delft Hydraulics: Delft, The Netherlands, 1982.
54. Delft Hydraulics. *Dune Erosion at the Hotels 'La Specia' en 'De Blanke Top' in Zeeuwsch-Vlaanderen*; Research Report M1811; Delft Hydraulics: Delft, The Netherlands, 1982.
55. Delft Hydraulics. *Scale Series on Dune Erosion, Large Scale Tests in the Delta Flume*; Research Report M1263, Part III; Delft Hydraulics: Delft, The Netherlands, 1984.
56. Delft Hydraulics. *Dune Erosion Due to Design Storm Surge at Walcheren, Golflinks en Vijgeter, Investigation of Stability and Effect of the Dune Revetments*; Research Report M1958; Delft Hydraulics: Delft, The Netherlands, 1984.
57. Delft Hydraulics. *Systematic Research on the Effectiveness of Dune Toe Revetments, Large Scale Model Investigations*; Research Report H298, Part I; Delft Hydraulics: Delft, The Netherlands, 1987.
58. Delft Hydraulics. *Model Study of Dune Erosion: Experimental Results of Small-Scale Pilot Tests*; Research Report H4265; Delft Hydraulics: Delft, The Netherlands, 2004. (In Dutch)
59. Delft Hydraulics. *Dune erosion, Large-Scale Model Tests and Dune Erosion Prediction Method*; Research Report H4357; Delft Hydraulics: Delft, The Netherlands, 2006.
60. Delft Hydraulics. *Technical Report Dune Erosion*; Research Report H4357; Delft Hydraulics: Delft, The Netherlands, 2007. (In Dutch)

61. Dette, H.H.; Peters, K.; Newe, J. *Large Wave Flume Experiments '96/97, Experiments on Beach and Dune Stability*; Rep. No. 830; Technical University of Leichtweiss Institute: Braunschweig, Germany, 1998.
62. Kraus, N.C.; Smith, J.M. *Supertank Laboratory Data Collection Project*; Technical Report CERC 94-3; U.S. Army Engineers Waterways Experiment Station, Coastal Engineer Research Center: Vicksburg, MS, USA, 1994.
63. van Thiel de Vries, J.S.M.; Van Gent, M.R.A.; Walstra, D.J.R.; Reniers, A.J.H.M. Analysis of dune erosion processes in large-scale flume experiments. *Coast. Eng.* **2008**, *55*, 1028–1040. [CrossRef]
64. Salvadori, G.; Tomasicchio, G.R.; D'Alessandro, F. Multivariate approach to design coastal and off-shore structures. *J. Coast. Res.* **2013**, *65*, 386–391. [CrossRef]
65. Salvadori, G.; Tomasicchio, G.R.; D'Alessandro, F. Practical guidelines for multivariate analysis and design in coastal engi-neering. *Coast. Eng.* **2014**, *88*, 1–14. [CrossRef]
66. Salvadori, G.; Durante, F.; Tomasicchio, G.R.; D'Alessandro, F. Practical guidelines for the multivariate assessment of the structural risk in coastal and off-shore engineering. *Coast. Eng.* **2015**, *95*, 77–83. [CrossRef]
67. Salvadori, G.; Tomasicchio, G.R.; D'Alessandro, F.; Lusito, L.; Francone, A. Multivariate sea storm hindcasting and design: The isotropic buoy-ungauged generator procedure. *Sci. Rep.* **2020**, *10*, 20517. [CrossRef]
68. Figlus, J.; Kobayashi, N.; Gralher, C.; Iranzo, V. Wave overtopping and overwash of dunes. *J. Waterw. Port. Coast. Ocean Eng.* **2011**, *137*, 26–33. [CrossRef]
69. Hancock, M.W.; Kobayashi, N. Wave overtopping and sediment transport over dunes. In Proceedings of the 24th Conference on Coastal Engineering, Kobe, Japan, 23–28 October 1994; pp. 2028–2042.
70. Kobayashi, N.; Buck, M.; Payo, A.; Johnson, B.D. Berm and dune erosion during a storm. *J. Waterw. Port. Coast. Ocean Eng.* **2009**, *135*, 1. [CrossRef]
71. Kobayashi, N.; Tega, Y.; Hancock, M.W. Wave reflection and overwash of dunes. *J. Waterw. Port. Coast. Ocean Eng.* **1996**, *122*, 150–153. [CrossRef]
72. Williams, J.; Buscombe, D.; Masselink, G.; Turner, I.; Swinkels, C. Barrier dynamics experiment (BARDEX): Aims, design and procedures. *Coast. Eng.* **2012**, *63*, 3–12. [CrossRef]
73. Blenkinsopp, C.; Matias, A.; Howe, D.; Castelle, B.; Marieu, V.; Turner, I. Wave runup and overwash on a prototype-scale sand barrier. *Coast. Eng.* **2016**, *113*, 88–103. [CrossRef]
74. Matias, A.; Masselink, G.; Castelle, B.; Blenkinsopp, C.E.; Kroon, A. Measurements of morphodynamic and hydrodynamic overwash processes in a large-scale wave flume. *Coast. Eng.* **2016**, *113*, 33–46. [CrossRef]
75. Kobayashi, N.; Gralher, C.; Do, K. Effects of woody plants on dune erosion and overwash. *J. Waterw. Port. Coast. Ocean Eng.* **2013**, *139*, 466–472. [CrossRef]
76. Figlus, J.; Sigren, J.M.; Armitage, A.R.; Tyler, R.C. Erosion of vegetated coastal dunes. In Proceedings of the 34th Conference on Coastal Engineering, Seoul, Korea, 15–20 June 2014; Volume 1. sediment.20.
77. Silva, R.; Martínez, M.L.; Odériz, I.; Mendoza, E.; Feagin, R.A. Response of vegetated dune-beach systems to storm conditions. *Coast. Eng.* **2016**, *109*, 53–62. [CrossRef]
78. Ranasinghe, R.; Duong, T.; Uhlenbrook, S.; Roelvink, D.; Stive, M. Climate-change impact assessment for inlet-interrupted coastlines. *Nat. Clim. Chang.* **2013**, *3*, 83–87. [CrossRef]
79. Tavakol, A.; Rahmani, V.; Harrington, J. Probability of compound climate extremes in a changing climate: A copula-based study of hot, dry, and windy events in the central United States. *Environ. Res. Lett.* **2020**, *15*, 104058. [CrossRef]
80. Gann, G.D.; McDonald, T.; Walder, B.; Aronson, J.; Nelson, C.R.; Jonson, J.; Hallett, J.G.; Eisenberg, C.; Guariguata, M.R.; Liu, J.; et al. International Principles and Standards for the Practice of Ecological Restoration. *Restor. Ecol.* **2019**, *27*, S1–S46. [CrossRef]
81. Owen, M. *Design of Seawalls Allowing for Wave Overtopping*; HR Wallingford HR Report: Wallingford, UK, 1980; 96p.
82. Takahashi, S. *Design of Vertical Breakwaters*; Reference Document 34; Port and Harbor Research Institute: Kanagawa, Japan, 2002; 106p.
83. van der Meer, J.W. Conceptual design of rubble mound breakwaters. In *Advances in Coastal and Ocean Engineering*; Liu, P.L.-F., Ed.; Delft Hydraulics: Delft, The Netherlands, 1993; Publication No. 483; pp. 221–315.
84. van der Meer, J.W.; Bruce, T. New physical insights and design formulas on wave overtopping at sloping and vertical struc-tures. *J. Waterw. Port. Coast. Ocean Eng.* **2013**, *140*, 04014025. [CrossRef]
85. van der Meer, J.W.; Allsop, N.W.H.; Bruce, T.; De Rouck, J.; Kortenhaus, A.; Pullen, T.; Schuttrumpf, H.; Troch, P.; Zanuttigh, B. *EurOtop: Manual on Wave Overtopping of Sea Defences and Related Structures-An Overtopping Manual Largely Based on European Research, but for Worldwide Application*, 2nd ed.; 2016; p. 252. Available online: https://www.kennisbank-waterbouw.nl/DesignCodes/EurOtop2016.pdf (accessed on 24 March 2022).
86. Takagi, H.; Goda, Y. A reliability design method of caisson breakwaters with optimal wave heights. *Coast. Eng. J.* **2000**, *42*, 357–387.
87. van der Meer, J.W. Deterministic and probabilistic design of breakwater armor layers. *J. Waterw. PortCoast. Ocean. Eng.* **1988**, *114*, 66–80. [CrossRef]
88. Tomasicchio, G.R.; D'Alessandro, F.; Barbaro, G.; Malara, G. General longshore transport model. *Coast. Eng.* **2013**, *71*, 28–36. [CrossRef]
89. Smith, E.R.; D'Alessandro, F.; Tomasicchio, G.R.; Gailani, J.Z. Nearshore placement of a sand dredged mound. *Coast. Eng.* **2017**, *126*, 1–10. [CrossRef]

90. Tomasicchio, G.R.; D'Alessandro, F.; Frega, F.; Francone, A.; Ligorio, F. Recent improvements for estimation of longshore transport. *Ital. J. Eng. Geol. Environ.* **2018**, *1*, 179–187.
91. Corbella, S.; Stretch, D.D. Multivariate return periods of sea storms for coastal erosion risk assessment. *Nat. Hazards Earth Syst. Sci.* **2012**, *12*, 2699–2708. [CrossRef]
92. Pantusa, D.; D'Alessandro, F.; Riefolo, L.; Principato, F.; Tomasicchio, G.R. Application of a Coastal Vulnerability Index. A case study along the Apulian coastline, Italy. *Water* **2018**, *10*, 1218. [CrossRef]
93. Barbaro, G.; Foti, G.; Barillà, G.C.; Frega, F. Beach and dune erosion: Causes and interventions, case study: Kaulon archeological site. *J. Mar. Sci. Eng.* **2022**, *10*, 14. [CrossRef]
94. Foti, G.; Barbaro, G.; Barillà, G.C.; Frega, F. Effects of Anthropogenic Pressures on Dune Systems-Case Study: Calabria (Italy). *J. Mar. Sci. Eng.* **2022**, *10*, 10. [CrossRef]
95. Barbaro, G.; Bombino, G.; Foti, G.; Borrello, M.M.; Puntorieri, P. Shoreline evolution near river mouth: Case study of Petrace River (Calabria, Italy). *Reg. Stud. Mar. Sci.* **2019**, *29*, 100619. [CrossRef]
96. Foti, G.; Barbaro, G.; Bombino, G.; Fiamma, V.; Puntorieri, P.; Minniti, F.; Pezzimenti, C. Shoreline changes near river mouth: Case study of Sant'Agata River (Reggio Calabria, Italy). *Eur. J. Remote Sens.* **2019**, *52*, 102–112. [CrossRef]
97. Tomasicchio, G.R.; Lusito, L.; D'Alessandro, F.; Frega, F.; Francone, A.; De Bartolo, S. A direct scaling analysis for the sea level rise. *Stoch. Environ. Res. Risk Assess.* **2018**, *32*, 3397–3408. [CrossRef]
98. de Boer, R. Building-with-Nature Solutions for Hurricane Flood Risk Reduction in Galveston Bay-Texas. Master's Thesis, Delft University of Technology, Delft, The Netherlands, 2015. 221p.
99. de Vriend, H.J.; van Koningsveld, M.; Aarninkhof, S.G.; de Vries, M.B.; Baptist, M.J. Sustainable hydraulic engineering through building with nature. *J. Hydro-Environ. Res.* **2015**, *9*, 159–171. [CrossRef]
100. Firth, L.B.; Thompson, R.C.; Bohn, K.; Abbiati, M.; Airoldi, L.; Bouma, T.J.; Bozzeda, F.; Ceccherelli, V.U.; Colangelo, M.A.; Evans, A.; et al. Between a rock and a hard place: Environmental and engineering consider-ations when designing coastal defence structures. *Coast. Eng.* **2014**, *87*, 122–135. [CrossRef]
101. Miller, J.K.; Mahon, A.M.; Herrington, T.O. *Development of the Stevens Dynamic Underwater Coastal Surveying (DUCKS) System*; Technical Report IT-DL-09-9-2878; Stevens Institute of Technology, Coastal Protection Technical Assistance Service, Davidson Laboratory: Hoboken, NJ, USA, 2009; 15p.
102. Foti, G.; Barbaro, G.; Barillà, G.C.; Mancuso, P.; Puntorieri, P. Shoreline Evolutionary Trends Along Calabrian Coasts: Causes and Classification. *Front. Mar. Sci.* **2022**, *9*, 846914. [CrossRef]
103. Sutton-Grier, A.E.; Wowk, K.; Bamford, H. Future of our coasts: The potential for natural and hybrid infrastructure to enhance the resilience of our coastal communities, economies and ecosystems. *Environ. Sci. Policy* **2015**, *51*, 137–148. [CrossRef]
104. Boers, M. *Technisch Rapport Duinwaterkeringen en Hybride Keringen 2011*; Deltares: Delft, The Netherlands, 2011; 140p. (In Dutch)
105. Stronkhorst, J.; Lagendijk, O. Toekomstbestendige ver-harde zeeweringen. In *Opdracht Van Rijkswaterstaat, Water-Dienst, Rapport 1206188-000-VEB-0009*; Deltares: Delft, The Netherlands, 2012; 81p. (In Dutch)
106. Wamsley, T.V.; Waters, J.P.; King, D.B. Performance of Experimental Low Volume Beach Fill and Clay Core Dune Shore Protection Project. *J. Coast. Res.* **2011**, *59*, 202–210. [CrossRef]
107. West, N.A. Conceptual Design and Physical Model Tests of A Levee-in-Dune Hurricane Barrier. Master's Thesis, Texas A&M University, College Station, TX, USA, 2014.
108. Spalding, M.D.; Ruffo, S.; Lacambra, C.; Meliane, I.; Hale, L.Z.; Shepard, C.C.; Beck, M.W. The role of ecosystems in coastal protection: Adapting to climate change and coastal hazards. *Ocean Coast. Manag.* **2014**, *90*, 50–57. [CrossRef]
109. Maza, M. Nature-based solutions for coastal defence: Key aspects in the modelling of flow-ecosystem interactions. In Proceedings of the Virtual Conference on Costal Engineering, Online, 3 July 2020.
110. Toimil, A.; Losada, I.J.; Nicholls, R.J.; Dalrymple, R.A.; Stive, M.J. Addressing the challenges of climate change risks and ad-aptation in coastal areas: A review. *Coast. Eng.* **2020**, *156*, 103611. [CrossRef]
111. Birds Directive, 2009. Directive 2009/147/EC on the Conservation of Wild Birds. 2009. Available online: http://ec.europa.eu/environment/nature/legislation/birdsdirective/index_en.htm (accessed on 10 December 2021).
112. EIA. Council Directive 97/11/EC of 3 March 1997 Amending Directive 85/337/ EEC on the Assessment of the Effects of Certain Public and Private Projects on the Environment. 1997. Available online: http://ec.europa.eu/environment/eia/full-legal-text/9711.htm (accessed on 21 January 2022).
113. SEA. Directive 2001/42/EC on the Assessment of the Effects of Certain Plans and Programmes on the Environment (EN:NOT). 2001. Available online: https://eur-lex.europa.eu/legal-content/EN/NIM/?uri=CELEX:32001L0042&qid=1487172800769 (accessed on 25 January 2022).
114. Cowles, H.C. The Ecological Relations of the Vegetation on the Sand Dunes of Lake Michigan. *Bot. Gaz.* **1899**, *27*, 167–202. [CrossRef]
115. Hesp, P.A. A review of biological and geomorphological processes involved in the initiation and development of incipient foredunes. *Proc. R. Soc. Edinb. Sect. B Biol. Sci.* **1989**, *96*, 181–201. [CrossRef]
116. Olson, J.S. Lake Michigan dune development 2. Plants as agents and tools in geomorphology. *J. Geol.* **1958**, *66*, 345–351. [CrossRef]
117. Fernández-Montblanc, T.; Duo, E.; Ciavola, P. Dune reconstruction and revegetation as a potential measure to decrease coastal erosion and flooding under extreme storm conditions. *Ocean Coast. Manag.* **2020**, *188*, 105075. [CrossRef]

118. Bryant, D.B.; Bryant, M.A.; Sharp, J.A.; Bell, G.L.; Moore, C. The response of vegetated dunes to wave attack. *Coast. Eng.* **2019**, *152*, 103506. [CrossRef]
119. Feagin, R.A.; Furman, M.; Salgado, K.; Martinez, M.L.; Innocenti, R.A.; Eubanks, K.; Figlus, J.; Huff, T.P.; Sigren, J.; Silva, R. The role of beach and sand dune vegetation in mediating wave run up erosion. *Estuar. Coast. Shelf Sci.* **2019**, *219*, 97–106. [CrossRef]

MDPI
St. Alban-Anlage 66
4052 Basel
Switzerland
Tel. +41 61 683 77 34
Fax +41 61 302 89 18
www.mdpi.com

Journal of Marine Science and Engineering Editorial Office
E-mail: jmse@mdpi.com
www.mdpi.com/journal/jmse

www.ingramcontent.com/pod-product-compliance
Lightning Source LLC
Chambersburg PA
CBHW061831210926
51850CB00006B/857

* 9 7 8 3 0 3 6 5 5 0 2 7 5 *